# The Generalized Relative Gol'dberg Order and Type: Some Remarks on Functions of Complex Variables

Authored by

**Tanmay Biswas**

*Rajbari, Rabindrapally, R. N. Tagore Road*
*P.O.-Krishnagar, Dist.-Nadia, West Bengal*
*India*

&

**Chinmay Biswas**

*Department of Mathematics*
*Nabadwip Vidyasagar College*
*Nabadwip, West Bengal 741302*
*India*

# The Generalized Relative Gol'dberg Order and Type: Some Remarks on Functions of Complex Variables

Authors: Tanmay Biswas & Chinmay Biswas

ISBN (Online):  978-981-4998-03-1

ISBN (Print): 978-981-4998-04-8

ISBN (Paperback): 978-981-4998-05-5

need for a court order if at any point you breach any terms of this License Agreement. In no event will any delay or failure by Bentham Science Publishers in enforcing your compliance with this License Agreement constitute a waiver of any of its rights.

3. You acknowledge that you have read this License Agreement, and agree to be bound by its terms and conditions. To the extent that any other terms and conditions presented on any website of Bentham Science Publishers conflict with, or are inconsistent with, the terms and conditions set out in this License Agreement, you acknowledge that the terms and conditions set out in this License Agreement shall prevail.

**Bentham Science Publishers Ltd.**
Executive Suite Y - 2
PO Box 7917, Saif Zone
Sharjah, U.A.E.
Email: subscriptions@benthamscience.net

**BENTHAM SCIENCE**

# CONTENTS

# PREFACE

The main object of this book is to discuss the generalized comparative growth analysis of entire functions of n-complex variables, which covers the important branch of complex analysis, especially the theory of analytic functions of several variables. Our book contains eight chapters.

Chapter 1 contains the introductory parts and some preliminary definitions. In chapter 2, we have developed some results related to generalized Gol'dberg order (α, β) and generalized Gol'dberg type (α, β) of entire functions of several complex variables. In chapter 3, we have proved some results about generalized relative Gol'dbergorder (α, β) of entire functions of several complex variables. In chapter 4, some inequalities using generalized relative Gol'dberg order (α, β) and generalized relative Gol'dberg lower order (α, β) of entire functions of several complex variables are established. In chapter 5, we have improved some relation connecting to generalized relative Gol'dberg type (α, β) and generalized relative Gol'dberg weak type (α, β) of entire functions of several complex variables. In chapter 6, we have derived some inequalities using generalized relative Gol'dberg type (α, β) and generalized relative Gol'dberg weak type (α, β) of entire functions of several complex variables. In chapter 7, we have discussed generalized relative Gol'dberg order (α, β) and generalized relative Gol'dberg type (α, β) based growth measure of entire functions of several complex variables. And finally, in chapter 8, we mainly focus on sum and product theorems depending on the generalized relative Gol'dberg order (α, β) and generalized relative Gol'dberg type (α, β).

To improve our results, we took help from many publications of different authors and we are thankful to them and cited their publications in the bibliography. We think this book will be very helpful for research scholars and students. We are also thankful to the Bentham Science publishers to give us the opportunity to publish this monograph.

## CONSENT FOR PUBLICATION

Not applicable.

## CONFLICT OF INTEREST

The author declares no conflict of interest, financial or otherwise.

**Tanmay Biswas**
Rajbari, Rabindrapally, R. N. Tagore Road
P.O.-Krishnagar, Dist.-Nadia, West Bengal
India

&

**Chinmay Biswas**
Department of Mathematics
Nabadwip Vidyasagar College
Nabadwip, West Bengal 741302
India

# ACKNOWLEDGEMENT

We are very much thankful to the authors of different publications as many new ideas are abstracted from them. The authors are highly thankful to Mr. Obaid Sadiq, Ms. Asma Ahmed, and the Bentham Science publishers for providing the opportunity to publish this book. Authors also express gratefulness to their family members for their continuous help, inspirations, encouragement, and sacrifices without which the book could not be executed. Finally, the main target of this book will not be achieved unless it is used by students, research scholars, and authors in their future works. The authors will remain ever grateful to those who helped by giving constructive suggestions for this work. The authors are also responsible for any possible errors and shortcomings, if any in the book, despite the best attempt to make it immaculate.

# Chapter 1

# Introduction, definitions and notations

**Abstract:** In this chapter, we discussed about the introductory parts connected to the entire functions of $n$ complex variables. In this connection, we add some preliminary definitions related to different Gol'dberg growth indicators such as Gol'dberg order, Gol'dberg type etc.

**Keywords:** Entire functions, several complex variables, different growth indicators.

**Mathematics Subject Classification (2010) :** 32A15.

## 1.1  Introduction, definitions and notations.

The present chapter consists of some preliminary definitions in connection to the entire function $f(z)$ of $n$ complex variables. Let $\mathbb{C}^n$ and $R^n$ respectively denote the complex and real $n$-space. Also let us indicate the point $(z_1, z_2, \cdots, z_n)$, $(m_1, m_2, \cdots, m_n)$ of $\mathbb{C}^n$ or $I^n$ by their corresponding unsuffixed symbols $z, m$ respectively where $I$ denotes the set of non-negative integers. The modulus of $z$, denoted by $|z|$, is defined as $|z| = (|z_1|^2 + \cdots + |z_n|^2)^{\frac{1}{2}}$. If the coordinates of the vector $m$ are non-negative integers, then $z^m$ will denote $z_1^{m_1} \cdots z_n^{m_n}$ and $\|m\| = m_1 + \cdots + m_n$ .

If $D \subseteq \mathbb{C}^n$ ($\mathbb{C}^n$ denote the $n$-dimensional complex space) be an arbitrary bounded complex $n$-circular domain with center at the origin of coordinates then for any entire function $f(z)$ of $n$ complex variables and $R > 0$, $M_{f,D}(R)$ may be define as $M_{f,D}(R) = \sup_{z \in D_R} |f(z)|$ where a point $z \in D_R$ if and only if $\frac{z}{R} \in D$. If $f(z)$ is non-constant, then $M_{f,D}(R)$ is strictly increasing and its inverse $M_{f,D}^{-1} : (|f(0)|, \infty) \to (0, \infty)$ exists such that $\lim_{R \to \infty} M_{f,D}^{-1}(R) = \infty$.

Considering this, the Gol'dberg order and Gol'dberg lower order (cf. [1, 2]) of an entire function $f(z)$ with respect to any bounded complete $n$-circular domain $D$ with center at all the origin $\mathbb{C}^n$ are given by

$$\begin{aligned} \rho_D(f) \\ \lambda_D(f) \end{aligned} = \lim_{R \to \infty} \begin{aligned} \sup \\ \inf \end{aligned} \frac{\log \log M_{f,D}(R)}{\log R}.$$

It is well known that $\rho_D(f)$ is independent of the choice of the domain $D$, and therefore we write $\rho(f)$ instead of $\rho_D(f)$ (respectively $\lambda(f)$ instead of $\lambda_D(f)$) (cf. [1, 2]).

For any bounded complete $n$-circular domain $D$, an entire function of $n$ complex variables for which Gol'dberg order and Gol'dberg lower order are the same is said to be of regular growth. Functions which are not of regular growth are said to be of irregular growth.

To compare the relative growth of two entire functions of $n$ complex variables having same non-zero finite Gol'dberg order, one may introduce the definition of Gol'dberg type and Gol'dberg lower type in the following manner:

**Definition 1.1.1** *(cf. [1, 2]) The Gol'dberg type and Gol'dberg lower type respectively denoted by $\sigma_D(f)$ and $\overline{\sigma}_D(f)$ of an entire function $f(z)$ of $n$ complex variables with respect to any bounded complete $n$-circular domain $D$ with center at all the origin $\mathbb{C}^n$ are defined as follows:*

$$\begin{matrix} \sigma_D(f) \\ \overline{\sigma}_D(f) \end{matrix} = \lim_{R \to \infty} \begin{matrix} \sup \\ \inf \end{matrix} \frac{\log M_{f,D}(R)}{(R)^{\rho(f)}}, \ 0 < \rho(f) < \infty.$$

Analogously to determine the relative growth of two entire functions of $n$ complex variables having same non-zero finite Gol'dberg lower order, one may introduce the definition of Gol'dberg weak type in the following way:

**Definition 1.1.2** *The Gol'dberg weak type denoted by $\tau_D(f)$ of an entire function $f(z)$ of $n$ complex variables with respect to any bounded complete $n$-circular domain $D$ with center at all the origin $\mathbb{C}^n$ is defined as follows:*

$$\tau_D(f) = \liminf_{R \to \infty} \frac{\log M_{f,D}(R)}{(R)^{\lambda(f)}}, \ 0 < \lambda(f) < \infty.$$

*Also one may define the Gol'dberg upper weak type denoted by $\overline{\tau}_D(f)$ in the following manner :*

$$\overline{\tau}_D(f) = \limsup_{R \to \infty} \frac{\log M_{f,D}(R)}{(R)^{\lambda(f)}}, \ 0 < \lambda(f) < \infty.$$

Gol'dberg has shown that [2] Gol'dberg type depends on the domain $D$. Hence all the growth indicators define in Definition 1.1.1 and Definition 1.1.2 are also depend on $D$.

In the sequel the following two notations are used:

$$\log^{[k]} R = \log(\log^{[k-1]} R) \text{ for } k = 1, 2, 3, \cdots ;$$
$$\log^{[0]} R = R$$

and

$$\exp^{[k]} R = \exp(\exp^{[k-1]} R) \text{ for } k = 1, 2, 3, \cdots ;$$
$$\exp^{[0]} R = R.$$

Taking this into account the, one can give the definitions of generalized Gol'dberg order $\rho_D^{(l)}(f)$ and generalized Gol'dberg lower order $\lambda_D^{(l)}(f)$ in the following way:

**Definition 1.1.3** *The generalized Gol'dberg order $\rho_D^{(l)}(f)$ and generalized Gol'dberg lower order $\lambda_D^{(l)}(f)$ of an entire function $f(z)$ of $n$ complex variables with respect to any bounded complete $n$-circular domain $D$ with center at all the origin $\mathbb{C}^n$ are defined as follows:*

$$\begin{matrix} \rho_D^{(l)}(f) \\ \lambda_D^{(l)}(f) \end{matrix} = \lim_{R \to \infty} \begin{matrix} \sup \\ \inf \end{matrix} \frac{\log^{[l]} M_{f,D}(R)}{\log R},$$

*where $l$ is any positive integer such that $l \geq 2$.*

In the line of Gol'dberg (cf. [1, 2]), one can easily verify that $\rho_D^{(l)}(f)$ and $\lambda_D^{(l)}(f)$ are independent of the choice of the domain $D$, and therefore we write $\rho^{(l)}(f)$ instead of $\rho_D^{(l)}(f)$ and $\lambda^{(l)}(f)$ instead of $\lambda_D^{(l)}(f)$.

This definition extended the Gol'dberg order $\rho(f)$ and Gol'dberg lower order $\lambda(f)$ of an entire function $f(z)$ of $n$ complex variables with respect to any bounded complete $n$-circular domain $D$ since this correspond to the particular case $\rho^{(2)}(f) = \rho(f)$ and $\lambda^{(2)}(f) = \lambda(f)$.

However, an entire function $f(z)$ for which $\rho^{(l)}(f)$ and $\lambda^{(l)}(f)$ are the same is called a function of regular generalized Gol'dberg growth. Otherwise, $f(z)$ is said to be irregular generalized Gol'dberg growth.

The following two definitions are the natural consequences of the above study:

**Definition 1.1.4** *The generalized Gol'dberg type $\sigma_f^{[l]}$ and generalized Gol'dberg lower type $\overline{\sigma}_f^{[l]}$ of an entire function $f(z)$ of $n$ complex variables with respect to any bounded complete $n$-circular domain $D$ with center at all the origin $\mathbb{C}^n$ are defined as*

$$\begin{matrix} \sigma_D^{(l)}(f) \\ \overline{\sigma}_D^{(l)}(f) \end{matrix} = \lim_{R \to \infty} \begin{matrix} \sup \\ \inf \end{matrix} \frac{\log^{[l-1]} M_{f,D}(R)}{R^{\rho^{(l)}(f)}}, \quad 0 < \rho^{(l)}(f) < \infty,$$

*where $l$ is any positive integer such that $l \geq 2$. Moreover, when $l = 2$ then $\sigma_D^{(2)}(f)$ and $\overline{\sigma}_D^{(2)}(f)$ are correspondingly denoted as $\sigma_D(f)$ and $\overline{\sigma}_D(f)$.*

Similarly, extending the notion of Gol'dberg weak type, one can define generalized Gol'dberg weak type in the following manner:

**Definition 1.1.5** *The generalized Gol'dberg weak type $\tau_D^{(l)}(f)$ for any positive integer $l \geq 2$ of an entire function $f(z)$ of $n$ complex variables with respect to any bounded complete $n$-circular domain $D$ with center at all the origin $\mathbb{C}^n$ having finite positive generalized Gol'dberg lower order $\lambda^{(l)}(f)$ are defined by*

$$\tau_D^{(l)}(f) = \liminf_{R \to \infty} \frac{\log^{[l-1]} M_{f,D}(R)}{R^{\lambda^{(l)}(f)}}, \quad 0 < \lambda^{(l)}(f) < \infty.$$

*Also one may define the generalized Gol'dberg upper weak type denoted by $\overline{\tau}_D^{(l)}(f)$ in the following way:*

$$\overline{\tau}_D^{(l)}(f) = \limsup_{R \to \infty} \frac{\log^{[l-1]} M_{f,D}(R)}{R^{\lambda^{(l)}(f)}}, \quad 0 < \lambda^{(l)}(f) < \infty.$$

For $l = 2$, the above definition reduces to the classical Definition 1.1.2.

Since Gol'dberg has shown that [2] Gol'dberg type depends on the domain $D$. Hence all the growth indicators define in Definition 1.1.4 and Definition 1.1.5 are also depend on $D$.

However, extending the notion of Gol'dberg order, Datta et al. [3] defined the concept of $(p, q)$-th Gol'dberg order and $(p, q)$-th Gol'dberg lower order of an entire function $f(z)$ for any bounded complete $n$-circular domain $D$ with center at all the origin $\mathbb{C}^n$ where $p$ and $q$ are any positive integers with $p \geq q \geq 1$ in the following way:

$$\begin{aligned} \rho_D^{(p,q)}(f) \\ \lambda_D^{(p,q)}(f) \end{aligned} = \lim_{R \to \infty} \begin{aligned} \sup \\ \inf \end{aligned} \frac{\log^{[p]} M_{f,D}(R)}{\log^{[q]} R}.$$

Further in the line of Gol'dberg (cf. [1, 2]), one can easily verify that $\rho_D^{(p,q)}(f)$ and $\lambda_D^{(p,q)}(f)$ are independent of the choice of the domain $D$, and therefore one can write $\rho^{(p,q)}(f)$ and $\lambda^{(p,q)}(f)$ instead of $\rho_D^{(p,q)}(f)$ and $\lambda_D^{(p,q)}(f)$ respectively.

These definitions extended the generalized Gol'dberg order $\rho^{(l)}(f)$ and generalized Gol'dberg lower order $\lambda^{(l)}(f)$ of an entire function $f(z)$ for any bounded complete $n$-circular domain $D$ for each integer $l \geq 2$ since these correspond to the particular case $\rho^{(l)}(f) = \rho^{(l,1)}(f)$ and $\lambda^{(l)}(f) = \lambda^{(l,1)}(f)$ respectively. Clearly $\rho^{(2,1)}(f) = \rho(f)$ and $\lambda^{(2,1)}(f) = \lambda(f)$.

However, an entire function $f(z)$ for which $\rho^{(p,q)}(f)$ and $\lambda^{(p,q)}(f)$ are the same is called a function of regular $(p, q)$ Gol'dberg growth. Otherwise, $f(z)$ is said to be irregular $(p, q)$ Gol'dberg growth.

In this connection let us recall that if $0 < \rho^{(p,q)}(f) < \infty$, then the following properties hold where $p, q$ and $m$ are any positive integers

$$\begin{cases} \rho^{(p-m,q)}(f) = \infty & \text{for} \quad m < p, \\ \rho^{(p,q-m)}(f) = 0 & \text{for} \quad m < q, \\ \rho^{(p+m,q+m)}(f) = 1 & \text{for} \quad m = 1, 2, \cdots \end{cases}.$$

Similarly for $0 < \lambda^{(p,q)}(f) < \infty$, one can easily verify that

$$\begin{cases} \lambda^{(p-m,q)}(f) = \infty & \text{for} \quad m < p, \\ \lambda^{(p,q-m)}(f) = 0 & \text{for} \quad m < q, \\ \lambda^{(p+m,q+m)}(f) = 1 & \text{for} \quad m = 1, 2, \cdots \end{cases}.$$

Recalling that for any pair of integer numbers $a, b$ the Kronecker function is defined by $\delta_{a,b} = 1$ for $a = b$ and $\delta_{a,b} = 0$ for $a \neq b$, the aforementioned properties provide the following definition.

**Definition 1.1.6** *For any bounded complete $n$-circular domain $D$, an entire function $f(z)$ of $n$ complex variables is said to have index-pair $(1,1)$ if $0 < \rho^{(1,1)}(f) < \infty$. Otherwise, $f(z)$ is said to have index-pair $(p, q) \neq (1, 1)$, $p \geq q \geq 1$, if $\delta_{p-q,0} < \rho^{(p,q)}(f) < \infty$ and $\rho^{(p-1,q-1)}(f) \notin R^+$.*

**Definition 1.1.7** *For any bounded complete n-circular domain $D$, an entire function $f(z)$ of $n$ complex variables is said to have lower index-pair $(1,1)$ if $0 < \lambda^{(1,1)}(f) < \infty$. Otherwise, $f(z)$ is said to have lower index-pair $(p,q) \neq (1,1)$, $p \geq q \geq 1$, if $\delta_{p-q,0} < \lambda^{(p,q)}(f) < \infty$ and $\lambda^{(p-1,q-1)}(f) \notin R^+$.*

For any bounded complete $n$-circular domain $D$, an entire function $f(z)$ of $n$ complex variables of index-pair $((p,q)$ is said to be of regular $(p,q)$ Gol'dberg growth if its $(p,q)$-th Gol'dberg order coincides with its $(p,q)$-th Gol'dberg lower order, otherwise $f(z)$ is said to be of irregular $(p,q)$ Gol'dberg growth.

To compare the relative growth of two entire functions having same non-zero finite $(p,q)$-th Gol'dberg order, one may introduce the definition of $(p,q)$-th Gol'dberg type and $(p,q)$-th Gol'dberg lower type in the following manner:

**Definition 1.1.8** *Let $0 < \rho^{(p,q)}(f) < \infty$. The $(p,q)$-th Gol'dberg type and $(p,q)$-th Gol'dberg lower type respectively denoted by $\sigma_D^{(p,q)}(f)$ and $\overline{\sigma}_D^{(p,q)}(f)$ of an entire function $f(z)$ of $n$ complex variables with respect to any bounded complete $n$-circular domain $D$ with center at all the origin $\mathbb{C}^n$ are defined as follows:*

$$
\begin{aligned}
\sigma_D^{(p,q)}(f) \\
\overline{\sigma}_D^{(p,q)}(f)
\end{aligned}
= \lim_{R \to \infty} \begin{matrix} \sup \\ \inf \end{matrix} \frac{\log^{[p-1]} M_{f,D}(R)}{\left[\log^{[q-1]} R\right]^{\rho^{(p,q)}(f)}},
$$

*where $p$ and $q$ are any positive integers with $p \geq q \geq 1$*

Analogously to determine the relative growth of two entire functions of $n$ complex variables having same non-zero finite $(p,q)$-th Gol'dberg lower order, one may introduce the definition of $(p,q)$-th Gol'dberg weak type in the following way:

**Definition 1.1.9** *Let $p$ and $q$ are any positive integers with $p \geq q \geq 1$ The $(p,q)$-th Gol'dberg weak type denoted by $\tau_D^{(p,q)}(f)$ of an entire function $f(z)$ of $n$ complex variables with respect to any bounded complete $n$-circular domain $D$ with center at all the origin $\mathbb{C}^n$ is defined as follows:*

$$
\tau_D^{(p,q)}(f) = \liminf_{R \to \infty} \frac{\log^{[p-1]} M_{f,D}(R)}{\left[\log^{[q-1]} R\right]^{\lambda^{(p,q)}(f)}}, \quad 0 < \lambda^{(p,q)}(f) < \infty.
$$

*Also one may define $(p,q)$-th Gol'dberg upper weak type denoted by $\overline{\tau}_D^{(p,q)}(f)$ in the following way:*

$$
\overline{\tau}_D^{(p,q)}(f) = \limsup_{R \to \infty} \frac{\log^{[p-1]} M_{f,D}(R)}{\left[\log^{[q-1]} R\right]^{\lambda^{(p,q)}(f)}}, \quad 0 < \lambda^{(p,q)}(f) < \infty.
$$

Definition 1.1.8 and Definition 1.1.9 are extended the generalized Gol'dberg type $\sigma_D^{(l)}(f)$, generalized Gol'dberg lower type $\overline{\sigma}_D^{(l)}(f)$, generalized Gol'dberg weak order $\tau_D^{(l)}(f)$ and generalized Gol'dberg upper weak order $\overline{\tau}_D^{(l)}(f)$ of an entire function $f(z)$ of $n$ complex variables with respect to any bounded complete $n$-circular domain $D$ for each integer $l \geq 2$ since these correspond to the particular case $\sigma_D^{(l)}(f) = \sigma_D^{(l,1)}(f)$, $\overline{\sigma}_D^{(l)}(f) = \overline{\sigma}_D^{(l,1)}(f)$, $\tau_D^{(l)}(f) = \tau_D^{(l,1)}(f)$ and $\overline{\tau}_D^{(l)}(f) = \overline{\tau}_D^{(l,1)}(f)$. Clearly $\sigma_D^{(2,1)}(f) = \sigma_D(f)$, $\overline{\sigma}_D^{(2,1)}(f) = \overline{\sigma}_D(f)$, $\tau_D^{(2,1)}(f) = \tau_D(f)$ and $\overline{\tau}_D^{(2,1)}(f) = \overline{\tau}_D(f)$.

Since Gol'dberg has shown that [2] Gol'dberg type depends on the domain $D$, therefore in general all the growth indicators defined in Definition 1.1.8 and Definition 1.1.9 are also depending on $D$.

Extending the notion $(p, q)$-th order of entire function of single variable, Shen et al. [4] introduced the new concept of $(p, q)$-$\varphi$ order of an entire function where $p \geq q$. Later on, combining the definition of $(p, q)$-th Gol'dberg order and $(p, q)$-$\varphi$ order of entire function of single variable, Biswas et al. [5] introduced the definitions of $(p, q)$-$\varphi$ Gol'dberg order and $(p, q)$-$\varphi$ Gol'dberg lower order of an entire function $f(z)$ of $n$ complex variables which are as follows:

**Definition 1.1.10** *(cf. [5]) Let $\varphi(R) : [0, +\infty) \rightarrow (0, +\infty)$ be a non-decreasing unbounded function. Then the $(p, q)$-$\varphi$ Gol'dberg order $\rho_D^{(p,q)}(f, \varphi)$ and $(p, q)$-$\varphi$ Gol'dberg lower order $\lambda_D^{(p,q)}(f, \varphi)$ of an entire function $f(z)$ of $n$ complex variables are defined as*

$$\begin{array}{c} \rho_D^{(p,q)}(f, \varphi) \\ \lambda_D^{(p,q)}(f, \varphi) \end{array} = \lim_{R \to \infty} \begin{array}{c} \sup \\ \inf \end{array} \frac{\log^{[p]} M_{f,D}(R)}{\log^{[q]} \varphi(R)}.$$

However an entire function $f(z)$ for which $\rho_D^{(p,q)}(f, \varphi)$ and $\lambda_D^{(p,q)}(f, \varphi)$ are the same is called a function of regular $(p, q)$-$\varphi$ Gol'dberg growth. Otherwise, $f(z)$ is said to be irregular $(p, q)$-$\varphi$ Gol'dberg growth.

**Remark 1.1.1** *(cf. [5]) If $\lim_{R \to +\infty} \frac{\log^{[q]} \varphi(\alpha R)}{\log^{[q]} \varphi(R)} = 1$ for all $\alpha > 0$ where $\varphi(R) : [0, +\infty) \rightarrow (0, +\infty)$ is any non-decreasing unbounded function, then $\rho_D^{(p,q)}(f, \varphi)$ and $\lambda_D^{(p,q)}(f, \varphi)$ are independent of the choice of the domain $D$.*

If $\varphi(R) = R$ and $p \geq q$, then Definition 1.1.10 coincides with the definition of $(p, q)$-th Gol'dberg order and $(p, q)$-th Gol'dberg lower order introduced by Datta et al. [3].

Concerning this we just recall the following definition:

**Definition 1.1.11** *(cf. [5]) An entire function $f(z)$ of $n$ complex variables is said to have index-pair $(p, q)$-$\varphi$ if $b < \rho^{(p,q)}(f, \varphi) < \infty$ and $\rho^{(p-1,q-1)}(f, \varphi)$ is not a nonzero finite number, where $b = 1$ if $p = q$ and $b = 0$ for otherwise. Moreover if $0 < \rho^{(p,q)}(f, \varphi) < \infty$, then for any positive integer $a$*

$$\begin{cases} \rho^{(p-a,q)}(f, \varphi) = \infty & for \quad a < p, \\ \rho^{(p,q-a)}(f, \varphi) = 0 & for \quad a < q, \\ \rho^{(p+a,q+a)}(f, \varphi) = 1 & for \quad a = 1, 2, \cdots \end{cases}.$$

*Similarly for $0 < \lambda^{(p,q)}(f, \varphi) < \infty$,*

$$
\begin{cases}
\lambda^{(p-a,q)}(f,\varphi) = \infty & \text{for} \quad a < p, \\
\lambda^{(p,q-a)}(f,\varphi) = 0 & \text{for} \quad a < q, \\
\lambda^{(p+a,q+a)}(f,\varphi) = 1 & \text{for} \quad a = 1,2,\cdots
\end{cases}.
$$

Consequently for $\varphi(R) = R$, Definition 1.1.11 reduces to the the definition of index-pair $(p,q)$ of an entire function $f(z)$ of $n$ complex variables.

Now, for the development of such growth indicators, one may introduce $(p,q)$-$\varphi$ Gol'dberg type $\sigma_D^{(p,q)}(f,\varphi)$ and $(p,q)$-$\varphi$ Gol'dberg weak type $\tau_D^{(p,q)}(f,\varphi)$ in the following way:

**Definition 1.1.12** *Let $\varphi(R) : [0, +\infty) \to (0, +\infty)$ be a non-decreasing unbounded function. Let $f(z)$ be an entire functions of $n$ complex variables such that $0 < \rho_D^{(p,q)}(f,\varphi) < \infty$. Then the $(p,q)$-$\varphi$ Gol'dberg type $\sigma_D^{(p,q)}(f,\varphi)$ and the $(p,q)$-$\varphi$ Gol'dberg lower type $\overline{\sigma}_D^{(p,q)}(f,\varphi)$ of $f(z)$ are defined as:*

$$
\begin{matrix}
\sigma_D^{(p,q)}(f,\varphi) \\
\overline{\sigma}_D^{(p,q)}(f,\varphi)
\end{matrix}
= \lim_{R \to \infty}
\begin{matrix}
\sup \\
\inf
\end{matrix}
\frac{\log^{[p-1]}(M_{f,D}(R))}{\left(\log^{[q-1]}\varphi(R)\right)^{\rho_D^{(p,q)}(f,\varphi)}}.
$$

**Definition 1.1.13** *Let $\varphi(R) : [0, +\infty) \to (0, +\infty)$ be a non-decreasing unbounded function. Let $f(z)$ be an entire functions of $n$ complex variables such that $0 < \lambda_D^{(p,q)}(f,\varphi) < \infty$. Then the $(p,q)$-$\varphi$ Gol'dberg weak type $\tau_D^{(p,q)}(f,\varphi)$ and $(p,q)$-$\varphi$ Gol'dberg upper weak type $\overline{\tau}_D^{(p,q)}(f,\varphi)$ of $f(z)$ are defined as:*

$$
\begin{matrix}
\overline{\tau}_D^{(p,q)}(f,\varphi) \\
\tau_D^{(p,q)}(f,\varphi)
\end{matrix}
= \lim_{R \to \infty}
\begin{matrix}
\sup \\
\inf
\end{matrix}
\frac{\log^{[p-1]}(M_{f,D}(R))}{\left(\log^{[q-1]}\varphi(R)\right)^{\lambda_D^{(p,q)}(f,\varphi)}}.
$$

Gol'dberg has shown that (see [2]) Gol'dberg type depends on the domain $D$, so in general all the growth indicators defined in Definition 1.1.12 and Definition 1.1.13 also depend on $D$.

For any two entire functions $f(z)$ and $g(z)$ of $n$ complex variables and for any bounded complete $n$-circular domain $D$ with center at all the origin $\mathbb{C}^n$, Mondal et al. [6] introduced the concept relative Gol'dberg order which is as follows:

$$
\rho_{g,D}(f) = \inf\{\mu > 0 : M_{f,D}(R) < M_{g,D}(R^{\mu}) \text{ for all } R > R_0(\mu) > 0\}
$$
$$
= \limsup_{R \to \infty} \frac{\log M_{g,D}^{-1}(M_{f,D}(R))}{\log R}.
$$

In [6], Mandal et al. also proved that the relative Gol'dberg order of $f(z)$ with respect to $g(z)$ is independent of the choice of the domain $D$. So the relative Gol'dberg order of $f(z)$ with respect to $g(z)$ may be denoted as $\rho_g(f)$ instead of $\rho_{g,D}(f)$.

Likewise, one can define the relative Gol'dberg lower order $\lambda_{g,D}(f)$ in the following manner:

$$\lambda_{g,D}(f) = \liminf_{R\to\infty} \frac{\log M_{g,D}^{-1}(M_{f,D}(R))}{\log R}.$$

In the line of Mandal et al. (cf. [6]), one can also verify that $\lambda_{g,D}(f)$ is independent of the choice of the domain $D$, and therefore one can write $\lambda_g(f)$ instead of $\lambda_{g,D}(f)$.

An entire function $f(z)$ of $n$ complex variables for which $\rho_g(f)$ and $\lambda_g(f)$ are the same is called a function of regular relative Gol'dberg growth with respect to an entire function $g(z)$ of $n$ complex variables. Otherwise, $f(z)$ is said to be irregular relative Gol'dberg growth with respect to $g(z)$.

To compare the relative growth of two entire functions of $n$ complex variables having same non-zero finite relative Gol'dberg order with respect to another entire function of $n$ complex variables, one may introduce the definition of relative Gol'dberg type and relative Gol'dberg lower type in the following manner:

**Definition 1.1.14** *Let $f(z)$ and $g(z)$ be any two entire functions of $n$ complex variables with respect to any bounded complete $n$-circular domain $D$ with center at all the origin $\mathbb{C}^n$ and $0 < \rho_g(f) < \infty$. Then the relative Gol'dberg type of $f(z)$ with respect to $g(z)$ is defined as:*

$$\sigma_{g,D}(f) = \inf\left\{\mu > 0 : M_{f,D}(R) < M_{g,D}\left(\mu R^{\rho_g(f)}\right) \text{ for all sufficiently large values of } R\right\}$$
$$= \limsup_{R\to\infty} \frac{M_{g,D}^{-1}(M_{f,D}(R))}{R^{\rho_g(f)}}.$$

*Similarly, one can define the relative Gol'dberg lower type of $f(z)$ with respect to $g(z)$ denoted by $\overline{\sigma}_g(f)$ as follows :*

$$\overline{\sigma}_{g,D}(f) = \liminf_{R\to\infty} \frac{M_{g,D}^{-1}(M_{f,D}(R))}{R^{\rho_g(f)}}, \ 0 < \rho_g(f) < \infty.$$

Analogously, to determine the relative growth of two entire functions of $n$ complex variables having same non-zero finite relative Gol'dberg lower order with respect to another entire function of $n$ complex variables one may introduce the definition of relative Gol'dberg weak type and relative Gol'dberg upper type in the following manner:

**Definition 1.1.15** *Let $f(z)$ and $g(z)$ be any two entire functions of $n$ complex variables with respect to any bounded complete $n$-circular domain $D$ with center at all the origin $\mathbb{C}^n$ and $0 < \lambda_g(f) < \infty$. Then the relative Gol'dberg weak type and relative Gol'dberg upper weak type denoted by $\tau_{D,g}(f)$ and $\overline{\tau}_{D,g}(f)$ respectively of $f(z)$ with respect to $g(z)$ are defined as:*

$$\begin{matrix}\overline{\tau}_{g,D}(f) \\ \tau_{g,D}(f)\end{matrix} = \lim_{R\to\infty} \begin{matrix}\sup \\ \inf\end{matrix} \frac{M_{g,D}^{-1}(M_{f,D}(R))}{R^{\lambda_g(f)}}.$$

Since Gol'dberg has shown that (see [2]) Gol'dberg type depends on the domain $D$, therefore all the growth indicators define in Definition 1.1.14 and Definition 1.1.15 are also depend on $D$.

In this connection we state the following definition which will be needed in the sequel:

**Definition 1.1.16** *A non-constant entire function $f(z)$ of $n$ complex variables is said to have Property (G), if for any $\delta > 1$*

$$[M_{f,D}(R)]^2 \leq M_{f,D}(R^\delta).$$

Prajapati et al. [7] gave a more generalized concept of relative Gol'dberg order in the following way:

**Definition 1.1.17** *(cf. [7]) Let $l \geq 1$ is a positive integer and $f(z)$, $g(z)$ be any two entire functions of $n$ complex variables with respect to any bounded complete $n$-circular domain $D$ with center at all the origin $\mathbb{C}^n$. Then the generalized Gol'dberg relative order and generalized Gol'dberg relative lower order of $f(z)$ with respect to $g(z)$, denoted by $\rho_g^{(l)}(f)$ and $\lambda_g^{(l)}(f)$ respectively are defined by*

$$\begin{matrix} \rho_{g,D}^{(l)}(f) \\ \lambda_{g,D}^{(l)}(f) \end{matrix} = \lim_{R\to\infty} \begin{matrix} \sup \\ \inf \end{matrix} \frac{\log^{[l]} M_{g,D}^{-1}(M_{f,D}(R))}{\log R}.$$

In the line of Mandal et al. (cf. [6]), one can also verify that $\rho_{g,D}^{(l)}(f)$ and $\lambda_{g,D}^{(l)}(f)$ are independent of the choice of the domain $D$, and therefore one can write $\rho_g^{(l)}(f)$ and $\rho_g^{(l)}(f)$ instead of $\rho_{g,D}^{(l)}(f)$ and $\rho_{g,D}^{(l)}(f)$ respectively. Clearly $\rho_g^{(1)}(f) = \rho_g(f)$ and $\lambda_g^{(1)}(f) = \lambda_g(f)$.

An entire function $f(z)$ for which generalized Gol'dberg relative order and generalized Gol'dberg relative lower order with respect to another entire function $g(z)$ are the same is called a function of regular generalized Gol'dberg relative growth with respect to $g(z)$. Otherwise, $f(z)$ is said to be irregular generalized Gol'dberg relative growth.with respect to $g(z)$.

Further to compare the relative growth of two entire functions of $n$ complex variables having same non-zero finite generalized Gol'dberg relative order with respect to another entire function of $n$ complex variables, one may introduce the definition of generalized relative Gol'dberg type and generalized relative Gol'dberg lower type in the following way:

**Definition 1.1.18** *Let $f(z)$ and $g(z)$ be any two entire functions of $n$ complex variables with respect to any bounded complete $n$-circular domain $D$ with center at all the origin $\mathbb{C}^n$ and $0 < \rho_g^{(1)}(f) < \infty$. Then the generalized relative Gol'dberg type and generalized relative Gol'dberg lower type of $f(z)$ with respect to $g(z)$ are defined as:*

$$\begin{matrix} \sigma_{g,D}^{[l]}(f) \\ \overline{\sigma}_{g,D}^{[l]}(f) \end{matrix} = \lim_{R\to\infty} \begin{matrix} \sup \\ \inf \end{matrix} \frac{\log^{[l-1]} M_{g,D}^{-1}(M_{f,D}(R))}{R^{\rho_g^{[l]}(f)}},$$

*where $l \geq 1$.*

For $l = 1$, Definition 1.1.18 reduces to Definition 1.1.14.

Similarly to determine the relative growth of two entire functions of $n$ complex variables having same non-zero finite generalized relative Gol'dberg lower order with respect to another entire function of $n$ complex variables one may introduce the definition of generalized relative Gol'dberg weak type and generalized relative Gol'dberg upper type in the following manner:

**Definition 1.1.19** *Let $f(z)$ and $g(z)$ be any two entire functions of $n$ complex variables with respect to any bounded complete $n$-circular domain $D$ with center at all the origin $\mathbb{C}^n$ and $0 < \lambda_g^{(1)}(f) < \infty$. Then the generalized relative Gol'dberg weak type and generalized relative Gol'dberg upper weak type denoted by $\tau_{g,D}^{[l]}(f)$ and $\overline{\tau}_{g,D}^{[l]}(f)$ respectively of $f(z)$ with respect to $g(z)$ are defined as:*

$$\begin{array}{c} \overline{\tau}_{g,D}^{[l]}(f) \\ \tau_{g,D}^{[l]}(f) \end{array} = \lim_{R \to \infty} \begin{array}{c} \sup \\ \inf \end{array} \frac{\log^{[l-1]} M_{g,D}^{-1}(M_{f,D}(R))}{R^{\lambda_g^{[l]}(f)}}.$$

Definition 1.1.19 also reduces to Definition 1.1.15 for particular $l = 1$.

Since Gol'dberg has shown that [2] Gol'dberg type depends on the domain $D$, therefore all the growth indicators define in Definition 1.1.18 and Definition 1.1.19 are also depend on $D$.

In the case of relative Gol'dberg order, it was then natural for Biswas [8, 9, 10] to define the $(p, q)$-th relative Gol'dberg order of entire function of $n$ complex variables and for any bounded complete $n$-circular domain $D$ with center at the origin in $\mathbb{C}^n$ in the light of index-pair. Next definition avoids the restriction $p > q$ and gives the more natural particular case of Generalized relative Gol'dberg order.

**Definition 1.1.20** *(cf. [8, 9, 10]) Let $f(z)$ and $g(z)$ be any two entire functions of $n$ complex variables with index-pair $(m, q)$ and $(m, p)$, respectively, where $p, q, m$ are positive integers such that $m \geq q \geq 1$ and $m \geq p \geq 1$ and $D$ be any bounded complete $n$-circular domain with center at the origin in $\mathbb{C}^n$. Then the $(p, q)$-th relative Gol'dberg order and $(p, q)$-th relative Gol'dberg lower order of $f(z)$ with respect to $g(z)$ are defined as*

$$\begin{array}{c} \rho_{g,D}^{(p,q)}(f) \\ \lambda_{g,D}^{(p,q)}(f) \end{array} = \lim_{R \to \infty} \begin{array}{c} \sup \\ \inf \end{array} \frac{\log^{[p]} M_{g,D}^{-1}(M_{f,D}(R))}{\log^{[q]} R}.$$

In the line of Mandal et al. (cf. [6]), one may prove that $\rho_{g,D}^{(p,q)}(f)$ (respectively $\lambda_{g,D}^{(p,q)}(f)$) is independent of the choice of the domain $D$, and therefore one can write $\rho_g^{(p,q)}(f)$ (respectively $\lambda_g^{(p,q)}(f)$) instead of $\rho_{g,D}^{(p,q)}(f)$ (respectively $\lambda_{g,D}^{(p,q)}(f)$).

An entire function $f(z)$ of $n$ complex variables for which $\rho_g^{(p,q)}(f)$ and $\lambda_g^{(p,q)}(f)$ are the same is called a function of regular $(p, q)$ relative Gol'dberg growth with respect to an entire function $g(z)$ of $n$ complex variables. Otherwise, $f(z)$ is said to be irregular $(p, q)$ relative Gol'dberg growth with respect to $g(z)$.

In this connection, one may introduce the definition of relative index-pair of an entire function with respect to another entire function (both of $n$ complex variables) which is relevant in the sequel :

**Definition 1.1.21** *(cf. [8, 9, 10]) Let $f(z)$ and $g(z)$ be any two entire functions (both of $n$ complex variables) with index-pairs $(m,q)$ and $(m,p)$ respectively where $m \geq q \geq 1$, $m \geq p \geq 1$ and $D$ be any bounded complete $n$-circular domain. Then the entire function $f(z)$ is said to have relative index-pair $(p,q)$ with respect to another entire function $g(z)$, if $b < \rho_g^{(p,q)}(f) < \infty$ and $\rho_g^{(p-1,q-1)}(f)$ is not a nonzero finite number, where $b = 1$ if $p = q = m$ and $b = 0$ for otherwise. Moreover if $0 < \rho_g^{(p,q)}(f) < \infty$, then for any positive integer $a$*

$$
\begin{cases}
\rho_g^{(p-a,q)}(f) = \infty & for \quad a < p, \\
\rho_g^{(p,q-a)}(f) = 0 & for \quad a < q, \\
\rho_g^{(p+a,q+a)}(f) = 1 & for \quad a = 1, 2, \cdots .
\end{cases}
$$

*Similarly for $0 < \lambda_g^{(p,q)}(f) < \infty$, one can easily verify that*

$$
\begin{cases}
\lambda_g^{(p-a,q)}(f) = \infty & for \quad a < p, \\
\lambda_g^{(p,q-a)}(f) = 0 & for \quad a < q, \\
\lambda_g^{(p+a,q+a)}(f) = 1 & for \quad a = 1, 2, \cdots .
\end{cases}
$$

Next we give the definition of $(p,q)$-th relative Gol'dberg type and $(p,q)$-th relative Gol'dberg lower type in order to compare the relative growth of two entire functions of $n$ complex variables having same non-zero finite $(p,q)$-th relative Gol'dberg order with respect to another entire function of $n$ complex variables.

**Definition 1.1.22** *(cf. [8, 9, 10]) Let $f(z)$ and $g(z)$ be any two entire functions of $n$ complex variables with index-pair $(m,q)$ and $(m,p)$, respectively, where $p,q,m$ are positive integers such that $m \geq q \geq 1$ and $m \geq p \geq 1$ and $D$ be any bounded complete $n$-circular domain with center at the origin in $\mathbb{C}^n$. Then the $(p,q)$-th relative Gol'dberg type and $(p,q)$-th relative Gol'dberg lower type of $f(z)$ with respect to $g(z)$ are defined as*

$$
\begin{aligned}
\sigma_{g,D}^{(p,q)}(f) \\
\overline{\sigma}_{g,D}^{(p,q)}(f)
\end{aligned}
= \lim_{R \to \infty} \begin{aligned} \sup \\ \inf \end{aligned} \frac{\log^{[p-1]} M_{g,D}^{-1}(M_{f,D}(R))}{(\log^{[q-1]} R)^{\rho_g^{(p,q)}(f)}},
$$

*where $0 < \rho_g^{(p,q)}(f) < +\infty$.*

Analogously to determine the relative growth of two entire functions of $n$ complex variables having same non-zero finite $(p,q)$-th relative Gol'dberg lower order with respect to another entire function of $n$ complex variables, one may introduce the definition of $(p,q)$-th relative Gol'dberg weak type in the following way:

**Definition 1.1.23** *(cf. [8, 9, 10]) Let $f(z)$ and $g(z)$ be any two entire functions of $n$ complex variables with index-pair $(m,q)$ and $(m,p)$, respectively, where $p,q,m$ are positive integers such that $m \geq q \geq 1$ and $m \geq p \geq 1$ and $D$ be any bounded complete $n$-circular domain with center at the origin in $\mathbb{C}^n$. Then $(p,q)$-th relative Gol'dberg weak type denoted* by $\tau_{g,D}^{(p,q)}(f)$ of an entire function $f(z)$ with respect to another entire function $g(z)$ is defined as follows:

$$\tau_{g,D}^{(p,q)}(f) = \liminf_{R \to \infty} \frac{\log^{[p-1]} M_{g,D}^{-1}(M_{f,D}(R))}{(\log^{[q-1]} R)^{\lambda_g^{(p,q)}(f)}}, \ 0 < \lambda_g^{(p,q)}(f) < \infty.$$

Similarly the $(p,q)$-th relative Gol'dberg upper weak type denoted by $\overline{\tau}_{g,D}^{(p,q)}(f)$ of an entire function $f(z)$ with respect to another entire function $g(z)$ both of $n$ complex variables in the following manner :

$$\overline{\tau}_{g,D}^{(p,q)}(f) = \limsup_{R \to \infty} \frac{\log^{[p-1]} M_{g,D}^{-1}(M_{f,D}(R))}{(\log^{[q-1]} R)^{\lambda_g^{(p,q)}(f)}}, \ 0 < \lambda_g^{(p,q)}(f) < \infty.$$

Therefore, for any two entire functions $f(z)$ and $g(z)$ both of $n$ complex variables, we note that

$$\rho_g^{(p,q)}(f) \neq \lambda_g^{(p,q)}(f), \sigma_{g,D}^{(p,q)}(f) > 0 \Rightarrow \overline{\tau}_{g,D}^{(p,q)}(f) = \infty \text{ and}$$

$$\rho_g^{(p,q)}(f) \neq \lambda_g^{(p,q)}(f), \overline{\sigma}_{g,D}^{(p,q)}(f) > 0 \Rightarrow \tau_{g,D}^{(p,q)}(f) = \infty.$$

Since Gol'dberg has shown that [2] Gol'dberg type depends on the domain $D$. Hence in general all the growth indicators define in Definition 1.1.22 and Definition 1.1.23 are also depend on $D$.

Extending the notion of $(p,q)$-th relative Gol'dberg order Biswas et al. [11, 5] introduced the following definition:

**Definition 1.1.24** (cf. [11, 5]) Let $\varphi(R) : [0, +\infty) \to (0, +\infty)$ be a non-decreasing unbounded function. Also let $f(z)$ and $g(z)$ be any two entire functions of $n$ complex variables. The $(p,q)$-$\varphi$ relative Gol'dberg order and the $(p,q)$-$\varphi$ relative Gol'dberg lower order of of $f(z)$ with respect to $g(z)$ are defined as

$$\begin{array}{l} \rho_{g,D}^{(p,q)}(f,\varphi) \\ \lambda_{g,D}^{(p,q)}(f,\varphi) \end{array} = \lim_{R \to \infty} \begin{array}{l} \sup \\ \inf \end{array} \frac{\log^{[p]} M_{g,D}^{-1}(M_{f,D}R))}{\log^{[q]} \varphi((R)}.$$

An entire function $f(z)$ of $n$ complex variables for which $\rho_{g,D}^{(p,q)}(f,\varphi)$ and $\lambda_{g,D}^{(p,q)}(f,\varphi)$ are the same is called a function of regular relative $(p,q)$-$\varphi$ Gol'dberg growth with respect to an entire function $g(z)$ of $n$ complex variables. Otherwise, $f(z)$ is said to be irregular relative $(p,q)$-$\varphi$ Gol'dberg growth.with respect to $g(z)$.

Biswas et al. [5] have already shown that $\rho_{g,D}^{(p,q)}(f,\varphi)$ and $\lambda_{g,D}^{(p,q)}(f,\varphi)$ are independent of the choice of the domain $D$ when $\varphi(R) : [0, +\infty) \to (0, +\infty)$ is a nondecreasing unbounded function and satisfies $\lim_{R \to +\infty} \frac{\log^{[q]} \varphi(\alpha R)}{\log^{[q]} \varphi(R)} = 1$ for all $\alpha > 0$.

Now, for the development of such growth indicators, one may introduce $(p,q)$-$\varphi$ relative Gol'dberg type $\sigma_{g,D}^{(p,q)}(f,\varphi)$ and $(p,q)$-$\varphi$ relative Gol'dberg weak type $\tau_{g,D}^{(p,q)}(f,\varphi)$ in the following way:

**Definition 1.1.25** *(cf. [12])Let $\varphi(R) : [0, +\infty) \to (0, +\infty)$ be a non-decreasing unbounded function. Let $f(z)$ and $g(z)$ be any two entire functions of $n$ complex variables such that $0 < \rho_{g,D}^{(p,q)}(f, \varphi) < \infty$. Then the $(p,q)$-$\varphi$ relative Gol'dberg type $\sigma_{g,D}^{(p,q)}(f, \varphi)$ and the $(p,q)$-$\varphi$ relative Gol'dberg lower type $\overline{\sigma}_{g,D}^{(p,q)}(f, \varphi)$ of $f(z)$ with respect to $g(z)$ are defined as:*

$$
\begin{matrix} \sigma_{g,D}^{(p,q)}(f, \varphi) \\ \overline{\sigma}_{g,D}^{(p,q)}(f, \varphi) \end{matrix} = \lim_{R \to \infty} \begin{matrix} \sup \\ \inf \end{matrix} \frac{\log^{[p-1]} M_{g,D}^{-1}(M_{f,D}(R))}{\left( \log^{[q-1]} \varphi(R) \right)^{\rho_{g,D}^{(p,q)}(f,\varphi)}}.
$$

**Definition 1.1.26** *(cf. [12]) Let $\varphi(R) : [0, +\infty) \to (0, +\infty)$ be a non-decreasing unbounded function. Let $f(z)$ and $g(z)$ be any two entire functions of $n$ complex variables such that $0 < \lambda_{g,D}^{(p,q)}(f, \varphi) < \infty$. Then the $(p,q)$-$\varphi$ relative Gol'dberg weak type $\tau_{g,D}^{(p,q)}(f, \varphi)$ and $(p,q)$-$\varphi$ relative Gol'dberg upper weak type $\overline{\tau}_{g,D}^{(p,q)}(f, \varphi)$ of $f(z)$ with respect to $g(z)$ are defined as:*

$$
\begin{matrix} \overline{\tau}_{g,D}^{(p,q)}(f, \varphi) \\ \tau_{g,D}^{(p,q)}(f, \varphi) \end{matrix} = \lim_{R \to \infty} \begin{matrix} \sup \\ \inf \end{matrix} \frac{\log^{[p-1]} M_{g,D}^{-1}(M_{f,D}(R))}{\left( \log^{[q-1]} \varphi(R) \right)^{\lambda_{g,D}^{(p,q)}(f,\varphi)}}.
$$

As Gol'dberg has shown that (see [2]) Gol'dberg type depends on the domain $D$, therefore in general all the growth indicators defined in Definition 1.1.25 and Definition 1.1.26 also depend on $D$.

During the past decades, several authors {cf. [1] to [22]} made closed investigations on the growth properties of entire functions of several complex variables using different growth indicator such as Gol'dberg order, $(p,q)$-th Gol'dberg order, relative Gol'dberg order, etc. In this book we wish to establish some basic growth properties of entire functions of several complex variables on the basis of their generalized relative Gol'dberg order $(\alpha, \beta)$ and generalized relative Gol'dberg type $(\alpha, \beta)$ where $\alpha$ and $\beta$ continuous non-negative functions defined on $(-\infty, +\infty)$.

Considering this, in the next section we wish to introduce the definition generalized Gol'dberg order $(\alpha, \beta)$ and generalized Gol'dberg type $(\alpha, \beta)$ of entire functions of several complex variables where $\alpha$ and $\beta$ continuous non-negative on $(-\infty, +\infty)$ functions and establish some related growth properties.

## References

[1] B. A. Fuks: Introduction to the theory of analytic functions of several complex variables, American Mathematical Soci., Providence, R. I., 1963.

[2] A. A. Gol'dberg: Elementary remarks on the formulas defining order and type of functions of several variables,Dokl. Akad. Nauk Arm. SSR, 29 (1959), 145-151 (Russian).

[3] S. K. Datta and A. R. Maji: Study of Growth properties on the basis of generalised Gol'dberg order of composite entire functions of several complex variables,International J. of Math.Sci.& Engg. Appls, 5(V) (2011), 297-311.

[4] X. Shen, J. Tu and H.Y. Xu: Complex oscillation of a second-order linear differential equation with entire coefficients of $[p, q] - \varphi$ order, Adv. Difference Equ. 2014, 2014:200, 14 pages.

[5] T. Biswas and R. Biswas: Sum and product theorems relating to (p,q)-$\varphi$ relative Gol'dberg order and (p,q)-$\varphi$ relative Gol'dberg lower order of entire functions of several variables,Uzbek Math. J., 2018(4) (2018), 160-169.

[6] B. C. Mondal and C. Roy: Relative gol'dberg order of an entire function of several variables, Bull Cal. Math. Soc., 102(4) (2010), 371-380.

[7] B. Prajapati and A. Rastogi: Some results o $p^{th}$ gol'dberg relative order, International Journal of Applied Mathematics and Statistical Sciences, 5(2) (2016), 147-154.

[8] T. Biswas: Some growth analysis of entire functions of several variables on the basis of their (p,q)-th relative Gol'dberg order and (p,q)-th relative Gol'dberg type, Palest. J. Math., 9(1) (2020), 149-158.

[9] T. Biswas: Sum and product theorems relating to relative (p,q)-th Gol'dberg order, relative (p,q)-th Gol'dberg type and relative (p,q)-th Gol'dberg weak type of entire functions of several variables, J. Interdiscip. Math., 22(1) (2019), 53-63.

[10] T. Biswas: Some results relating to (p,q)-th relative Gol'dberg order and (p,q)-relative Gol'dberg type of entire functions of several variables, J. Fract. Calc. Appl., 10(2) (2019), 249-272.

[11] T. Biswas and R. Biswas: Some growth properties of entire functions of several complex variables on the basis of their (p,q)-$\varphi$ relative Gol'dberg order and (p,q)-$\varphi$ relative Gol'dberg lower order, Electron. J. Math. Anal. Appl., 8(1) (2020), 229-236.

[12] T. Biswas and R. Biswas: Some growth estimations based on (p,q)-$\varphi$ relative Gol'dberg type and (p,q)-$\varphi$ relative Gol'dberg weak type of entire functions of several complex variables, Korean J. Math., 28(3) (2020), 489-507.

[13] D. Banerjee and S. Sarkar: A note on (p,q)$^{th}$ relative Gol'dberg order of entire functions of several variables. Bull. Allahabad Math. Soc., 34(1) (2019), 25-37.

[14] D. Banerjee and S. Sarkar: On (p,q)$^{th}$ Gol'dberg order and (p,q)$^{th}$ Gol'dberg type of an entire function of several complex variables represented by multiple dirichlet series, South East Asian J. Math. Math. Sci., 15(1) (2019), 15-24.

[15] D. Banerjee and S. Sarkar: (p,q)$^{th}$ relative Gol'dberg order of entire functions of several variables. International J. of Math. Sci. & Engg. Appls., 11(III) (2017), 185-201.

[16] T. Biswas and R. Biswas: On some (p,q)-$\varphi$ relative Gol'dberg type and (p,q)-$\varphi$ relative Gol'dberg weak type based growth properties of entire functions of several complex variables, Ital. J. Pure Appl. Math., N. 44 (2020), 403-414.

[17] S.K. Datta and A.R. Maji: Some study of the comparative growth rates on the basis of generalised relative Gol'dberg order of composite entire functions of several complex variables, International J.of Math. Sci. & Engg. Appls, 5(V) ( 2011), 335-344.

[18] S.K. Datta and A.R. Maji: Some study of the comparative growth properties on the basis of relative Gol'dberg order of composite entire functions of several complex variables, Int. J. Contemp. Math. Sci., 6(42) (2011), 2075-2082.

[19] A. Feruj: Gol'dberg order and Gol'dberg type of entire functions represented by multiple Dirichlet series, Ganit J. Bangladesh Math. Soc., 29, (2009), 63-70.

[20] C. Roy: Some properties of entire functions in one and several complex variables, Ph.D. Thesis ( 2010), University of Calcutta.

[21] P. K. Sarkar: On Gol'dberg order and Gol'dberg type of an entire function of several complex variables represented by multiple Dirichlet series, Indian J. of pure and App. Math. 13(10) (1982), 1221-1229.

[22] U. V. Singh and A. Rastogi: On Gol'dberg qth Order and Gol'dberg qth Type of an Entire Function Represented by Multiple Dirichlet Series, International Journal of Mathematics and its Applications, 3(3-D) (2015), 51-56.

# Chapter 2

# Generalized Gol'dberg order $(\alpha, \beta)$ and generalized Gol'dberg type $(\alpha, \beta)$ of entire functions of several complex variables

**Abstract:** In this chapter, first we introduce the definitions of generalized Gol'dberg order $(\alpha, \beta)$, generalized hyper Gol'dberg order $(\alpha, \beta)$ generalized logarithmic Gol'dberg order $(\alpha, \beta)$, generalized Gol'dberg type $(\alpha, \beta)$ and generalized Gol'dberg weak type $(\alpha, \beta)$ of entire functions of several complex variables and then using these growth indicators, we discuss of some related growth properties of entire functions of n complex variables, where $\alpha, \beta$ are continuous non-negative functions defined on $(-\infty, +\infty)$.

**Keywords:** Increasing function, generalized Gol'dberg order $(\alpha, \beta)$, generalized hyper Gol'dberg order $(\alpha, \beta)$, generalized logarithmic Gol'dberg order $(\alpha, \beta)$, generalized Gol'dberg type $(\alpha, \beta)$, generalized Gol'dberg weak type $(\alpha, \beta)$.

**Mathematics Subject Classification (2010) :** 32A15.

## 2.1   Introduction.

The Gol'dberg order and Gol'dberg type of an entire function $f(z)$ with respect to any bounded complete $n$-circular domain $D$ with center at all the origin $\mathbb{C}^n$ which are generally used in computational purpose are classical. Datta et al. [1] defined the concept of $(p, q)$-th Gol'dberg order of an entire function $f(z)$ for any bounded complete $n$-circular domain $D$ with center at all the origin $\mathbb{C}^n$ where $p$ and $q$ are any positive integers with $p \geq q \geq 1$. Extending this notion, here in this chapter we wish to introduce the definitions of generalized Gol'dberg order $(\alpha, \beta)$ and generalized Gol'dberg type $(\alpha, \beta)$ of an entire functions of several complex variables and establish some related growth properties of entire functions of several complex variables.

**Tanmay Biswas & Chinmay Biswas**

## 2.2   Preliminary remarks and definitions.

Throughout the book we assume $L$ be a class of continuous non-negative functions $\alpha$ defined on $(-\infty, +\infty)$ such that $\alpha(x) = \alpha(x_0) \geq 0$ for $x \leq x_0$ with $\alpha(x) \uparrow +\infty$ as $x \to +\infty$. For any $\alpha \in L$, we say that $\alpha \in L^0$, if $\alpha(cx) = (1 + o(1))\alpha(x)$ as $x_0 \leq x \to +\infty$ for each $c \in (0, +\infty)$. Clearly, $L^0 \subset L$.

Further we assume that throughout the book, unless specified later, $\alpha, \alpha_1, \alpha_2, \gamma, \beta, \beta_1$ and $\beta_2$ always denote the functions belonging to $L^0$. Now considering this, we introduce the definition of the generalized Gol'dberg order $(\alpha, \beta)$ and generalized Gol'dberg lower order $(\alpha, \beta)$ of an entire function $f(z)$ with respect to any bounded complete $n$-circular domain $D$ with center at all the origin $\mathbb{C}^n$ which are as follows:

**Definition 2.2.1** *The generalized Gol'dberg order $(\alpha, \beta)$ and generalized Gol'dberg lower order $(\alpha, \beta)$ of an entire function $f(z)$ with respect to any bounded complete n-circular domain $D$ with center at all the origin $\mathbb{C}^n$ are defined as:*

$$
\begin{array}{c}
\rho_D^{(\alpha,\beta)}[f] \\
\lambda_D^{(\alpha,\beta)}[f]
\end{array}
= \lim_{R \to \infty} \begin{array}{c} \sup \\ \inf \end{array} \frac{\alpha(M_{f,D}(R))}{\beta(R)}.
$$

Definition of $(p, q)$-th Gol'dberg order is a special case of Definition 2.2.1 for $\alpha(R) = \log^{[p]} R$ and $\beta(R) = \log^{[q]} R$.

The function $f(z)$ is said to be of regular generalized Gol'dberg $(\alpha, \beta)$ growth when generalized Gol'dberg order $(\alpha, \beta)$ and generalized Gol'dberg lower order $(\alpha, \beta)$ of $f(z)$ are the same. Functions which are not of regular generalized Gol'dberg $(\alpha, \beta)$ growth are said to be of irregular generalized Gol'dberg $(\alpha, \beta)$ growth.

Now in order to refine the growth scale namely the generalized Gol'dberg order $(\alpha, \beta)$, we introduce the definitions of another growth indicators, called generalized Gol'dberg type $(\alpha, \beta)$ and generalized Gol'dberg lower type $(\alpha, \beta)$ respectively of an entire function $f(z)$ with respect to any bounded complete $n$-circular domain $D$ with center at all the origin $\mathbb{C}^n$ which are as follows:

**Definition 2.2.2** *The generalized Gol'dberg type $(\alpha, \beta)$ and generalized Gol'dberg lower type $(\alpha, \beta)$ of an entire function $f(z)$ with respect to any bounded complete n-circular domain $D$ with center at all the origin $\mathbb{C}^n$ having finite positive generalized Gol'dberg order $(\alpha, \beta)$ $\left(0 < \rho_D^{(\alpha,\beta)}[f] < \infty\right)$ are defined as :*

$$
\begin{array}{c}
\sigma_D^{(\alpha,\beta)}[f] \\
\overline{\sigma}_D^{(\alpha,\beta)}[f]
\end{array}
= \lim_{R \to +\infty} \begin{array}{c} \sup \\ \inf \end{array} \frac{\exp(\alpha(M_{f,D}(R)))}{(\exp(\beta(r)))^{\rho_D^{(\alpha,\beta)}[f]}}.
$$

*It is obvious that $0 \leq \overline{\sigma}_D^{(\alpha,\beta)}[f] \leq \sigma_D^{(\alpha,\beta)}[f] \leq \infty$.*

Analogously to determine the relative growth of two entire functions of $n$ complex variables having same non-zero finite generalized Gol'dberg lower order $(\alpha, \beta)$, one may introduce the definition of generalized Gol'dberg weak type $(\alpha, \beta)$ and generalized

Gol'dberg upper weak type $(\alpha, \beta)$ of an entire function $f(z)$ with respect to any bounded complete $n$-circular domain $D$ with center at all the origin $\mathbb{C}^n$ having finite positive generalized Gol'dberg lower order $(\alpha, \beta)$, $\lambda_D^{(\alpha,\beta)}[f]$ in the following way:

**Definition 2.2.3** *The generalized Gol'dberg upper weak type* $(\alpha, \beta)$ *denoted by* $\overline{\tau}_D^{(\alpha,\beta)}[f]$ *and generalized Gol'dberg weak type* $(\alpha, \beta)$ *denoted by* $\tau_D^{(\alpha,\beta)}[f]$ *of an entire function* $f(z)$ *with respect to any bounded complete $n$-circular domain $D$ with center at all the origin* $\mathbb{C}^n$ *having finite positive generalized Gol'dberg lower order* $(\alpha, \beta)$ $\left(0 < \lambda_D^{(\alpha,\beta)}[f] < \infty\right)$ *are defined as :*

$$\begin{aligned}\overline{\tau}_D^{(\alpha,\beta)}[f] \\ \tau_D^{(\alpha,\beta)}[f]\end{aligned} = \lim_{R \to +\infty} \begin{array}{c}\sup \\ \inf\end{array} \frac{\exp(\alpha(M_{f,D}(R)))}{(\exp(\beta(r)))^{\lambda_D^{(\alpha,\beta)}[f]}}.$$

*It is obvious that* $0 \leq \tau_D^{(\alpha,\beta)}[f] \leq \overline{\tau}_D^{(\alpha,\beta)}[f] \leq \infty.$

**Remark 2.2.1** *As Gol'dberg has shown that (see [2]) Gol'dberg type depends on the domain $D$, so in general all the growth indicators defined in Definition 2.2.2 and Definition 2.2.3 also depend on $D$.*

Now one may give the following definitions of generalized hyper Gol'dberg order $(\alpha, \beta)$ and generalized logarithmic Gol'dberg order $(\alpha, \beta)$ of an entire function $f(z)$ with respect to any bounded complete $n$-circular domain $D$ with center at all the origin $\mathbb{C}^n$ in the following way:

**Definition 2.2.4** *The generalized hyper Gol'dberg order* $(\alpha, \beta)$ *and generalized hyper Gol'dberg lower order* $(\alpha, \beta)$ *of an entire function* $f(z)$ *with respect to any bounded complete $n$-circular domain $D$ with center at all the origin* $\mathbb{C}^n$ *are defined as:*

$$\begin{aligned}\overline{\rho}_D^{(\alpha,\beta)}[f] \\ \overline{\lambda}_D^{(\alpha,\beta)}[f]\end{aligned} = \lim_{R \to \infty} \begin{array}{c}\sup \\ \inf\end{array} \frac{\alpha(\log(M_{f,D}(R)))}{\beta(R)}.$$

**Definition 2.2.5** *The generalized logarithmic Gol'dberg order* $(\alpha, \beta)$ *and generalized logarithmic Gol'dberg lower order* $(\alpha, \beta)$ *of an entire function* $f(z)$ *with respect to any bounded complete $n$-circular domain $D$ with center at all the origin* $\mathbb{C}^n$ *are defined as:*

$$\begin{aligned}\underline{\rho}_D^{(\alpha,\beta)}[f] \\ \underline{\lambda}_D^{(\alpha,\beta)}[f]\end{aligned} = \lim_{R \to \infty} \begin{array}{c}\sup \\ \inf\end{array} \frac{\alpha(M_{f,D}(R))}{\beta(\log R)}.$$

## 2.3   Main Results.

In this section we state the main results of this chapter.

**Theorem 2.3.1** *Let $f(z)$ be any entire function of $n$ complex variables. Then generalized Gol'dberg order $(\alpha, \beta)$ and generalized Gol'dberg lower order $(\alpha, \beta)$ of $f(z)$ are independent of the choice of the domain $D$.*

**Proof.** Let us consider $D_1$ and $D_2$ ne any two bounded complete $n$-circular domains. Then there exist two real numbers $a$, $b > 0$ such that $aD_1 \subset D_2 \subset bD_1$ and therefore

$$M_{f,aD_1}(R) \leq M_{f,D_2}(R) \leq M_{f,bD_1}(R). \tag{1}$$

Now for any $c > 0$ and any $D$, we get that

$$M_{f,cD}(R) = M_{f,D}(cR).$$

Therefore

$$\limsup_{R \to \infty} \frac{\alpha(M_{f,cD}(R))}{\beta(R)} = \limsup_{R \to \infty} \frac{\alpha(M_{f,D}(cR))}{\beta(R)} = \limsup_{\frac{R}{c} \to \infty} \frac{\alpha(M_{f,D}(R))}{\beta\left(\frac{R}{c}\right)}$$

$$= \limsup_{\frac{R}{c} \to \infty} \frac{\alpha(M_{f,D}(R))}{\beta(R)} \cdot \lim_{R \to \infty} \frac{\beta(cR)}{\beta(R)}$$

$$= \limsup_{R \to \infty} \frac{\alpha(M_{f,D}(R))}{\beta(R)}.$$

Hence by (1) we obtain that

$$\limsup_{R \to \infty} \frac{\alpha(M_{f,D_1}(R))}{\beta(R)} = \limsup_{R \to \infty} \frac{\alpha(M_{f,aD_1}(R))}{\beta(R)}$$

$$\leq \limsup_{R \to \infty} \frac{\alpha(M_{f,D_2}(R))}{\beta(R)}$$

$$\leq \limsup_{R \to \infty} \frac{\alpha(M_{f,bD_1}(R))}{\beta(R)}$$

$$\leq \limsup_{R \to \infty} \frac{\alpha(M_{f,D_1}(R))}{\beta(R)}.$$

Thus

$$\limsup_{R \to \infty} \frac{\alpha(M_{f,D_1}(R))}{\beta(R)} = \limsup_{R \to \infty} \frac{\alpha(M_{f,D_2}(R))}{\beta(R)}.$$

Similarly one can easily verify that

$$\liminf_{R \to \infty} \frac{\alpha(M_{f,D_1}(R))}{\beta(R)} = \liminf_{R \to \infty} \frac{\alpha(M_{f,D_2}(R))}{\beta(R)}.$$

Hence the theorem follows. ∎

Since $\rho_D^{(\alpha,\beta)}[f]$ and $\lambda_D^{(\alpha,\beta)}[f]$ are independent of the choice of the domain $D$, so after this we shall always write $\rho^{(\alpha,\beta)}[f]$ and $\lambda^{(\alpha,\beta)}[f]$ instead of $\rho_D^{(\alpha,\beta)}[f]$ and $\lambda_D^{(\alpha,\beta)}[f]$ respectively.

**Remark 2.3.1** *In the line of Theorem 2.3.2, one can easily verify that generalized hyper Gol'dberg order $(\alpha,\beta)$ (respectively generalized hyper lower Gol'dberg order $(\alpha,\beta)$) and generalized logarithmic Gol'dberg order $(\alpha,\beta)$ (respectively generalized logarithmic lower*

*Gol'dberg order $(\alpha, \beta))$ are independent of the choice of the domain. So after this we shall always write, $\overline{\rho}^{(\alpha,\beta)}[f]$ $\left(\overline{\lambda}^{(\alpha,\beta)}[f]\right)$ and $\underline{\rho}^{(\alpha,\beta)}[f]$ $\left(\underline{\lambda}^{(\alpha,\beta)}[f]\right)$ instead of $\overline{\rho}_D^{(\alpha,\beta)}[f]$ $\left(\overline{\lambda}_D^{(\alpha,\beta)}[f]\right)$ and $\underline{\rho}_D^{(\alpha,\beta)}[f]$ $\left(\underline{\lambda}_D^{(\alpha,\beta)}[f]\right)$ respectively.*

**Theorem 2.3.2** *Let $f(z)$ and $g(z)$ be any two entire functions of $n$ complex variables. Also let $0 < \lambda^{(\alpha_1,\beta_1)}[f] \le \rho^{(\alpha_1,\beta_1)}[f] < \infty$ and $0 < \lambda^{(\alpha_2,\beta_2)}[g] \le \rho^{(\alpha_2,\beta_2)}[g] < \infty$. Then*

$$\frac{\lambda^{(\alpha_1,\beta_1)}[f]}{\rho^{(\alpha_2,\beta_2)}[g]} \le \liminf_{R \to \infty} \frac{\alpha_1(M_{f,D}(R))}{\alpha_2(M_{g,D}(\beta_2^{-1}(\beta_1(R))))} \le \frac{\lambda^{(\alpha_1,\beta_1)}[f]}{\lambda^{(\alpha_2,\beta_2)}[g]} \le$$

$$\limsup_{R \to \infty} \frac{\alpha_1(M_{f,D}(R))}{\alpha_2(M_{g,D}(\beta_2^{-1}(\beta_1(R))))} \le \frac{\rho^{(\alpha_1,\beta_1)}[f]}{\lambda^{(\alpha_2,\beta_2)}[g]}.$$

**Proof.** From the definition of $\rho^{(\alpha,\beta)}[f]$ and $\lambda^{(\alpha,\beta)}[f]$ we have for arbitrary positive $\varepsilon$ and for all large values of $R$,

$$\alpha_1(M_{f,D}(R)) \geqslant \left(\lambda^{(\alpha_1,\beta_1)}[f] - \varepsilon\right)\beta_1(R) \tag{2}$$

and

$$\alpha_2(M_{g,D}(\beta_2^{-1}(\beta_1(R)))) \le \left(\rho^{(\alpha_2,\beta_2)}[g] + \varepsilon\right)\beta_1(R). \tag{3}$$

Now from (2) and (3) it follows for all large values of $R$,

$$\frac{\alpha_1(M_{f,D}(R))}{\alpha_2(M_{g,D}(\beta_2^{-1}(\beta_1(R))))} \geqslant \frac{\lambda^{(\alpha_1,\beta_1)}[f] - \varepsilon}{\rho^{(\alpha_2,\beta_2)}[g] + \varepsilon}.$$

As $\varepsilon(> 0)$ is arbitrary, we obtain that

$$\liminf_{r \to \infty} \frac{\alpha_1(M_{f,D}(R))}{\alpha_2(M_{g,D}(\beta_2^{-1}(\beta_1(R))))} \geqslant \frac{\lambda^{(\alpha_1,\beta_1)}[f]}{\rho^{(\alpha_2,\beta_2)}[g]}. \tag{4}$$

Again for a sequence of values of $R$ tending to infinity,

$$\alpha_1(M_{f,D}(R)) \le \left(\lambda^{(\alpha_1,\beta_1)}[f] + \varepsilon\right)\beta_1(R) \tag{5}$$

and for all large values of $R$,

$$\alpha_2(M_{g,D}(\beta_2^{-1}(\beta_1(R)))) \geqslant \left(\lambda^{(\alpha_2,\beta_2)}[g] - \varepsilon\right)\beta_1(R). \tag{6}$$

Combining (5) and (6) we get for a sequence of values of $R$ tending to infinity,

$$\frac{\alpha_1(M_{f,D}(R))}{\alpha_2(M_{g,D}(\beta_2^{-1}(\beta_1(R))))} \le \frac{\lambda^{(\alpha_1,\beta_1)}[f] + \varepsilon}{\lambda^{(\alpha_2,\beta_2)}[g] - \varepsilon}.$$

Since $\varepsilon(> 0)$ is arbitrary it follows that

$$\liminf_{R \to \infty} \frac{\alpha_1(M_{f,D}(R))}{\alpha_2(M_{g,D}(\beta_2^{-1}(\beta_1(R))))} \le \frac{\lambda^{(\alpha_1,\beta_1)}[f]}{\lambda^{(\alpha_2,\beta_2)}[g]}. \tag{7}$$

Also for a sequence of values of $R$ tending to infinity ,

$$\alpha_2(M_{g,D}\left(\beta_2^{-1}(\beta_1(R))\right)) \leq \left(\lambda^{(\alpha_2,\beta_2)}[g] + \varepsilon\right)\beta_1(R).\qquad(8)$$

Now from (2) and (8) we obtain for a sequence of values of $R$ tending to infinity ,

$$\frac{\alpha_1(M_{f,D}(R))}{\alpha_2(M_{g,D}\left(\beta_2^{-1}(\beta_1(R))\right))} \geq \frac{\lambda^{(\alpha_1,\beta_1)}[f] - \varepsilon}{\lambda^{(\alpha_2,\beta_2)}[g] + \varepsilon}.$$

As $\varepsilon\,(>0)$ is arbitrary, we get from above that

$$\limsup_{R\to\infty}\frac{\alpha_1(M_{f,D}(R))}{\alpha_2(M_{g,D}\left(\beta_2^{-1}(\beta_1(R))\right))} \geq \frac{\lambda^{(\alpha_1,\beta_1)}[f]}{\lambda^{(\alpha_2,\beta_2)}[g]}.\qquad(9)$$

Also for all large values of $R$ ,

$$\alpha_1(M_{f,D}(R)) \leq \left(\rho^{(\alpha_1,\beta_1)}[f] + \varepsilon\right)\beta_1(R).\qquad(10)$$

So from (6) and (10) it follows for all large values of $R$,

$$\frac{\alpha_1(M_{f,D}(R))}{\alpha_2(M_{g,D}\left(\beta_2^{-1}(\beta_1(R))\right))} \leq \frac{\rho^{(\alpha_1,\beta_1)}[f] + \varepsilon}{\lambda^{(\alpha_2,\beta_2)}[g] - \varepsilon}.$$

Since $\varepsilon\,(>0)$ is arbitrary we obtain that

$$\limsup_{R\to\infty}\frac{\alpha_1(M_{f,D}(R))}{\alpha_2(M_{g,D}\left(\beta_2^{-1}(\beta_1(R))\right))} \leq \frac{\rho^{(\alpha_1,\beta_1)}[f]}{\lambda^{(\alpha_2,\beta_2)}[g]}.\qquad(11)$$

Thus the theorem follows from $(4),(7),(9)$ and $(11)$ . ∎

**Theorem 2.3.3** *Let $f(z)$ and $g(z)$ be any two entire functions of $n$ complex variables. Also let $0 < \lambda^{(\alpha_1,\beta_1)}[f] \leq \rho^{(\alpha_1,\beta_1)}[f] < \infty$ and $0 < \rho^{(\alpha_2,\beta_2)}[g] < \infty$. Then*

$$\liminf_{R\to\infty}\frac{\alpha_1(M_{f,D}(R))}{\alpha_2(M_{g,D}\left(\beta_2^{-1}(\beta_1(R))\right))} \leq \frac{\rho^{(\alpha_1,\beta_1)}[f]}{\rho^{(\alpha_2,\beta_2)}[g]} \leq \limsup_{R\to\infty}\frac{\alpha_1(M_{f,D}(R))}{\alpha_2(M_{g,D}\left(\beta_2^{-1}(\beta_1(R))\right))}.$$

**Proof.** From the definition of $\rho^{(\alpha,\beta)}[f]$ we get for a sequence of values of $R$ tending to infinity ,

$$\alpha_2(M_{g,D}\left(\beta_2^{-1}(\beta_1(R))\right)) \geq \left(\rho^{(\alpha_2,\beta_2)}[g] - \varepsilon\right)\beta_1(R).\qquad(12)$$

Now from (10) and (12) it follows for a sequence of values of $R$ tending to infinity ,

$$\frac{\alpha_1(M_{f,D}(R))}{\alpha_2(M_{g,D}\left(\beta_2^{-1}(\beta_1(R))\right))} \leq \frac{\rho^{(\alpha_1,\beta_1)}[f] + \varepsilon}{\rho^{(\alpha_2,\beta_2)}[g] - \varepsilon}.$$

As $\varepsilon\,(>0)$ is arbitrary we obtain that

$$\liminf_{R\to\infty}\frac{\alpha_1(M_{f,D}(R))}{\alpha_2(M_{g,D}\left(\beta_2^{-1}(\beta_1(R))\right))} \leq \frac{\rho^{(\alpha_1,\beta_1)}[f]}{\rho^{(\alpha_2,\beta_2)}[g]}.\qquad(13)$$

Again for a sequence of values of $R$ tending to infinity ,

$$\alpha_1(M_{f,D}(R)) \geqslant \left(\rho^{(\alpha_1,\beta_1)}[f] - \varepsilon\right)\beta_1(R). \tag{14}$$

So combining (3) and (14) we get for a sequence of values of $R$ tending to infinity ,

$$\frac{\alpha_1(M_{f,D}(R))}{\alpha_2(M_{g,D}\left(\beta_2^{-1}(\beta_1(R))\right))} \geqslant \frac{\rho^{(\alpha_1,\beta_1)}[f] - \varepsilon}{\rho^{(\alpha_2,\beta_2)}[g] + \varepsilon}.$$

Since $\varepsilon\,(>0)$ is arbitrary it follows that

$$\limsup_{R\to\infty}\frac{\alpha_1(M_{f,D}(R))}{\alpha_2(M_{g,D}\left(\beta_2^{-1}(\beta_1(R))\right))} \geqslant \frac{\rho^{(\alpha_1,\beta_1)}[f]}{\rho^{(\alpha_2,\beta_2)}[g]}. \tag{15}$$

Thus the theorem follows from (13) and (15) . ∎

The following theorem is a natural consequence of Theorem 2.3.2 and Theorem 2.3.3.

**Theorem 2.3.4** *Let $f(z)$ and $g(z)$ be any two entire functions of $n$ complex variables. Also let $0 < \lambda^{(\alpha_1,\beta_1)}[f] \leq \rho^{(\alpha_1,\beta_1)}[f] < \infty$ and $0 < \lambda^{(\alpha_2,\beta_2)}[g] \leq \rho^{(\alpha_2,\beta_2)}[g] < \infty$. Then*

$$\liminf_{R\to\infty}\frac{\alpha_1(M_{f,D}(R))}{\alpha_2(M_{g,D}\left(\beta_2^{-1}(\beta_1(R))\right))} \leq \min\left\{\frac{\lambda^{(\alpha_1,\beta_1)}[f]}{\lambda^{(\alpha_2,\beta_2)}[g]}, \frac{\rho^{(\alpha_1,\beta_1)}[f]}{\rho^{(\alpha_2,\beta_2)}[g]}\right\} \leq$$

$$\max\left\{\frac{\lambda^{(\alpha_1,\beta_1)}[f]}{\lambda^{(\alpha_2,\beta_2)}[g]}, \frac{\rho^{(\alpha_1,\beta_1)}[f]}{\rho^{(\alpha_2,\beta_2)}[g]}\right\} \leq \limsup_{R\to\infty}\frac{\alpha_1(M_{f,D}(R))}{\alpha_2(M_{g,D}\left(\beta_2^{-1}(\beta_1(R))\right))}.$$

The proof is omitted.

We may now state the following three theorems without proof based on generalized hyper Gol'dberg order $(\alpha, \beta)$ of an entire function of $n$ complex variables.

**Theorem 2.3.5** *Let $f(z)$ and $g(z)$ be any two entire functions of $n$ complex variables. Also let $0 < \overline{\lambda}^{(\alpha_1,\beta_1)}[f] \leq \overline{\rho}^{(\alpha_1,\beta_1)}[f] < \infty$ and $0 < \overline{\lambda}^{(\alpha_2,\beta_2)}[g] \leq \overline{\rho}^{(\alpha_2,\beta_2)}[g] < \infty$. Then*

$$\frac{\overline{\lambda}^{(\alpha_1,\beta_1)}[f]}{\overline{\rho}^{(\alpha_2,\beta_2)}[g]} \leq \liminf_{R\to\infty}\frac{\alpha_1(\log(M_{f,D}(R)))}{\alpha_2(\log(M_{g,D}\left(\beta_2^{-1}(\beta_1(R))\right)))} \leq \frac{\overline{\lambda}^{(\alpha_1,\beta_1)}[f]}{\overline{\lambda}^{(\alpha_2,\beta_2)}[g]} \leq$$

$$\limsup_{R\to\infty}\frac{\alpha_1(\log(M_{f,D}(R)))}{\alpha_2(\log(M_{g,D}\left(\beta_2^{-1}(\beta_1(R))\right)))} \leq \frac{\overline{\rho}^{(\alpha_1,\beta_1)}[f]}{\overline{\lambda}^{(\alpha_2,\beta_2)}[g]}.$$

**Theorem 2.3.6** *Let $f(z)$ and $g(z)$ be any two entire functions of $n$ complex variables. Also let $0 < \overline{\lambda}^{(\alpha_1,\beta_1)}[f] \leq \overline{\rho}^{(\alpha_1,\beta_1)}[f] < \infty$ and $0 < \overline{\rho}^{(\alpha_2,\beta_2)}[g] < \infty$. Then*

$$\liminf_{R\to\infty}\frac{\alpha_1(\log(M_{f,D}(R)))}{\alpha_2(\log(M_{g,D}\left(\beta_2^{-1}(\beta_1(R))\right)))} \leq \frac{\overline{\rho}^{(\alpha_1,\beta_1)}[f]}{\overline{\rho}^{(\alpha_2,\beta_2)}[g]} \leq \limsup_{R\to\infty}\frac{\alpha_1(\log(M_{f,D}(R)))}{\alpha_2(\log(M_{g,D}\left(\beta_2^{-1}(\beta_1(R))\right)))}.$$

**Theorem 2.3.7** *Let $f(z)$ and $g(z)$ be any two entire functions of $n$ complex variables. Also let $0 < \overline{\lambda}^{(\alpha_1,\beta_1)}[f] \leq \overline{\rho}^{(\alpha_1,\beta_1)}[f] < \infty$ and $0 < \overline{\lambda}^{(\alpha_2,\beta_2)}[g] \leq \overline{\rho}^{(\alpha_2,\beta_2)}[g] < \infty$. Then*

$$\liminf_{R \to \infty} \frac{\alpha_1(\log(M_{f,D}(R)))}{\alpha_2(\log(M_{g,D}(\beta_2^{-1}(\beta_1(R)))))} \leq \min \left\{ \frac{\overline{\lambda}^{(\alpha_1,\beta_1)}[f]}{\overline{\lambda}^{(\alpha_2,\beta_2)}[g]}, \frac{\overline{\rho}^{(\alpha_1,\beta_1)}[f]}{\overline{\rho}^{(\alpha_2,\beta_2)}[g]} \right\} \leq$$

$$\max \left\{ \frac{\overline{\lambda}^{(\alpha_1,\beta_1)}[f]}{\overline{\lambda}^{(\alpha_2,\beta_2)}[g]}, \frac{\overline{\rho}^{(\alpha_1,\beta_1)}[f]}{\overline{\rho}^{(\alpha_2,\beta_2)}[g]} \right\} \leq \limsup_{R \to \infty} \frac{\alpha_1(\log(M_{f,D}(R)))}{\alpha_2(\log(M_{g,D}(\beta_2^{-1}(\beta_1(R)))))}.$$

Using the concept of generalized logarithmic Gol'dberg order $(\alpha,\beta)$ of an entire function of $n$ complex variables, one may prove the following theorems. We omit the details.

**Theorem 2.3.8** *Let $f(z)$ and $g(z)$ be any two entire functions of $n$ complex variables. Also let $0 < \underline{\lambda}^{(\alpha_1,\beta_1)}[f] \leq \underline{\rho}^{(\alpha_1,\beta_1)}[f] < \infty$ and $0 < \underline{\lambda}^{(\alpha_2,\beta_2)}[g] \leq \underline{\rho}^{(\alpha_2,\beta_2)}[g] < \infty$. Then*

$$\frac{\underline{\lambda}^{(\alpha_1,\beta_1)}[f]}{\underline{\rho}^{(\alpha_2,\beta_2)}[g]} \leq \liminf_{R \to \infty} \frac{\alpha_1(M_{f,D}(R))}{\alpha_2(M_{g,D}(\exp(\beta_2^{-1}(\beta_1(\log R)))))} \leq \frac{\underline{\lambda}^{(\alpha_1,\beta_1)}[f]}{\underline{\lambda}^{(\alpha_2,\beta_2)}[g]} \leq$$

$$\limsup_{R \to \infty} \frac{\alpha_1(M_{f,D}(R))}{\alpha_2(M_{g,D}(\exp(\beta_2^{-1}(\beta_1(\log R)))))} \leq \frac{\underline{\rho}^{(\alpha_1,\beta_1)}[f]}{\underline{\lambda}^{(\alpha_2,\beta_2)}[g]}.$$

**Theorem 2.3.9** *Let $f(z)$ and $g(z)$ be any two entire functions of $n$ complex variables. Also let $0 < \underline{\lambda}^{(\alpha_1,\beta_1)}[f] \leq \underline{\rho}^{(\alpha_1,\beta_1)}[f] < \infty$ and $0 < \underline{\rho}^{(\alpha_2,\beta_2)}[g] < \infty$. Then*

$$\liminf_{R \to \infty} \frac{\alpha_1(M_{f,D}(R))}{\alpha_2(M_{g,D}(\exp(\beta_2^{-1}(\beta_1(\log R)))))} \leq \frac{\underline{\rho}^{(\alpha_1,\beta_1)}[f]}{\underline{\rho}^{(\alpha_2,\beta_2)}[g]} \leq \limsup_{R \to \infty} \frac{\alpha_1(M_{f,D}(R))}{\alpha_2(M_{g,D}(\exp(\beta_2^{-1}(\beta_1(\log R)))))}.$$

**Theorem 2.3.10** *Let $f(z)$ and $g(z)$ be any two entire functions of $n$ complex variables. Also let $0 < \underline{\lambda}^{(\alpha_1,\beta_1)}[f] \leq \underline{\rho}^{(\alpha_1,\beta_1)}[f] < \infty$ and $0 < \underline{\lambda}^{(\alpha_2,\beta_2)}[g] \leq \underline{\rho}^{(\alpha_2,\beta_2)}[g] < \infty$. Then*

$$\liminf_{R \to \infty} \frac{\alpha_1(M_{f,D}(R))}{\alpha_2(M_{g,D}(\exp(\beta_2^{-1}(\beta_1(\log R)))))} \leq \min \left\{ \frac{\underline{\lambda}^{(\alpha_1,\beta_1)}[f]}{\underline{\lambda}^{(\alpha_2,\beta_2)}[g]}, \frac{\underline{\rho}^{(\alpha_1,\beta_1)}[f]}{\underline{\rho}^{(\alpha_2,\beta_2)}[g]} \right\} \leq$$

$$\max \left\{ \frac{\underline{\lambda}^{(\alpha_1,\beta_1)}[f]}{\underline{\lambda}^{(\alpha_2,\beta_2)}[g]}, \frac{\underline{\rho}^{(\alpha_1,\beta_1)}[f]}{\underline{\rho}^{(\alpha_2,\beta_2)}[g]} \right\} \leq \limsup_{R \to \infty} \frac{\alpha_1(M_{f,D}(R))}{\alpha_2(M_{g,D}(\exp(\beta_2^{-1}(\beta_1(\log R)))))}.$$

**Theorem 2.3.11** *Let $f(z)$ and $g(z)$ be any two entire functions of $n$ complex variables. Also let $0 < \overline{\sigma}_D^{(\alpha_1,\beta_1)}[f] \leq \sigma_D^{(\alpha_1,\beta_1)}[f] < \infty$, $0 < \overline{\sigma}_D^{(\alpha_2,\beta_2)}[g] \leq \sigma_D^{(\alpha_2,\beta_2)}[g] < \infty$ and $\rho^{(\alpha_1,\beta_1)}[f] = \rho^{(\alpha_2,\beta_2)}[g]$. Then*

$$\frac{\overline{\sigma}_D^{(\alpha_1,\beta_1)}[f]}{\sigma_D^{(\alpha_2,\beta_2)}[g]} \leq \liminf_{R \to \infty} \frac{\exp(\alpha_1(M_{f,D}(R)))}{\exp(\alpha_2(M_{g,D}(\beta_2^{-1}(\beta_1(R)))))} \leq \frac{\overline{\sigma}_D^{(\alpha_1,\beta_1)}[f]}{\overline{\sigma}_D^{(\alpha_2,\beta_2)}[g]}$$

$$\leq \limsup_{R \to \infty} \frac{\exp(\alpha_1(M_{f,D}(R)))}{\exp(\alpha_2(M_{g,D}(\beta_2^{-1}(\beta_1(R)))))} \leq \frac{\sigma_D^{(\alpha_1,\beta_1)}[f]}{\overline{\sigma}_D^{(\alpha_2,\beta_2)}[g]}.$$

**Proof.** From the definition of $\sigma_D^{(\alpha_2,\beta_2)}[g]$ and $\overline{\sigma}_D^{(\alpha_1,\beta_1)}[f]$, we have for arbitrary positive $\varepsilon$ and for all large values of $R$ that

$$\exp(\alpha_1(M_{f,D}(R))) \geqslant \left(\overline{\sigma}_D^{(\alpha_1,\beta_1)}[f] - \varepsilon\right)(\exp \beta_1(R))^{\rho^{(\alpha_1,\beta_1)}[f]} \tag{16}$$

and

$$\exp(\alpha_2(M_{g,D}\left(\beta_2^{-1}(\beta_1(R))\right))) \leq \left(\sigma_D^{(\alpha_2,\beta_2)}[g] + \varepsilon\right)(\exp \beta_1(R))^{\rho^{(\alpha_2,\beta_2)}[g]}. \tag{17}$$

Now from (16), (17) and the condition $\rho^{(\alpha_1,\beta_1)}[f] = \rho^{(\alpha_2,\beta_2)}[g]$, it follows for all large values of $R$ that

$$\frac{\exp(\alpha_1(M_{f,D}(R)))}{\exp(\alpha_2(M_{g,D}\left(\beta_2^{-1}(\beta_1(R))\right)))} \geqslant \frac{\overline{\sigma}_D^{(\alpha_1,\beta_1)}[f] - \varepsilon}{\sigma_D^{(\alpha_2,\beta_2)}[g] + \varepsilon}.$$

As $\varepsilon (> 0)$ is arbitrary , we obtain that

$$\liminf_{R \to \infty} \frac{\exp(\alpha_1(M_{f,D}(R)))}{\exp(\alpha_2(M_{g,D}\left(\beta_2^{-1}(\beta_1(R))\right)))} \geqslant \frac{\overline{\sigma}_D^{(\alpha_1,\beta_1)}[f]}{\sigma_D^{(\alpha_2,\beta_2)}[g]}. \tag{18}$$

Again for a sequence of values of $R$ tending to infinity ,

$$\exp(\alpha_1(M_{f,D}(R))) \leq \left(\overline{\sigma}_D^{(\alpha_1,\beta_1)}[f] + \varepsilon\right)(\exp \beta_1(R))^{\rho^{(\alpha_1,\beta_1)}[f]} \tag{19}$$

and for all sufficiently large values of $R$ ,

$$\exp(\alpha_2(M_{g,D}\left(\beta_2^{-1}(\beta_1(R))\right))) \geqslant \left(\overline{\sigma}_D^{(\alpha_2,\beta_2)}[g] - \varepsilon\right)(\exp \beta_1(R))^{\rho^{(\alpha_2,\beta_2)}[g]}. \tag{20}$$

Combining the condition $\rho^{(\alpha_1,\beta_1)}[f] = \rho^{(\alpha_2,\beta_2)}[g]$, (19) and (20) we get for a sequence of values of $R$ tending to infinity that

$$\frac{\exp(\alpha_1(M_{f,D}(R)))}{\exp(\alpha_2(M_{g,D}\left(\beta_2^{-1}(\beta_1(R))\right)))} \leq \frac{\overline{\sigma}_D^{(\alpha_1,\beta_1)}[f] + \varepsilon}{\overline{\sigma}_D^{(\alpha_2,\beta_2)}[g] - \varepsilon}.$$

Since $\varepsilon (> 0)$ is arbitrary, it follows that

$$\liminf_{R \to \infty} \frac{\exp(\alpha_1(M_{f,D}(R)))}{\exp(\alpha_2(M_{g,D}\left(\beta_2^{-1}(\beta_1(R))\right)))} \leq \frac{\overline{\sigma}_D^{(\alpha_1,\beta_1)}[f]}{\overline{\sigma}_D^{(\alpha_2,\beta_2)}[g]}. \tag{21}$$

Also for a sequence of values of $R$ tending to infinity that

$$\exp(\alpha_2(M_{g,D}\left(\beta_2^{-1}(\beta_1(R))\right))) \leq \left(\overline{\sigma}_D^{(\alpha_2,\beta_2)}[g] + \varepsilon\right)(\exp \beta_1(R))^{\rho^{(\alpha_2,\beta_2)}[g]}. \tag{22}$$

Now from (16), (22) and the condition $\rho^{(\alpha_1,\beta_1)}[f] = \rho^{(\alpha_2,\beta_2)}[g]$, we obtain for a sequence of values of $R$ tending to infinity that

$$\frac{\exp(\alpha_1(M_{f,D}(R)))}{\exp(\alpha_2(M_{g,D}\left(\beta_2^{-1}(\beta_1(R))\right)))} \geqslant \frac{\overline{\sigma}_D^{(\alpha_1,\beta_1)}[f] - \varepsilon}{\overline{\sigma}_D^{(\alpha_2,\beta_2)}[g] + \varepsilon}.$$

As $\varepsilon\,(>0)$ is arbitrary, we get from above that

$$\limsup_{R\to\infty}\frac{\exp(\alpha_1(M_{f,D}\,(R)))}{\exp(\alpha_2(M_{g,D}\,(\beta_2^{-1}(\beta_1(R)))))}\geq\frac{\overline{\sigma}_D^{(\alpha_1,\beta_1)}\,[f]}{\overline{\sigma}_D^{(\alpha_2,\beta_2)}\,[g]}. \qquad (23)$$

Also for all sufficiently large values of $R$ ,

$$\exp(\alpha_1(M_{f,D}\,(R)))\leq\left(\sigma_D^{(\alpha_1,\beta_1)}\,[f]+\varepsilon\right)(\exp\beta_1(R))^{\rho^{(\alpha_1,\beta_1)}[f]}. \qquad (24)$$

As the condition $\rho^{(\alpha_1,\beta_1)}\,[f]=\rho^{(\alpha_2,\beta_2)}\,[g]$ , it follows from (20) and (24) for all large values of $R$ that

$$\frac{\exp(\alpha_1(M_{f,D}\,(R)))}{\exp(\alpha_2(M_{g,D}\,(\beta_2^{-1}(\beta_1(R)))))}\leq\frac{\sigma_D^{(\alpha_1,\beta_1)}\,[f]+\varepsilon}{\overline{\sigma}_D^{(\alpha_2,\beta_2)}\,[g]-\varepsilon}.$$

Since $\varepsilon\,(>0)$ is arbitrary, we obtain that

$$\limsup_{R\to\infty}\frac{\exp(\alpha_1(M_{f,D}\,(R)))}{\exp(\alpha_2(M_{g,D}\,(\beta_2^{-1}(\beta_1(R)))))}\leq\frac{\sigma_D^{(\alpha_1,\beta_1)}\,[f]}{\overline{\sigma}_D^{(\alpha_2,\beta_2)}\,[g]}. \qquad (25)$$

Thus the theorem follows from $(18),(21),(23)$ and $(25)$ . ∎

**Theorem 2.3.12** *Let $f\,(z)$ and $g\,(z)$ be any two entire functions of $n$ complex variables. Also let $0<\sigma_D^{(\alpha_1,\beta_1)}\,[f]<\infty,\;0<\sigma_D^{(\alpha_2,\beta_2)}\,[g]<\infty$ and $\rho^{(\alpha_1,\beta_1)}\,[f]=\rho^{(\alpha_2,\beta_2)}\,[g]$. Then*

$$\liminf_{R\to\infty}\frac{\exp(\alpha_1(M_{f,D}\,(R)))}{\exp(\alpha_2(M_{g,D}\,(\beta_2^{-1}(\beta_1(R)))))}\leq\frac{\sigma_D^{(\alpha_1,\beta_1)}\,[f]}{\sigma_D^{(\alpha_2,\beta_2)}\,[g]}\leq\limsup_{R\to\infty}\frac{\exp(\alpha_1(M_{f,D}\,(R)))}{\exp(\alpha_2(M_{g,D}\,(\beta_2^{-1}(\beta_1(R)))))}.$$

**Proof.** From the definition of $\sigma_D^{(\alpha_2,\beta_2)}\,[g]$, we get for a sequence of values of $R$ tending to infinity that

$$\exp(\alpha_2(M_{g,D}\,(\beta_2^{-1}(\beta_1(R)))))\geqslant\left(\sigma_D^{(\alpha_2,\beta_2)}\,[g]-\varepsilon\right)(\exp\beta_1(R))^{\rho^{(\alpha_2,\beta_2)}[g]}. \qquad (26)$$

Now from (24), (26) and the condition $\rho^{(\alpha_1,\beta_1)}\,[f]=\rho^{(\alpha_2,\beta_2)}\,[g]$ , it follows for a sequence of values of $R$ tending to infinity that

$$\frac{\exp(\alpha_1(M_{f,D}\,(R)))}{\exp(\alpha_2(M_{g,D}\,(\beta_2^{-1}(\beta_1(R)))))}\leq\frac{\sigma_D^{(\alpha_1,\beta_1)}\,[f]+\varepsilon}{\sigma_D^{(\alpha_2,\beta_2)}\,[g]-\varepsilon}.$$

As $\varepsilon\,(>0)$ is arbitrary, we obtain that

$$\liminf_{R\to\infty}\frac{\exp(\alpha_1(M_{f,D}\,(R)))}{\exp(\alpha_2(M_{g,D}\,(\beta_2^{-1}(\beta_1(R)))))}\leq\frac{\sigma_D^{(\alpha_1,\beta_1)}\,[f]}{\sigma_D^{(\alpha_2,\beta_2)}\,[g]}. \qquad (27)$$

Again for a sequence of values of $R$ tending to infinity that

$$\exp(\alpha_1(M_{f,D}\,(R)))\geqslant\left(\sigma_D^{(\alpha_1,\beta_1)}\,[f]-\varepsilon\right)(\exp\beta_1(R))^{\rho^{(\alpha_1,\beta_1)}[f]}. \qquad (28)$$

So combining the condition $\rho^{(\alpha_1,\beta_1)}[f] = \rho^{(\alpha_2,\beta_2)}[g]$, (17) and (28) we get for a sequence of values of $R$ tending to infinity,

$$\frac{\exp(\alpha_1(M_{f,D}(R)))}{\exp(\alpha_2(M_{g,D}(\beta_2^{-1}(\beta_1(R)))))} \geqslant \frac{\sigma_D^{(\alpha_1,\beta_1)}[f] - \varepsilon}{\sigma_D^{(\alpha_2,\beta_2)}[g] + \varepsilon}.$$

Since $\varepsilon (> 0)$ is arbitrary, it follows that

$$\limsup_{R\to\infty}\frac{\exp(\alpha_1(M_{f,D}(R)))}{\exp(\alpha_2(M_{g,D}(\beta_2^{-1}(\beta_1(R)))))} \geqslant \frac{\sigma_D^{(\alpha_1,\beta_1)}[f]}{\sigma_D^{(\alpha_2,\beta_2)}[g]}. \tag{29}$$

Thus the theorem follows from (27) and (29). ∎

The following theorem is a natural consequence of Theorem 2.3.11 and Theorem 2.3.12.

**Theorem 2.3.13** *Let* $f(z)$ *and* $g(z)$ *be any two entire functions of* $n$ *complex variables. Also let* $0 < \overline{\sigma}_D^{(\alpha_1,\beta_1)}[f] \leq \sigma_D^{(\alpha_1,\beta_1)}[f] < \infty$, $0 < \overline{\sigma}_D^{(\alpha_2,\beta_2)}[g] \leq \sigma_D^{(\alpha_2,\beta_2)}[g] < \infty$ *and* $\rho^{(\alpha_1,\beta_1)}[f] = \rho^{(\alpha_2,\beta_2)}[g]$. *Then*

$$\liminf_{R\to\infty}\frac{\exp(\alpha_1(M_{f,D}(R)))}{\exp(\alpha_2(M_{g,D}(\beta_2^{-1}(\beta_1(R)))))} \leq \min\left\{\frac{\overline{\sigma}_D^{(\alpha_1,\beta_1)}[f]}{\overline{\sigma}_D^{(\alpha_2,\beta_2)}[g]}, \frac{\sigma_D^{(\alpha_1,\beta_1)}[f]}{\sigma_D^{(\alpha_2,\beta_2)}[g]}\right\}$$

$$\leq \max\left\{\frac{\overline{\sigma}_D^{(\alpha_1,\beta_1)}[f]}{\overline{\sigma}_D^{(\alpha_2,\beta_2)}[g]}, \frac{\sigma_D^{(\alpha_1,\beta_1)}[f]}{\sigma_D^{(\alpha_2,\beta_2)}[g]}\right\} \leq \limsup_{R\to\infty}\frac{\exp(\alpha_1(M_{f,D}(R)))}{\exp(\alpha_2(M_{g,D}(\beta_2^{-1}(\beta_1(R)))))}.$$

The proof is omitted.

Now in the line of Theorem 2.3.11, Theorem 2.3.12 and Theorem 2.3.13 respectively one can easily prove the following three theorems using the notion of generalized Gol'dberg weak type $(\alpha, \beta)$ of an entire function with respect to any bounded complete $n$-circular domain $D$ with center at all the origin $\mathbb{C}^n$ and therefore their proofs are omitted.

**Theorem 2.3.14** *Let* $f(z)$ *and* $g(z)$ *be any two entire functions of* $n$ *complex variables. Also let* $0 < \tau_D^{(\alpha_1,\beta_1)}[f] \leq \overline{\tau}_D^{(\alpha_1,\beta_1)}[f] < \infty$, $0 < \tau_D^{(\alpha_2,\beta_2)}[g] \leq \overline{\tau}_D^{(\alpha_2,\beta_2)}[g] < \infty$ *and* $\lambda^{(\alpha_1,\beta_1)}[f] = \lambda^{(\alpha_2,\beta_2)}[g]$. *Then*

$$\frac{\tau_D^{(\alpha_1,\beta_1)}[f]}{\overline{\tau}_D^{(\alpha_2,\beta_2)}[g]} \leq \liminf_{R\to\infty}\frac{\exp(\alpha_1(M_{f,D}(R)))}{\exp(\alpha_2(M_{g,D}(\beta_2^{-1}(\beta_1(R)))))} \leq \frac{\tau_D^{(\alpha_1,\beta_1)}[f]}{\tau_D^{(\alpha_2,\beta_2)}[g]}$$

$$\leq \limsup_{R\to\infty}\frac{\exp(\alpha_1(M_{f,D}(R)))}{\exp(\alpha_2(M_{g,D}(\beta_2^{-1}(\beta_1(R)))))} \leq \frac{\overline{\tau}_D^{(\alpha_1,\beta_1)}[f]}{\tau_D^{(\alpha_2,\beta_2)}[g]}.$$

**Theorem 2.3.15** *Let* $f(z)$ *and* $g(z)$ *be any two entire functions of* $n$ *complex variables. Also let* $0 < \overline{\tau}_D^{(\alpha_1,\beta_1)}[f] < \infty$, $0 < \overline{\tau}_D^{(\alpha_2,\beta_2)}[g] < \infty$ *and* $\lambda^{(\alpha_1,\beta_1)}[f] = \lambda^{(\alpha_2,\beta_2)}[g]$. *Then*

$$\liminf_{R\to\infty}\frac{\exp(\alpha_1(M_{f,D}(R)))}{\exp(\alpha_2(M_{g,D}(\beta_2^{-1}(\beta_1(R)))))} \leq \frac{\overline{\tau}_D^{(\alpha_1,\beta_1)}[f]}{\overline{\tau}_D^{(\alpha_2,\beta_2)}[g]} \leq \limsup_{R\to\infty}\frac{\exp(\alpha_1(M_{f,D}(R)))}{\exp(\alpha_2(M_{g,D}(\beta_2^{-1}(\beta_1(R)))))}.$$

**Theorem 2.3.16** *Let $f(z)$ and $g(z)$ be any two entire functions of $n$ complex variables. Also let $0 < \tau_D^{(\alpha_1,\beta_1)}[f] \leq \overline{\tau}_D^{(\alpha_1,\beta_1)}[f] < \infty$, $0 < \tau_D^{(\alpha_2,\beta_2)}[g] \leq \overline{\tau}_D^{(\alpha_2,\beta_2)}[g] < \infty$ and $\lambda^{(\alpha_1,\beta_1)}[f] = \lambda^{(\alpha_2,\beta_2)}[g]$. Then*

$$\liminf_{R\to\infty} \frac{\exp(\alpha_1(M_{f,D}(R)))}{\exp(\alpha_2(M_{g,D}(\beta_2^{-1}(\beta_1(R)))))} \leq \min\left\{\frac{\tau_D^{(\alpha_1,\beta_1)}[f]}{\tau_D^{(\alpha_2,\beta_2)}[g]}, \frac{\overline{\tau}_D^{(\alpha_1,\beta_1)}[f]}{\overline{\tau}_D^{(\alpha_2,\beta_2)}[g]}\right\}$$

$$\leq \max\left\{\frac{\tau_D^{(\alpha_1,\beta_1)}[f]}{\tau_D^{(\alpha_2,\beta_2)}[g]}, \frac{\overline{\tau}_D^{(\alpha_1,\beta_1)}[f]}{\overline{\tau}_D^{(\alpha_2,\beta_2)}[g]}\right\} \leq \limsup_{R\to\infty} \frac{\exp(\alpha_1(M_{f,D}(R)))}{\exp(\alpha_2(M_{g,D}(\beta_2^{-1}(\beta_1(R)))))}.$$

We may now state the following theorems without their proofs based on generalized Gol'dberg type $(\alpha, \beta)$ and generalized Gol'dberg weak type $(\alpha, \beta)$ of an entire function with respect to any bounded complete $n$-circular domain $D$ with center at all the origin $\mathbb{C}^n$.

**Theorem 2.3.17** *Let $f(z)$ and $g(z)$ be any two entire functions of $n$ complex variables. Also let $0 < \overline{\sigma}_D^{(\alpha_1,\beta_1)}[f] \leq \sigma_D^{(\alpha_1,\beta_1)}[f] < \infty$, $0 < \tau_D^{(\alpha_2,\beta_2)}[g] \leq \overline{\tau}_D^{(\alpha_2,\beta_2)}[g] < \infty$ and $\rho^{(\alpha_1,\beta_1)}[f] = \lambda^{(\alpha_2,\beta_2)}[g]$. Then*

$$\frac{\overline{\sigma}_D^{(\alpha_1,\beta_1)}[f]}{\overline{\tau}_D^{(\alpha_2,\beta_2)}[g]} \leq \liminf_{R\to\infty} \frac{\exp(\alpha_1(M_{f,D}(R)))}{\exp(\alpha_2(M_{g,D}(\beta_2^{-1}(\beta_1(R)))))} \leq \frac{\overline{\sigma}_D^{(\alpha_1,\beta_1)}[f]}{\tau_D^{(\alpha_2,\beta_2)}[g]}$$

$$\leq \limsup_{R\to\infty} \frac{\exp(\alpha_1(M_{f,D}(R)))}{\exp(\alpha_2(M_{g,D}(\beta_2^{-1}(\beta_1(R)))))} \leq \frac{\sigma_D^{(\alpha_1,\beta_1)}[f]}{\tau_D^{(\alpha_2,\beta_2)}[g]}.$$

**Theorem 2.3.18** *Let $f(z)$ and $g(z)$ be any two entire functions of $n$ complex variables. Also let $0 < \sigma_D^{(\alpha_1,\beta_1)}[f] < \infty$, $0 < \overline{\tau}_D^{(\alpha_2,\beta_2)}[g] < \infty$ and $\rho^{(\alpha_1,\beta_1)}[f] = \lambda^{(\alpha_2,\beta_2)}[g]$. Then*

$$\liminf_{R\to\infty} \frac{\exp(\alpha_1(M_{f,D}(R)))}{\exp(\alpha_2(M_{g,D}(\beta_2^{-1}(\beta_1(R)))))} \leq \frac{\sigma_D^{(\alpha_1,\beta_1)}[f]}{\overline{\tau}_D^{(\alpha_2,\beta_2)}[g]} \leq \limsup_{R\to\infty} \frac{\exp(\alpha_1(M_{f,D}(R)))}{\exp(\alpha_2(M_{g,D}(\beta_2^{-1}(\beta_1(R)))))}.$$

**Theorem 2.3.19** *Let $f(z)$ and $g(z)$ be any two entire functions of $n$ complex variables. Also let $0 < \overline{\sigma}_D^{(\alpha_1,\beta_1)}[f] \leq \sigma_D^{(\alpha_1,\beta_1)}[f] < \infty$, $0 < \tau_D^{(\alpha_2,\beta_2)}[g] \leq \overline{\tau}_D^{(\alpha_2,\beta_2)}[g] < \infty$ and $\rho^{(\alpha_1,\beta_1)}[f] = \lambda^{(\alpha_2,\beta_2)}[g]$. Then*

$$\liminf_{R\to\infty} \frac{\exp(\alpha_1(M_{f,D}(R)))}{\exp(\alpha_2(M_{g,D}(\beta_2^{-1}(\beta_1(R)))))} \leq \min\left\{\frac{\overline{\sigma}_D^{(\alpha_1,\beta_1)}[f]}{\tau_D^{(\alpha_2,\beta_2)}[g]}, \frac{\sigma_D^{(\alpha_1,\beta_1)}[f]}{\overline{\tau}_D^{(\alpha_2,\beta_2)}[g]}\right\}$$

$$\leq \max\left\{\frac{\overline{\sigma}_D^{(\alpha_1,\beta_1)}[f]}{\tau_D^{(\alpha_2,\beta_2)}[g]}, \frac{\sigma_D^{(\alpha_1,\beta_1)}[f]}{\overline{\tau}_D^{(\alpha_2,\beta_2)}[g]}\right\} \leq \limsup_{R\to\infty} \frac{\exp(\alpha_1(M_{f,D}(R)))}{\exp(\alpha_2(M_{g,D}(\beta_2^{-1}(\beta_1(R)))))}.$$

**Theorem 2.3.20** *Let $f(z)$ and $g(z)$ be any two entire functions of $n$ complex variables. Also let $0 < \tau_D^{(\alpha_1,\beta_1)}[f] \leq \overline{\tau}_D^{(\alpha_1,\beta_1)}[f] < \infty$, $0 < \overline{\sigma}_D^{(\alpha_2,\beta_2)}[g] \leq \sigma_D^{(\alpha_2,\beta_2)}[g] < \infty$ and $\lambda^{(\alpha_1,\beta_1)}[f] = \rho^{(\alpha_2,\beta_2)}[g]$. Then*

$$\frac{\tau_D^{(\alpha_1,\beta_1)}[f]}{\sigma_D^{(\alpha_2,\beta_2)}[g]} \leq \liminf_{R\to\infty}\frac{\exp(\alpha_1(M_{f,D}(R)))}{\exp(\alpha_2(M_{g,D}(\beta_2^{-1}(\beta_1(R)))))} \leq \frac{\tau_D^{(\alpha_1,\beta_1)}[f]}{\sigma_D^{(\alpha_2,\beta_2)}[g]}$$

$$\leq \limsup_{R\to\infty}\frac{\exp(\alpha_1(M_{f,D}(R)))}{\exp(\alpha_2(M_{g,D}(\beta_2^{-1}(\beta_1(R)))))} \leq \frac{\overline{\tau}_D^{(\alpha_1,\beta_1)}[f]}{\overline{\sigma}_D^{(\alpha_2,\beta_2)}[g]}.$$

**Theorem 2.3.21** *Let $f(z)$ and $g(z)$ be any two entire functions of $n$ complex variables. Also let $0 < \overline{\tau}_D^{(\alpha_1,\beta_1)}[f] < \infty$, $0 < \sigma_D^{(\alpha_2,\beta_2)}[g] < \infty$ and $\lambda^{(\alpha_1,\beta_1)}[f] = \rho^{(\alpha_2,\beta_2)}[g]$. Then*

$$\liminf_{R\to\infty}\frac{\exp(\alpha_1(M_{f,D}(R)))}{\exp(\alpha_2(M_{g,D}(\beta_2^{-1}(\beta_1(R)))))} \leq \frac{\overline{\tau}_D^{(\alpha_1,\beta_1)}[f]}{\sigma_D^{(\alpha_2,\beta_2)}[g]} \leq \limsup_{R\to\infty}\frac{\exp(\alpha_1(M_{f,D}(R)))}{\exp(\alpha_2(M_{g,D}(\beta_2^{-1}(\beta_1(R)))))}.$$

**Theorem 2.3.22** *Let $f(z)$ and $g(z)$ be any two entire functions of $n$ complex variables. Also let $0 < \tau_D^{(\alpha_1,\beta_1)}[f] \leq \overline{\tau}_D^{(\alpha_1,\beta_1)}[f] < \infty$, $0 < \overline{\sigma}_D^{(\alpha_2,\beta_2)}[g] \leq \sigma_D^{(\alpha_2,\beta_2)}[g] < \infty$ and $\lambda^{(\alpha_1,\beta_1)}[f] = \rho^{(\alpha_2,\beta_2)}[g]$. Then*

$$\liminf_{R\to\infty}\frac{\exp(\alpha_1(M_{f,D}(R)))}{\exp(\alpha_2(M_{g,D}(\beta_2^{-1}(\beta_1(R)))))} \leq \min\left\{\frac{\tau_D^{(\alpha_1,\beta_1)}[f]}{\overline{\sigma}_D^{(\alpha_2,\beta_2)}[g]}, \frac{\overline{\tau}_D^{(\alpha_1,\beta_1)}[f]}{\sigma_D^{(\alpha_2,\beta_2)}[g]}\right\}$$

$$\leq \max\left\{\frac{\tau_D^{(\alpha_1,\beta_1)}[f]}{\overline{\sigma}_D^{(\alpha_2,\beta_2)}[g]}, \frac{\overline{\tau}_D^{(\alpha_1,\beta_1)}[f]}{\sigma_D^{(\alpha_2,\beta_2)}[g]}\right\} \leq \limsup_{R\to\infty}\frac{\exp(\alpha_1(M_{f,D}(R)))}{\exp(\alpha_2(M_{g,D}(\beta_2^{-1}(\beta_1(R)))))}.$$

# 2.4   Conclusion.

The main aim of this chapter is actually to extend and to modify the notion of Gol'dberg order, Gol'dberg type and Gol'dberg weak type to generalized Gol'dberg order $(\alpha,\beta)$, generalized Gol'dberg type $(\alpha,\beta)$ and generalized Gol'dberg weak type $(\alpha,\beta)$ of higher dimensions in case of an entire functions of of $n$ complex variables for any bounded complete $n$-circular domain $D$ with center at all the origin $\mathbb{C}^n$. Here we see that the previous definitions of $(p,q)$-th Gol'dberg order, $(p,q)$-th Gol'dberg type and $(p,q)$-th Gol'dberg weak type are easily generated as particular cases of the present definitions, e.g. if $\alpha(R) = \log^{[p]} R$ and $\beta(R) = \log^{[q]} R$.

However in the case of generalized Gol'dberg order $(\alpha,\beta)$, it therefore seems reasonable to define suitably the generalized relative Gol'dberg order $(\alpha,\beta)$ of an entire function for any bounded complete $n$-circular domain $D$ with center at all the origin $\mathbb{C}^n$. In the next section we wish to introduce the definition of the generalized relative Gol'dberg order $(\alpha,\beta)$ of entire functions of $n$ complex variables with respect to another entire functions of $n$ complex variables where $\alpha$ and $\beta$ continuous non-negative on $(-\infty, +\infty)$ function. Then we wish to establish their integral representations and study some comparative growth properties.

# References

[1] S. K. Datta and A. R. Maji: Study of Growth properties on the basis of generalised Gol'dberg order of composite entire functions of several complex variables,International J. of Math.Sci.& Engg.Appls, 5(V) (2011), 297-311.

[2] A. A. Gol'dberg: Elementary remarks on the formulas defining order and type of functions of several variables, Dokl. Akad. Nauk Arm. SSR, 29 (1959), 145-151 (Russian).

# Chapter 3

# Generalized relative Gol'dberg order $(\alpha, \beta)$ of entire functions of several complex variables

**Abstract:** The aim of the chapter is to introduce the concepts of generalized relative Gol'dberg order $(\alpha, \beta)$, generalized relative hyper Gol'dberg order $(\alpha, \beta)$, and generalized relative logarithmic Gol'dberg order $(\alpha, \beta)$ of an entire function of several complex variables with respect to another entire function of several complex variables, where $\alpha, \beta$ are continuous non-negative functions defined on $(-\infty, +\infty)$. Then we discuss some growth analysis of entire functions of several complex variables. Also we established some integral representations of the above growth indicators.

**Keywords:** Entire functions of several complex variables, increasing function, Generalized relative Gol'dberg order $(\alpha, \beta)$, generalized relative hyper Gol'dberg order $(\alpha, \beta)$, generalized relative logarithmic Gol'dberg order $(\alpha, \beta)$, generalized relative logarithmic Gol'dberg lower order $(\alpha, \beta)$.

**Mathematics Subject Classification (2010) :** 32A15.

## 3.1   Introduction.

The Gol'dberg order and Gol'dberg type of an entire function $f(z)$ with respect to any bounded complete $n$-circular domain $D$ with center at all the origin $\mathbb{C}^n$ which are generally used in computational purpose are classical. Mondal et al. [1] defined the concept of relative Gol'dberg order between two entire functions $f(z)$ and $g(z)$ for any bounded complete $n$-circular domain $D$ with center at all the origin $\mathbb{C}^n$. Extending this notion, here in this chapter we wish to introduce the definition of generalized relative Gol'dberg order $(\alpha, \beta)$ and generalized relative Gol'dberg lower order $(\alpha, \beta)$ between two entire functions of several complex variables and establish some related growth properties with their integral representations.

# 3.2   Preliminary remarks and definitions.

First we introduce the definitions of the generalized relative Gol'dberg order $(\alpha, \beta)$ and generalized relative Gol'dberg lower order $(\alpha, \beta)$ of an entire function in $\mathbb{C}^n$ with respect to another entire function of several variables in the following way:

**Definition 3.2.1** *Let $f(z)$ and $g(z)$ be any two entire functions of $n$ complex variables. The generalized relative Gol'dberg order $(\alpha, \beta)$ of $f(z)$ with respect to $g(z)$ is defined by:*

$$\rho_D^{(\alpha,\beta)} [f]_g = \limsup_{R \to \infty} \frac{\alpha(M_{g,D}^{-1}(M_{f,D}(R)))}{\beta(R)}.$$

**Definition 3.2.2** *Let $f(z)$ and $g(z)$ be any two entire functions of $n$ complex variables. The growth indicator $\rho_D^{(\alpha,\beta)} [f]_g$ is alternatively defined as : The integral*

$$\int_{R_0}^{\infty} \frac{\exp(\alpha(M_{g,D}^{-1}(M_{f,D}(R))))}{(\exp \beta(R))^{k+1}} dR \, (R_0 > 0)$$

*converges for $k > \rho_D^{(\alpha,\beta)} [f]_g$ and diverges for $k < \rho_D^{(\alpha,\beta)} [f]_g$.*

**Definition 3.2.3** *Let $f(z)$ and $g(z)$ be any two entire functions of $n$ complex variables. The generalized relative Gol'dberg lower order $(\alpha, \beta)$ of $f(z)$ with respect to $g(z)$ is defined as:*

$$\lambda_D^{(\alpha,\beta)} [f]_g = \liminf_{R \to \infty} \frac{\alpha(M_{g,D}^{-1}(M_{f,D}(R)))}{\beta(R)}.$$

**Definition 3.2.4** *Let $f(z)$ and $g(z)$ be any two entire functions of $n$ complex variables. The growth indicator $\lambda_D^{(\alpha,\beta)} [f]_g$ is alternatively defined as : The integral*

$$\int_{R_0}^{\infty} \frac{\exp(\alpha(M_{g,D}^{-1}(M_{f,D}(R))))}{(\exp \beta(R))^{k+1}} dR \, (R_0 > 0)$$

*converges for $k > \lambda_D^{(\alpha,\beta)} [f]_g$ and diverges for $k < \lambda_D^{(\alpha,\beta)} [f]_g$.*

An entire function $f(z)$ of $n$ complex variables for which $\rho_D^{(\alpha,\beta)} [f]_g$ and $\lambda_D^{(\alpha,\beta)} [f]_g$ are the same is called a function of regular generalized relative Gol'dberg $(\alpha, \beta)$ growth with respect to an entire function $g(z)$ of $n$ complex variables. Otherwise, $f(z)$ is said to be irregular generalized relative Gol'dberg $(\alpha, \beta)$ growth with respect to $g(z)$.

Now a question may arise about the equivalence of the definitions of generalized relative Gol'dberg order $(\alpha, \beta)$ and generalized relative Gol'dberg lower order $(\alpha, \beta)$ with their integral representations. In the next section we would like to establish such equivalence of Definition 3.2.1 and Definition 3.2.3 with Definition 3.2.2 and Definition 3.2.4 respectively and also investigate some growth properties related to generalized relative Gol'dberg order $(\alpha, \beta)$ and generalized relative Gol'dberg lower order $(\alpha, \beta)$ of an entire functions of $n$ complex variables with respect to another entire function of $n$ complex variables.

## 3.3 Lemma.

In this section we present a lemma which will be needed in the sequel.

**Lemma 3.3.1** *Let the integral* $\int\limits_{R_0}^{\infty} \frac{\exp(\alpha(M_{g,D}^{-1}(M_{f,D}(R))))}{(\exp \beta(R))^{k+1}} dR$ $(R_0 > 0)$ *converges for* $0 < k < \infty$. *Then*

$$\lim_{R \to \infty} \frac{\exp(\alpha(M_{g,D}^{-1}(M_{f,D}(R))))}{(\exp \beta(R))^k} = 0.$$

**Proof.** Since the integral $\int\limits_{R_0}^{\infty} \frac{\exp(\alpha(M_{g,D}^{-1}(M_{f,D}(R))))}{(\exp \beta(R))^{k+1}} dR$ is convergent for $0 < k < \infty$, given $\varepsilon$ $(> 0)$ there exists a number $\Re = \Re(\varepsilon)$ such that

$$\int\limits_{R_0}^{\infty} \frac{\exp(\alpha(M_{g,D}^{-1}(M_{f,D}(R))))}{(\exp \beta(R))^{k+1}} dR < \varepsilon \text{ for } R_0 > \Re.$$

i.e., for $R_0 > \Re$,

$$\int\limits_{R_0}^{R_0+R} \frac{\exp(\alpha(M_{g,D}^{-1}(M_{f,D}(R))))}{(\exp \beta(R))^{k+1}} dR < \varepsilon.$$

Since $\exp(\alpha(M_{g,D}^{-1}(M_{f,D}(R))))$ is an increasing function of $R$, so

$$\int\limits_{R_0}^{R_0+R} \frac{\exp(\alpha(M_{g,D}^{-1}(M_{f,D}(R))))}{(\exp \beta(R))^{k+1}} dR \geq \frac{\exp(\alpha(M_{g,D}^{-1}(M_{f,D}(R))))}{(\exp \beta(R_0))^{k+1}} \cdot (\exp \beta(R_0))$$

i.e., for all large values of $R$,

$$\int\limits_{R_0}^{R_0+R} \frac{\exp(\alpha(M_{g,D}^{-1}(M_{f,D}(R))))}{(\exp \beta(R))^{k+1}} dR \geq \frac{\exp(\alpha(M_{g,D}^{-1}(M_{f,D}(R))))}{(\exp \beta(R_0))^k}$$

$$i.e., \quad \frac{\exp(\alpha(M_{g,D}^{-1}(M_{f,D}(R))))}{(\exp \beta(R_0))^k} < \varepsilon \text{ for } R_0 > \Re,$$

from which it follows that

$$\lim_{R \to \infty} \frac{\exp(\alpha(M_{g,D}^{-1}(M_{f,D}(R))))}{(\exp \beta(R))^k} = 0.$$

This proves the lemma. ∎

## 3.4  Main Results.

In this section we present the main results of the chapter.

**Theorem 3.4.1** *Let $f(z)$ and $g(z)$ be two entire functions of $n$ complex variables. Then generalized relative Gol'dberg order $(\alpha, \beta)$ and generalized relative Gol'dberg lower order $(\alpha, \beta)$ of $f(z)$ are independent of the choice of the domain $D$.*

**Proof.** Let us consider $D_1$ and $D_2$ ne any two bounded complete $n$-circular domains. Then there exist two real numbers $a$, $b > 0$ such that $aD_1 \subset D_2 \subset bD_1$ and therefore

$$M_{f,aD_1}(R) \le M_{f,D_2}(R) \le M_{f,bD_1}(R).$$

Hence for any bounded complete $n$-circular domain $D$

$$M_{g,D}^{-1}(M_{f,aD_1}(R)) \le M_{g,D}^{-1}(M_{f,D_2}(R)) \le M_{g,D}^{-1}(M_{f,bD_1}(R)). \tag{30}$$

Now for any $c > 0$ and any $D$, we get that

$$M_{f,cD}(R) = M_{f,D}(cR).$$

Therefore

$$
\begin{aligned}
\limsup_{R\to\infty} \frac{\alpha(M_{g,D}^{-1}(M_{f,cD}(R)))}{\beta(R)} &= \limsup_{R\to\infty} \frac{\alpha(M_{g,D}^{-1}(M_{f,D}(cR)))}{\beta(R)} \\
&= \limsup_{\frac{R}{c}\to\infty} \frac{\alpha(M_{g,D}^{-1}(M_{f,D}(R)))}{\beta\left(\frac{R}{c}\right)} \\
&= \limsup_{\frac{R}{c}\to\infty} \frac{\alpha(M_{g,D}^{-1}(M_{f,D}(R)))}{\beta(R)} \cdot \lim_{R\to\infty} \frac{\beta(cR)}{\beta(R)} \\
&= \limsup_{R\to\infty} \frac{\alpha(M_{g,D}^{-1}(M_{f,D}(R)))}{\beta(R)}.
\end{aligned}
$$

Hence by (30) we obtain that

$$
\begin{aligned}
\limsup_{R\to\infty} \frac{\alpha(M_{g,D}^{-1}(M_{f,D_1}(R)))}{\beta(R)} &= \limsup_{R\to\infty} \frac{\alpha(M_{g,D}^{-1}(M_{f,aD_1}(R)))}{\beta(R)} \\
&\le \limsup_{R\to\infty} \frac{\alpha(M_{g,D}^{-1}(M_{f,D_2}(R)))}{\beta(R)} \\
&\le \limsup_{R\to\infty} \frac{\alpha(M_{g,D}^{-1}(M_{f,bD_1}(R)))}{\beta(R)} \\
&\le \limsup_{R\to\infty} \frac{\alpha(M_{g,D}^{-1}(M_{f,D_1}(R)))}{\beta(R)}.
\end{aligned}
$$

Thus

$$\limsup_{R\to\infty}\frac{\alpha(M_{g,D}^{-1}(M_{f,D_1}(R)))}{\beta\left(R\right)}=\limsup_{R\to\infty}\frac{\alpha(M_{g,D}^{-1}(M_{f,D_2}(R)))}{\beta\left(R\right)}.$$

Similarly one can easily verify that

$$\liminf_{R\to\infty}\frac{\alpha(M_{g,D}^{-1}(M_{f,D_1}(R)))}{\beta\left(R\right)}=\liminf_{R\to\infty}\frac{\alpha(M_{g,D}^{-1}(M_{f,D_2}(R)))}{\beta\left(R\right)}.$$

Hence the theorem follows. ∎

Since $\rho_D^{(\alpha,\beta)}\left[f\right]_g$ and $\lambda_D^{(\alpha,\beta)}\left[f\right]_g$ are independent of the choice of the domain $D$, so after this we shall always write $\rho^{(\alpha,\beta)}\left[f\right]_g$ and $\lambda^{(\alpha,\beta)}\left[f\right]_g$ instead of $\rho_D^{(\alpha,\beta)}\left[f\right]_g$ and $\lambda_D^{(\alpha,\beta)}\left[f\right]_g$ respectively.

**Theorem 3.4.2** *Let $f\left(z\right)$ and $g\left(z\right)$ be any two entire functions of $n$ complex variables. Then Definition 3.2.1 and Definition 3.2.2 are equivalent.*

**Proof. Case 1.** $\rho^{(\alpha,\beta)}\left[f\right]_g=\infty$.

**Definition 3.2.1 $\Rightarrow$ Definition 3.2.2.**

As $\rho^{(\alpha,\beta)}\left[f\right]_g=\infty$, from Definition 3.2.1 we have for arbitrary positive $G$ and for a sequence of values of $R$ tending to infinity that

$$\rho_D^{(\alpha,\beta)}\left[f\right]_g=\limsup_{R\to\infty}\frac{\alpha(M_{g,D}^{-1}\left(M_{f,D}\left(R\right)\right))}{\beta\left(R\right)}.$$

$$\alpha(M_{g,D}^{-1}\left(M_{f,D}\left(R\right)\right))>G\beta\left(R\right)$$

$$i.e,\exp(\alpha(M_{g,D}^{-1}\left(M_{f,D}\left(R\right)\right)))>(\exp\beta\left(R\right))^{G}. \tag{31}$$

If possible let the integral $\int\limits_{R_0}^{\infty}\frac{\exp(\alpha(M_{g,D}^{-1}\left(M_{f,D}(R)\right)))}{(\exp\beta(R))^{G+1}}dR$ $(R_0>0)$ be converge. Then by Lemma 3.3.1,

$$\limsup_{R\to\infty}\frac{\exp(\alpha(M_{g,D}^{-1}\left(M_{f,D}\left(R\right)\right)))}{(\exp\beta\left(R\right))^{G}}=0.$$

So for all sufficiently large values of $R$,

$$\exp(\alpha(M_{g,D}^{-1}\left(M_{f,D}\left(R\right)\right)))<(\exp\beta\left(R\right))^{G}. \tag{32}$$

Now from (31) and (32) we arrive at a contradiction.

Hence $\int\limits_{R_0}^{\infty}\frac{\exp(\alpha(M_{g,D}^{-1}\left(M_{f,D}(R)\right)))}{(\exp\beta(R))^{G+1}}dR$ $(R_0>0)$ diverges whenever $G$ is finite, which is Definition 3.2.2.

**Definition 3.2.2 $\Rightarrow$ Definition 3.2.1.**

Suppose $G$ be any positive number. Since $\rho^{(\alpha,\beta)}\left[f\right]_g=\infty$, from Definition 3.2.2 the divergence of the integral $\int\limits_{R_0}^{\infty}\frac{\exp(\alpha(M_{g,D}^{-1}\left(M_{f,D}(R)\right)))}{(\exp\beta(R))^{G+1}}dR$ $(R_0>0)$ gives for arbitrary positive $\varepsilon$ and for a sequence of values of $R$ tending to infinity,

$$\exp(\alpha(M_{g,D}^{-1}\left(M_{f,D}\left(R\right)\right)))>(\exp\beta\left(R\right))^{G-\varepsilon}$$

$$i.e,\ \alpha(M_{g,D}^{-1}(M_{f,D}(R))) > (G-\varepsilon)\,\beta(R).$$

This gives that

$$\limsup_{R\to\infty}\frac{\alpha(M_{g,D}^{-1}(M_{f,D}(R)))}{\beta(R)} \geq (G-\varepsilon).$$

Since $G > 0$ is arbitrary, it follows that

$$\limsup_{R\to\infty}\frac{\alpha(M_{g,D}^{-1}(M_{f,D}(R)))}{\beta(R)} = \infty.$$

Thus Definition 3.2.1 follows.

**Case 2.** $0 \leq \rho^{(\alpha,\beta)}[f]_g < \infty.$

**Definition 3.2.1 $\Rightarrow$ Definition 3.2.2.**

**Subcase** $(I)$. $0 < \rho^{(\alpha,\beta)}[f]_g < \infty.$

If $0 < \rho^{(\alpha,\beta)}[f]_g < \infty$, then for arbitrary $\varepsilon\,(>0)$ and for all sufficiently large values of $R$,

$$\frac{\alpha(M_{g,D}^{-1}(M_{f,D}(R)))}{\beta(R)} < \rho^{(\alpha,\beta)}[f]_g + \varepsilon$$

$$i.e, \exp(\alpha(M_{g,D}^{-1}(M_{f,D}(R)))) < (\exp\beta(R))^{(\rho^{(\alpha,\beta)}[f]_g+\varepsilon)}$$

$$i.e,\ \frac{\exp(\alpha(M_{g,D}^{-1}(M_{f,D}(R))))}{(\exp\beta(R))^k} < \frac{(\exp\beta(R))^{(\rho^{(\alpha,\beta)}[f]_g+\varepsilon)}}{(\exp\beta(R))^k}$$

$$i.e,\ \frac{\exp(\alpha(M_{g,D}^{-1}(M_{f,D}(R))))}{(\exp\beta(R))^k} < \frac{1}{(\exp\beta(R))^{k-(\rho^{(\alpha,\beta)}[f]_g+\varepsilon)}}.$$

Therefore $\int_{R_0}^{\infty}\frac{\exp(\alpha(M_{g,D}^{-1}(M_{f,D}(R))))}{(\exp\beta(R))^{k+1}}dR$ $(R_0 > 0)$ converges if $k > \rho^{(\alpha,\beta)}[f]_g$ and diverges if $k < \rho^{(\alpha,\beta)}[f]_g.$

**Subcase** $(II)$.

When $\rho^{(\alpha,\beta)}[f]_g = 0$, Definition 3.2.1 gives for all sufficiently large values of $R$ that

$$\frac{\alpha(M_{g,D}^{-1}(M_{f,D}(R)))}{\beta(R)} \leq \varepsilon.$$

Then as before we obtain that $\int_{R_0}^{\infty}\frac{\exp(\alpha(M_{g,D}^{-1}(M_{f,D}(R))))}{(\exp\beta(R))^{k+1}}dR$ $(R_0 > 0)$ converges for $k > 0$ and diverges for $k < 0$.

Thus from Subcase $(I)$ and Subcase $(II)$ Definition 3.2.2 follows.

**Definition 3.2.2 $\Rightarrow$ Definition 3.2.1.**

By Definition 3.2.2, for arbitrary $\varepsilon\,(>0)$ the integral $\int_{R_0}^{\infty}\frac{\exp(\alpha(M_{g,D}^{-1}(M_{f,D}(R))))}{(\exp\beta(R))^{\rho^{(\alpha,\beta)}[f]_g+\varepsilon+1}}dR$ converges. Then by Lemma 3.3.1 we have

$$\limsup_{R\to\infty}\frac{\exp(\alpha(M_{g,D}^{-1}(M_{f,D}(R))))}{(\exp\beta(R))^{\rho^{(\alpha,\beta)}[f]_g+\varepsilon}} = 0$$

i.e, for all sufficiently large values of $R$,

$$\frac{\exp(\alpha(M_{g,D}^{-1}(M_{f,D}(R))))}{(\exp\beta(R))^{\rho^{(\alpha,\beta)}[f]_g+\varepsilon}} < \varepsilon_0$$

$$i.e, \quad \exp(\alpha(M_{g,D}^{-1}(M_{f,D}(R)))) < \varepsilon_0 \cdot (\exp\beta(R))^{\rho^{(\alpha,\beta)}[f]_g+\varepsilon}$$

$$i.e, \quad \alpha(M_{g,D}^{-1}(M_{f,D}(R))) < \log\varepsilon_0 + \left(\rho^{(\alpha,\beta)}[f]_g + \varepsilon\right)\beta(R)$$

$$i.e, \quad \frac{\alpha(M_{g,D}^{-1}(M_{f,D}(R)))}{\beta(R)} \leq \frac{\log\varepsilon_0}{\beta(R)} + \left(\rho^{(\alpha,\beta)}[f]_g + \varepsilon\right)$$

$$i.e, \quad \limsup_{R\to\infty}\frac{\alpha(M_{g,D}^{-1}(M_{f,D}(R)))}{\beta(R)} \leq \rho^{(\alpha,\beta)}[f]_g + \varepsilon.$$

Since $\varepsilon\,(>0)$ is arbitrary, it follows from above that

$$\limsup_{R\to\infty}\frac{\alpha(M_{g,D}^{-1}(M_{f,D}(R)))}{\beta(R)} \leq \rho^{(\alpha,\beta)}[f]_g. \tag{33}$$

Again by Definition 3.2.2 the divergence of the integral $\int\limits_{R_0}^{\infty}\frac{\exp(\alpha(M_{g,D}^{-1}(M_{f,D}(R))))}{(\exp\beta(R))^{\rho^{(\alpha,\beta)}[f]_g-\varepsilon+1}}dR$ implies that there exists a sequence of values of $R$ tending to infinity such that

$$\frac{\exp(\alpha(M_{g,D}^{-1}(M_{f,D}(R))))}{(\exp\beta(R))^{\rho^{(\alpha,\beta)}[f]_g-\varepsilon+1}} > \frac{1}{(\exp\beta(R))^{1+\varepsilon}}$$

$$i.e, \exp(\alpha(M_{g,D}^{-1}(M_{f,D}(R)))) > (\exp\beta(R))^{\rho^{(\alpha,\beta)}[f]_g-2\varepsilon}$$

$$i.e, \quad \alpha(M_{g,D}^{-1}(M_{f,D}(R))) > (\rho^{(\alpha,\beta)}[f]_g - 2\varepsilon)\beta(R)$$

$$i.e, \quad \frac{\alpha(M_{g,D}^{-1}(M_{f,D}(R)))}{\beta(R)} > (\rho^{(\alpha,\beta)}[f]_g - 2\varepsilon).$$

As $\varepsilon\,(>0)$ is arbitrary, we get that

$$\limsup_{R\to\infty}\frac{\alpha(M_{g,D}^{-1}(M_{f,D}(R)))}{\beta(R)} \geq \rho^{(\alpha,\beta)}[f]_g. \tag{34}$$

Thus from (33) and (34) it follows that

$$\limsup_{R\to\infty}\frac{\alpha(M_{g,D}^{-1}(M_{f,D}(R)))}{\beta(R)} = \rho^{(\alpha,\beta)}[f]_g.$$

Thus we obtain Definition 3.2.1.

Now combining Case 1 and Case 2, the theorem follows. ∎

In the line of Theorem 3.4.2 we may now state the following theorem without proof.

**Theorem 3.4.3** *Let $f(z)$ and $g(z)$ be any two entire functions of $n$ complex variables. Then Definition 3.2.3 and Definition 3.2.4 are equivalent.*

**Theorem 3.4.4** *Let $f(z)$, $g(z)$ and $h(z)$ be three entire functions of $n$ complex variables. Then for $M_{g,D}(R) \leq M_{h,D}(R)$ and all sufficiently large values of $R$,*

$$\rho^{(\alpha,\beta)}[f]_h \leq \rho^{(\alpha,\beta)}[f]_g \text{ and } \lambda^{(\alpha,\beta)}[f]_h \leq \lambda^{(\alpha,\beta)}[f]_g.$$

**Proof.** As $M_{g,D}(R) \leq M_{h,D}(R)$ and $M_{f,D}(R)$ is an increasing function of $R$, we get for all sufficiently large values of $R$ that

$$M_{h,D}^{-1}(r) \leq M_{g,D}^{-1}(r)$$

$$\text{i.e., } \limsup_{R \to \infty} \frac{\alpha(M_{h,D}^{-1}(M_{f,D}(R)))}{\beta(R)} \leq \limsup_{R \to \infty} \frac{\alpha(M_{g,D}^{-1}(M_{f,D}(R)))}{\beta(R)}$$

$$\text{i.e., } \rho^{(\alpha,\beta)}[f]_h \leq \rho^{(\alpha,\beta)}[f]_g.$$

Similarly one can prove that $\lambda^{(\alpha,\beta)}[f]_h \leq \lambda^{(\alpha,\beta)}[f]_g$.

This proves the theorem. ∎

**Theorem 3.4.5** *Let $f(z)$, $g(z)$, $h(z)$ and $k(z)$ be four entire functions of $n$ complex variables such that $0 < \lambda^{(\alpha_1,\beta_1)}[f]_h \leq \rho^{(\alpha_1,\beta_1)}[f]_h < \infty$ and $0 < \lambda^{(\alpha_2,\beta_2)}[g]_k \leq \rho^{(\alpha_2,\beta_2)}[g]_k < \infty$. Then*

$$\frac{\lambda^{(\alpha_1,\beta_1)}[f]_h}{\rho^{(\alpha_2,\beta_2)}[g]_k} \leq \liminf_{R \to \infty} \frac{\alpha_1(M_{h,D}^{-1}(M_{f,D}(R)))}{\alpha_2(M_{k,D}^{-1}(M_{g,D}(\beta_2^{-1}(\beta_1(R)))))} \leq \frac{\lambda^{(\alpha_1,\beta_1)}[f]_h}{\lambda^{(\alpha_2,\beta_2)}[g]_k} \leq$$

$$\limsup_{R \to \infty} \frac{\alpha_1(M_{h,D}^{-1}(M_{f,D}(R)))}{\alpha_2(M_{k,D}^{-1}(M_{g,D}(\beta_2^{-1}(\beta_1(R)))))} \leq \frac{\rho^{(\alpha_1,\beta_1)}[f]_h}{\lambda^{(\alpha_2,\beta_2)}[g]_k}.$$

**Proof.** From the definition of relative order and relative lower order we have for arbitrary positive $\varepsilon$ and for all large values of $R$,

$$\alpha_1(M_{h,D}^{-1}(M_{f,D}(R))) \geqslant \left(\lambda^{(\alpha_1,\beta_1)}[f]_h - \varepsilon\right)\beta(R) \tag{35}$$

and

$$\alpha_2(M_{k,D}^{-1}(M_{g,D}(\beta_2^{-1}(\beta_1(R))))) \leq \left(\rho^{(\alpha_2,\beta_2)}[g]_k + \varepsilon\right)\beta(R). \tag{36}$$

Now from (35) and (36) it follows for all large values of $R$,

$$\frac{\alpha_1(M_{h,D}^{-1}(M_{f,D}(R)))}{\alpha_2(M_{k,D}^{-1}(M_{g,D}(\beta_2^{-1}(\beta_1(R)))))} \geqslant \frac{\lambda^{(\alpha_1,\beta_1)}[f]_h - \varepsilon}{\rho^{(\alpha_2,\beta_2)}[g]_k + \varepsilon}.$$

As $\varepsilon(>0)$ is arbitrary, we obtain that

$$\liminf_{R \to \infty} \frac{\alpha_1(M_{h,D}^{-1}(M_{f,D}(R)))}{\alpha_2(M_{k,D}^{-1}(M_{g,D}(\beta_2^{-1}(\beta_1(R)))))} \geqslant \frac{\lambda^{(\alpha_1,\beta_1)}[f]_h}{\rho^{(\alpha_2,\beta_2)}[g]_k}. \tag{37}$$

Again for a sequence of values of $R$ tending to infinity,

$$\alpha_1(M_{h,D}^{-1}(M_{f,D}(R))) \leq \left(\lambda^{(\alpha_1,\beta_1)}[f]_h + \varepsilon\right)\beta(R) \tag{38}$$

and for all large values of $R$ ,

$$\alpha_2(M_{k,D}^{-1}(M_{g,D}\left(\beta_2^{-1}(\beta_1(R))\right))) \geqslant \left(\lambda^{(\alpha_2,\beta_2)}[g]_k - \varepsilon\right)\beta(R). \qquad (39)$$

So combining (38) and (39) we get for a sequence of values of $R$ tending to infinity ,

$$\frac{\alpha_1(M_{h,D}^{-1}(M_{f,D}(R)))}{\alpha_2(M_{k,D}^{-1}(M_{g,D}\left(\beta_2^{-1}(\beta_1(R))\right)))} \leq \frac{\lambda^{(\alpha_1,\beta_1)}[f]_h + \varepsilon}{\lambda^{(\alpha_2,\beta_2)}[g]_k - \varepsilon}.$$

Since $\varepsilon\,(>0)$ is arbitrary it follows that

$$\liminf_{R\to\infty}\frac{\alpha_1(M_{h,D}^{-1}(M_{f,D}(R)))}{\alpha_2(M_{k,D}^{-1}(M_{g,D}\left(\beta_2^{-1}(\beta_1(R))\right)))} \leq \frac{\lambda^{(\alpha_1,\beta_1)}[f]_h}{\lambda^{(\alpha_2,\beta_2)}[g]_k}. \qquad (40)$$

Also for a sequence of values of $R$ tending to infinity ,

$$\alpha_2(M_{k,D}^{-1}(M_{g,D}\left(\beta_2^{-1}(\beta_1(R))\right))) \leq \left(\lambda^{(\alpha_2,\beta_2)}[g]_k + \varepsilon\right)\beta(R). \qquad (41)$$

Now from (35) and (41) we obtain for a sequence of values of $R$ tending to infinity ,

$$\frac{\alpha_1(M_{h,D}^{-1}(M_{f,D}(R)))}{\alpha_2(M_{k,D}^{-1}(M_{g,D}\left(\beta_2^{-1}(\beta_1(R))\right)))} \geqslant \frac{\lambda^{(\alpha_1,\beta_1)}[f]_h - \varepsilon}{\lambda^{(\alpha_2,\beta_2)}[g]_k + \varepsilon}.$$

Choosing $\varepsilon\to 0$ we get from above that

$$\limsup_{R\to\infty}\frac{\alpha_1(M_{h,D}^{-1}(M_{f,D}(R)))}{\alpha_2(M_{k,D}^{-1}(M_{g,D}\left(\beta_2^{-1}(\beta_1(R))\right)))} \geqslant \frac{\lambda^{(\alpha_1,\beta_1)}[f]_h}{\lambda^{(\alpha_2,\beta_2)}[g]_k}. \qquad (42)$$

Also for all large values of $R$ ,

$$\alpha_1(M_{h,D}^{-1}(M_{f,D}(R))) \leq \left(\rho^{(\alpha_1,\beta_1)}[f]_h + \varepsilon\right)\beta(R). \qquad (43)$$

So from (39) and (43) it follows for all large values of $R$ ,

$$\frac{\alpha_1(M_{h,D}^{-1}(M_{f,D}(R)))}{\alpha_2(M_{k,D}^{-1}(M_{g,D}\left(\beta_2^{-1}(\beta_1(R))\right)))} \leq \frac{\rho^{(\alpha_1,\beta_1)}[f]_h + \varepsilon}{\lambda^{(\alpha_2,\beta_2)}[g]_k - \varepsilon}.$$

As $\varepsilon\,(>0)$ is arbitrary we obtain that

$$\limsup_{R\to\infty}\frac{\alpha_1(M_{h,D}^{-1}(M_{f,D}(R)))}{\alpha_2(M_{k,D}^{-1}(M_{g,D}\left(\beta_2^{-1}(\beta_1(R))\right)))} \leq \frac{\rho^{(\alpha_1,\beta_1)}[f]_h}{\lambda^{(\alpha_2,\beta_2)}[g]_k}. \qquad (44)$$

Thus the theorem follows from $(37),(40),(42)$ and $(44)$ . ∎

**Theorem 3.4.6** *Let* $f(z)$, $g(z)$, $h(z)$ *and* $k(z)$ *be four entire functions of* $n$ *complex variables such that* $0 < \lambda^{(\alpha_1,\beta_1)}[f]_h \leq \rho^{(\alpha_1,\beta_1)}[f]_h < \infty$ *and* $0 < \rho^{(\alpha_2,\beta_2)}[g]_k < \infty$. *Then*

$$\liminf_{R\to\infty}\frac{\alpha_1(M_{h,D}^{-1}(M_{f,D}(R)))}{\alpha_2(M_{k,D}^{-1}(M_{g,D}\left(\beta_2^{-1}(\beta_1(R))\right)))} \leq \frac{\rho^{(\alpha_1,\beta_1)}[f]_h}{\rho^{(\alpha_2,\beta_2)}[g]_k} \leq \limsup_{R\to\infty}\frac{\alpha_1(M_{h,D}^{-1}(M_{f,D}(R)))}{\alpha_2(M_{k,D}^{-1}(M_{g,D}\left(\beta_2^{-1}(\beta_1(R))\right)))}.$$

**Proof.** From the definition of relative order we get for a sequence of values of $R$ tending to infinity ,

$$\alpha_2(M_{k,D}^{-1}(M_{g,D}\left(\beta_2^{-1}(\beta_1(R))\right))) \geqslant \left(\rho^{(\alpha_2,\beta_2)}\left[g\right]_k - \varepsilon\right)\beta\left(R\right). \tag{45}$$

Now from (43) and (45) it follows for a sequence of values of $R$ tending to infinity ,

$$\frac{\alpha_1(M_{h,D}^{-1}(M_{f,D}\left(R\right)))}{\alpha_2(M_{k,D}^{-1}(M_{g,D}\left(\beta_2^{-1}(\beta_1(R)))))} \leq \frac{\rho^{(\alpha_1,\beta_1)}\left[f\right]_h + \varepsilon}{\rho^{(\alpha_2,\beta_2)}\left[g\right]_k - \varepsilon}.$$

As $\varepsilon\,(>0)$ is arbitrary we obtain that

$$\liminf_{R\to\infty}\frac{\alpha_1(M_{h,D}^{-1}(M_{f,D}\left(R\right)))}{\alpha_2(M_{k,D}^{-1}(M_{g,D}\left(\beta_2^{-1}(\beta_1(R)))))} \leq \frac{\rho^{(\alpha_1,\beta_1)}\left[f\right]_h}{\rho^{(\alpha_2,\beta_2)}\left[g\right]_k}. \tag{46}$$

Again , for a sequence of values of $R$ tending to infinity ,

$$\alpha_1(M_{h,D}^{-1}(M_{f,D}\left(R\right))) \geqslant \left(\rho^{(\alpha_1,\beta_1)}\left[f\right]_h - \varepsilon\right)\beta\left(R\right). \tag{47}$$

So combining (36) and (47) we get for a sequence of values of $R$ tending to infinity ,

$$\frac{\alpha_1(M_{h,D}^{-1}(M_{f,D}\left(R\right)))}{\alpha_2(M_{k,D}^{-1}(M_{g,D}\left(\beta_2^{-1}(\beta_1(R)))))} \geqslant \frac{\rho^{(\alpha_1,\beta_1)}\left[f\right]_h - \varepsilon}{\rho^{(\alpha_2,\beta_2)}\left[g\right]_k + \varepsilon}.$$

Since $\varepsilon\,(>0)$ is arbitrary it follows that

$$\limsup_{R\to\infty}\frac{\alpha_1(M_{h,D}^{-1}(M_{f,D}\left(R\right)))}{\alpha_2(M_{k,D}^{-1}(M_{g,D}\left(\beta_2^{-1}(\beta_1(R)))))} \geqslant \frac{\rho^{(\alpha_1,\beta_1)}\left[f\right]_h}{\rho^{(\alpha_2,\beta_2)}\left[g\right]_k}. \tag{48}$$

Thus the theorem follows from (46) and (48) . ∎

The following theorem is a natural consequence of Theorem 3.4.5 and Theorem 3.4.6.

**Theorem 3.4.7** *Let $f\,(z)$, $g(z)$, $h(z)$ and $k\,(z)$ be four entire functions of $n$ complex variables such that $0 < \lambda^{(\alpha_1,\beta_1)}\left[f\right]_h \leq \rho^{(\alpha_1,\beta_1)}\left[f\right]_h < \infty$ and $0 < \lambda^{(\alpha_2,\beta_2)}\left[g\right]_k \leq \rho^{(\alpha_2,\beta_2)}\left[g\right]_k < \infty$. Then*

$$\liminf_{R\to\infty}\frac{\alpha_1(M_{h,D}^{-1}(M_{f,D}\left(R\right)))}{\alpha_2(M_{k,D}^{-1}(M_{g,D}\left(\beta_2^{-1}(\beta_1(R)))))} \leq \min\left\{\frac{\lambda^{(\alpha_1,\beta_1)}\left[f\right]_h}{\lambda^{(\alpha_2,\beta_2)}\left[g\right]_k}, \frac{\rho^{(\alpha_1,\beta_1)}\left[f\right]_h}{\rho^{(\alpha_2,\beta_2)}\left[g\right]_k}\right\} \leq$$

$$\max\left\{\frac{\lambda^{(\alpha_1,\beta_1)}\left[f\right]_h}{\lambda^{(\alpha_2,\beta_2)}\left[g\right]_k}, \frac{\rho^{(\alpha_1,\beta_1)}\left[f\right]_h}{\rho^{(\alpha_2,\beta_2)}\left[g\right]_k}\right\} \leq \limsup_{R\to\infty}\frac{\alpha_1(M_{h,D}^{-1}(M_{f,D}\left(R\right)))}{\alpha_2(M_{k,D}^{-1}(M_{g,D}\left(\beta_2^{-1}(\beta_1(R)))))}.$$

Next we introduce the following two relative growth indicators which will also enable help our subsequent study.

**Definition 3.4.1** *Let $f(z)$ and $g(z)$ be any two entire functions of $n$ complex variables. The generalized relative hyper Gol'dberg order $(\alpha, \beta)$ and generalized relative hyper Gol'dberg lower order $(\alpha, \beta)$ of $f(z)$ with respect to $g(z)$ are defined by:*

$$
\begin{aligned}
\overline{\rho}_D^{(\alpha,\beta)}[f]_g \\
\overline{\lambda}_D^{(\alpha,\beta)}[f]_g
\end{aligned}
= \lim_{R \to \infty} \begin{array}{c} \sup \\ \inf \end{array} \frac{\alpha(\log(M_{g,D}^{-1}(M_{f,D}(R))))}{\beta(R)}.
$$

**Definition 3.4.2** *Let $f(z)$ and $g(z)$ be any two entire functions of $n$ complex variables. The generalized relative logarithmic Gol'dberg order $(\alpha, \beta)$ and generalized relative logarithmic Gol'dberg lower order $(\alpha, \beta)$ of $f(z)$ with respect to $g(z)$ are defined by:*

$$
\begin{aligned}
\underline{\rho}_D^{(\alpha,\beta)}[f]_g \\
\underline{\lambda}_D^{(\alpha,\beta)}[f]_g
\end{aligned}
= \lim_{R \to \infty} \begin{array}{c} \sup \\ \inf \end{array} \frac{\alpha(M_{g,D}^{-1}(M_{f,D}(R)))}{\beta(\log R)}.
$$

**Remark 3.4.1** *In the line of Theorem 2.3.1, one can easily verify that generalized relative hyper Gol'dberg order $(\alpha, \beta)$ (respectively generalized relative hyper lower Gol'dberg order $(\alpha, \beta)$) and generalized relative logarithmic Gol'dberg order $(\alpha, \beta)$ (respectively generalized relative logarithmic lower Gol'dberg order $(\alpha, \beta)$) are independent of the choice of the domain. So after this we shall always write, $\overline{\rho}^{(\alpha,\beta)}[f]_g \left(\overline{\lambda}^{(\alpha,\beta)}[f]_g\right)$ and $\underline{\rho}^{(\alpha,\beta)}[f]_g \left(\underline{\lambda}^{(\alpha,\beta)}[f]_g\right)$ instead of $\overline{\rho}_D^{(\alpha,\beta)}[f]_g \left(\overline{\lambda}_D^{(\alpha,\beta)}[f]_g\right)$ and $\underline{\rho}_D^{(\alpha,\beta)}[f]_g \left(\underline{\lambda}_D^{(\alpha,\beta)}[f]_g\right)$ respectively.*

We may now state the following three theorems without proof based on generalized relative hyper Gol'dberg order $(\alpha, \beta)$ of an entire function of $n$ complex variables.

**Theorem 3.4.8** *Let $f(z)$, $g(z)$, $h(z)$ and $k(z)$ be four entire functions of $n$ complex variables such that $0 < \overline{\lambda}^{(\alpha_1,\beta_1)}[f]_h \leq \overline{\rho}^{(\alpha_1,\beta_1)}[f]_h < \infty$ and $0 < \overline{\lambda}^{(\alpha_2,\beta_2)}[g]_k \leq \overline{\rho}^{(\alpha_2,\beta_2)}[g]_k < \infty$. Then*

$$
\frac{\overline{\lambda}^{(\alpha_1,\beta_1)}[f]_h}{\overline{\rho}^{(\alpha_2,\beta_2)}[g]_k} \leq \liminf_{R \to \infty} \frac{\alpha_1(\log(M_{h,D}^{-1}(M_{f,D}(R))))}{\alpha_2(\log(M_{k,D}^{-1}(M_{g,D}(\beta_2^{-1}(\beta_1(R))))))} \leq \frac{\overline{\lambda}^{(\alpha_1,\beta_1)}[f]_h}{\overline{\lambda}^{(\alpha_2,\beta_2)}[g]_k} \leq
$$

$$
\limsup_{R \to \infty} \frac{\alpha_1(\log(M_{h,D}^{-1}(M_{f,D}(R))))}{\alpha_2(\log(M_{k,D}^{-1}(M_{g,D}(\beta_2^{-1}(\beta_1(R))))))} \leq \frac{\overline{\rho}^{(\alpha_1,\beta_1)}[f]_h}{\overline{\lambda}^{(\alpha_2,\beta_2)}[g]_k}.
$$

**Theorem 3.4.9** *Let $f(z)$, $g(z)$, $h(z)$ and $k(z)$ be four entire functions of $n$ complex variables such that $0 < \overline{\lambda}^{(\alpha_1,\beta_1)}[f]_h \leq \overline{\rho}^{(\alpha_1,\beta_1)}[f]_h < \infty$ and $0 < \overline{\rho}^{(\alpha_2,\beta_2)}[g]_k < \infty$. Then*

$$
\liminf_{R \to \infty} \frac{\alpha_1(\log(M_{h,D}^{-1}(M_{f,D}(R))))}{\alpha_2(\log(M_{k,D}^{-1}(M_{g,D}(\beta_2^{-1}(\beta_1(R))))))} \leq \frac{\overline{\rho}^{(\alpha_1,\beta_1)}[f]_h}{\overline{\rho}^{(\alpha_2,\beta_2)}[g]_k}
$$

$$
\leq \limsup_{R \to \infty} \frac{\alpha_1(\log(M_{h,D}^{-1}(M_{f,D}(R))))}{\alpha_2(\log(M_{k,D}^{-1}(M_{g,D}(\beta_2^{-1}(\beta_1(R))))))}.
$$

**Theorem 3.4.10** *Let* $f(z)$, $g(z)$, $h(z)$ *and* $k(z)$ *be four entire functions of* $n$ *complex variables such that* $0 < \overline{\lambda}^{(\alpha_1,\beta_1)}[f]_h \leq \overline{\rho}^{(\alpha_1,\beta_1)}[f]_h < \infty$ *and* $0 < \overline{\lambda}^{(\alpha_2,\beta_2)}[g]_k \leq \overline{\rho}^{(\alpha_2,\beta_2)}[g]_k < \infty$. *Then*

$$\liminf_{R\to\infty} \frac{\alpha_1(\log(M_{h,D}^{-1}(M_{f,D}(R))))}{\alpha_2(\log(M_{k,D}^{-1}(M_{g,D}(\beta_2^{-1}(\beta_1(R))))))} \leq \min\left\{ \frac{\overline{\lambda}^{(\alpha_1,\beta_1)}[f]_h}{\overline{\lambda}^{(\alpha_2,\beta_2)}[g]_k}, \frac{\overline{\rho}^{(\alpha_1,\beta_1)}[f]_h}{\overline{\rho}^{(\alpha_2,\beta_2)}[g]_k} \right\} \leq$$

$$\max\left\{ \frac{\overline{\lambda}^{(\alpha_1,\beta_1)}[f]_h}{\overline{\lambda}^{(\alpha_2,\beta_2)}[g]_k}, \frac{\overline{\rho}^{(\alpha_1,\beta_1)}[f]_h}{\overline{\rho}^{(\alpha_2,\beta_2)}[g]_k} \right\} \leq \limsup_{R\to\infty} \frac{\alpha_1(\log(M_{h,D}^{-1}(M_{f,D}(R))))}{\alpha_2(\log(M_{k,D}^{-1}(M_{g,D}(\beta_2^{-1}(\beta_1(R))))))}.$$

Using the concept of generalized relative logarithmic Gol'dberg order $(\alpha, \beta)$ of an entire function of $n$ complex variables, one may prove the following theorems. We omit the details.

**Theorem 3.4.11** *Let* $f(z)$, $g(z)$, $h(z)$ *and* $k(z)$ *be four entire functions of* $n$ *complex variables such that* $0 < \underline{\lambda}^{(\alpha_1,\beta_1)}[f]_h \leq \underline{\rho}^{(\alpha_1,\beta_1)}[f]_h < \infty$ *and* $0 < \underline{\lambda}^{(\alpha_2,\beta_2)}[g]_k \leq \underline{\rho}^{(\alpha_2,\beta_2)}[g]_k < \infty$. *Then*

$$\frac{\underline{\lambda}^{(\alpha_1,\beta_1)}[f]_h}{\underline{\rho}^{(\alpha_2,\beta_2)}[g]_k} \leq \liminf_{R\to\infty} \frac{\alpha_1(M_{h,D}^{-1}(M_{f,D}(R)))}{\alpha_2(M_{k,D}^{-1}(M_{g,D}(\exp(\beta_2^{-1}(\beta_1(\log R))))))} \leq \frac{\underline{\lambda}^{(\alpha_1,\beta_1)}[f]_h}{\underline{\lambda}^{(\alpha_2,\beta_2)}[g]_k} \leq$$

$$\limsup_{R\to\infty} \frac{\alpha_1(M_{h,D}^{-1}(M_{f,D}(R)))}{\alpha_2(M_{k,D}^{-1}(M_{g,D}(\exp(\beta_2^{-1}(\beta_1(\log R))))))} \leq \frac{\underline{\rho}^{(\alpha_1,\beta_1)}[f]_h}{\underline{\lambda}^{(\alpha_2,\beta_2)}[g]_k}.$$

**Theorem 3.4.12** *Let* $f(z)$, $g(z)$, $h(z)$ *and* $k(z)$ *be four entire functions of* $n$ *complex variables such that* $0 < \underline{\lambda}^{(\alpha_1,\beta_1)}[f]_h \leq \underline{\rho}^{(\alpha_1,\beta_1)}[f]_h < \infty$ *and* $0 < \underline{\rho}^{(\alpha_2,\beta_2)}[g]_k < \infty$. *Then*

$$\liminf_{R\to\infty} \frac{\alpha_1(M_{h,D}^{-1}(M_{f,D}(R)))}{\alpha_2(M_{k,D}^{-1}(M_{g,D}(\exp(\beta_2^{-1}(\beta_1(\log R))))))} \leq \frac{\rho^{(\alpha_1,\beta_1)}[f]_h}{\rho^{(\alpha_2,\beta_2)}[g]_k}$$

$$\leq \limsup_{R\to\infty} \frac{\alpha_1(M_{h,D}^{-1}(M_{f,D}(R)))}{\alpha_2(M_{k,D}^{-1}(M_{g,D}\exp(\beta_2^{-1}(\beta_1(\log R)))))}.$$

**Theorem 3.4.13** *Let* $f(z)$, $g(z)$, $h(z)$ *and* $k(z)$ *be four entire functions of* $n$ *complex variables such that* $0 < \underline{\lambda}^{(\alpha_1,\beta_1)}[f]_h \leq \underline{\rho}^{(\alpha_1,\beta_1)}[f]_h < \infty$ *and* $0 < \underline{\lambda}^{(\alpha_2,\beta_2)}[g]_k \leq \underline{\rho}^{(\alpha_2,\beta_2)}[g]_k < \infty$. *Then*

$$\liminf_{R\to\infty} \frac{\alpha_1(M_{h,D}^{-1}(M_{f,D}(R)))}{\alpha_2(M_{k,D}^{-1}(M_{g,D}(\exp(\beta_2^{-1}(\beta_1(\log R))))))} \leq \min\left\{ \frac{\underline{\lambda}^{(\alpha_1,\beta_1)}[f]_h}{\underline{\lambda}^{(\alpha_2,\beta_2)}[g]_k}, \frac{\rho^{(\alpha_1,\beta_1)}[f]_h}{\rho^{(\alpha_2,\beta_2)}[g]_k} \right\} \leq$$

$$\max\left\{ \frac{\underline{\lambda}^{(\alpha_1,\beta_1)}[f]_h}{\underline{\lambda}^{(\alpha_2,\beta_2)}[g]_k}, \frac{\rho^{(\alpha_1,\beta_1)}[f]_h}{\rho^{(\alpha_2,\beta_2)}[g]_k} \right\} \leq \limsup_{R\to\infty} \frac{\alpha_1(M_{h,D}^{-1}(M_{f,D}(R)))}{\alpha_2(M_{k,D}^{-1}(M_{g,D}(\exp(\beta_2^{-1}(\beta_1(\log R))))))}.$$

# 3.5   Conclusion.

The main aim of this chapter is actually to extend and to modify the notion of generalized Gol'dberg order $(\alpha, \beta)$ and generalized Gol'dberg lower order $(\alpha, \beta)$ to generalized relative Gol'dberg order $(\alpha, \beta)$ and generalized relative Gol'dberg lower order $(\alpha, \beta)$ of higher dimensions in case of entire functions of $n$ complex variables and establish its integral representation. Here we see that the previous definitions are easily generated as particular cases of the present definitions, e.g. if $\alpha(R) = \log^{[p]} R$ and $\beta(R) = \log^{[q]} R$ respectively, then Definition 3.2.1 and Definition 3.2.3 reduce to Definition 1.1.20.

However the results of this chapter are basically inclined in proving some bounds involving the growth ratios of entire functions of $n$ complex variables. These mostly reflect the finiteness of the ratios as mentioned earlier. But we also may have a scope to establish some theorems in order to get more illuminating results. Keeping this in mind, the theories as well as the results of the next chapter have been tackled under some different conditions.

## References

[1] B. C. Mondal and C. Roy: Relative gol'dberg order of an entire function of several variables, Bull Cal. Math. Soc., 102(4) (2010), 371-380.

# Chapter 4

# Some inequalities using generalized relative Gol'dberg order $(\alpha, \beta)$ and generalized relative Gol'dberg lower order $(\alpha, \beta)$ of entire functions of several complex variables

**Abstract:** In this chapter, Some inequalities using generalized Gol'dberg order $(\alpha, \beta)$, generalized Gol'dberg lower order $(\alpha, \beta)$, generalized relative Gol'dberg order $(\alpha, \beta)$ and generalized relative Gol'dberg lower order $(\alpha, \beta)$ of entire functions of several complex variables are established, where $\alpha, \beta$ are continuous non-negative functions defined on $(-\infty, +\infty)$.

**Keywords:** Entire function, several complex variables, generalized Gol'dberg order $(\alpha, \beta)$, generalized relative Gol'dberg order $(\alpha, \beta)$, increasing function.

**Mathematics Subject Classification (2010) :** 32A15.

## 4.1 Introduction.

The relative Gol'dberg order of an entire function of $n$ complex variables gives a quantitative assessment of how different functions scale each other and until what extent they are self-similar in growth. In Chapter Two and Chapter Three, we give relevant notations and definitions of $\rho^{(\alpha,\beta)}[f]$, $\lambda^{(\alpha,\beta)}[f]$, $\rho^{(\alpha,\beta)}[f]_g$, $\lambda^{(\alpha,\beta)}[f]_g$ etc. In this chapter we discuss some growth rates of entire functions of $n$ complex variables on the basis of the generalized Gol'dberg order $(\alpha, \beta)$, generalized Gol'dberg lower order $(\alpha, \beta)$, generalized relative Gol'dberg order $(\alpha, \beta)$ and generalized relative Gol'dberg lower order $(\alpha, \beta)$ where $\alpha, \beta \in L_0$. Further we assume that throughout the present chapter $\alpha, \beta$ and $\gamma$ always denote the functions belonging to $L^0$.

## 4.2   Main Results.

In this section we present the main results of this chapter.

**Theorem 4.2.1** *Let $f(z)$ and $g(z)$ be two entire functions of $n$ complex variables such that $0 < \lambda^{(\gamma,\beta)}[f] \le \rho^{(\gamma,\beta)}[f] < \infty$ and $0 < \lambda^{(\gamma,\alpha)}[g] \le \rho^{(\gamma,\alpha)}[g] < \infty$ . Then*

$$\frac{\lambda^{(\gamma,\beta)}[f]}{\rho^{(\gamma,\alpha)}[g]} \le \lambda^{(\alpha,\beta)}[f]_g \le \min\left\{\frac{\lambda^{(\gamma,\beta)}[f]}{\lambda^{(\gamma,\alpha)}[g]}, \frac{\rho^{(\gamma,\beta)}[f]}{\rho^{(\gamma,\alpha)}[g]}\right\}$$

$$\le \max\left\{\frac{\lambda^{(\gamma,\beta)}[f]}{\lambda^{(\gamma,\alpha)}[g]}, \frac{\rho^{(\gamma,\beta)}[f]}{\rho^{(\gamma,\alpha)}[g]}\right\} \le \rho^{(\alpha,\beta)}[f]_g \le \frac{\rho^{(\gamma,\beta)}[f]}{\lambda^{(\gamma,\alpha)}[g]}.$$

**Proof.** From the definitions of $\rho^{(\gamma,\beta)}[f]$ and $\lambda^{(\gamma,\beta)}[f]$, we have for all sufficiently large values of $R$ that

$$M_{f,D}(R) \le \gamma^{-1}((\rho^{(\gamma,\beta)}[f]+\varepsilon)\,\beta(R)), \tag{49}$$

$$M_{f,D}(R) \ge \gamma^{-1}((\lambda^{(\gamma,\beta)}[f]-\varepsilon)\,\beta(R)) \tag{50}$$

and also for a sequence of values of $R$ tending to infinity we get that

$$M_{f,D}(R) \ge \gamma^{-1}((\rho^{(\gamma,\beta)}[f]-\varepsilon)\,\beta(R)), \tag{51}$$

$$M_{f,D}(R) \le \gamma^{-1}((\lambda^{(\gamma,\beta)}[f]+\varepsilon)\,\beta(R)). \tag{52}$$

Similarly from the definitions of $\rho^{(\gamma,\alpha)}[g]$ and $\lambda^{(\gamma,\alpha)}[g]$, it follows for all sufficiently large values of $R$ that

$$M_{g,D}(R) \le \gamma^{-1}((\rho^{(\gamma,\alpha)}[g]+\varepsilon)\,\alpha(R))$$

$$i.e.,\ \ R \le M_{g,D}^{-1}\left(\gamma^{-1}((\rho^{(\gamma,\alpha)}[g]+\varepsilon)\,\alpha(R))\right)$$

$$i.e.,\ \ M_{g,D}^{-1}(R) \ge \alpha^{-1}\left(\frac{\gamma(R)}{(\rho^{(\gamma,\alpha)}[g]+\varepsilon)}\right), \tag{53}$$

$$M_{g,D}(R) \ge \gamma^{-1}((\lambda^{(\gamma,\alpha)}[g]-\varepsilon)\,\alpha(R))$$

$$i.e.,\ \ M_{g,D}^{-1}(R) \le \alpha^{-1}\left(\frac{\gamma(R)}{(\lambda^{(\gamma,\alpha)}[g]-\varepsilon)}\right) \tag{54}$$

and for a sequence of values of $R$ tending to infinity we obtain that

$$M_{g,D}(R) \ge \gamma^{-1}((\rho^{(\gamma,\alpha)}[g]-\varepsilon)\,\alpha(R))$$

$$i.e.\ \ M_{g,D}^{-1}(R) \le \alpha^{-1}\left(\frac{\gamma(R)}{(\rho^{(\gamma,\alpha)}[g]-\varepsilon)}\right), \tag{55}$$

$$M_{g,D}(R) \le \gamma^{-1}((\lambda^{(\gamma,\alpha)}[g]+\varepsilon)\,\alpha(R))$$

$$i.e.,\ \ M_{g,D}^{-1}(R) \ge \alpha^{-1}\left(\frac{\gamma(R)}{(\lambda^{(\gamma,\alpha)}[g]+\varepsilon)}\right). \tag{56}$$

Now from (51) and in view of (53), for a sequence of values of $R$ tending to infinity we get that

$$\alpha(M_{g,D}^{-1}(M_{f,D}(R))) \geq \alpha(M_{g,D}^{-1}(\gamma^{-1}((\rho^{(\gamma,\beta)}[f] - \varepsilon)\beta(R))))$$

i.e., $\alpha(M_{g,D}^{-1}(M_{f,D}(R))) \geq \alpha\left(\alpha^{-1}\left(\dfrac{\gamma(\gamma^{-1}((\rho^{(\gamma,\beta)}[f] - \varepsilon)\beta(R)))}{(\rho^{(\gamma,\alpha)}[g] + \varepsilon)}\right)\right)$

$$= \dfrac{(\rho^{(\gamma,\beta)}[f] - \varepsilon)}{(\rho^{(\gamma,\alpha)}[g] + \varepsilon)}\beta(R)$$

i.e., $\dfrac{\alpha(M_{g,D}^{-1}(M_{f,D}(R)))}{\beta(R)} \geq \dfrac{(\rho^{(\gamma,\beta)}[f] - \varepsilon)}{(\rho^{(\gamma,\alpha)}[g] + \varepsilon)}.$

As $\varepsilon\,(> 0)$ is arbitrary, it follows that

$$\rho_{(\alpha,\beta)}[f] \geq \frac{\rho^{(\gamma,\beta)}[f]}{\rho^{(\gamma,\alpha)}[g]}. \tag{57}$$

Analogously, from (50) and in view of (56) it follows for a sequence of values of $R$ tending to infinity that

$$\alpha(M_{g,D}^{-1}(M_{f,D}(R))) \geq \alpha(M_{g,D}^{-1}(\gamma^{-1}((\lambda^{(\gamma,\alpha)}[g] - \varepsilon)\beta(R))))$$

i.e., $\alpha(M_{g,D}^{-1}(M_{f,D}(R))) \geq \alpha\left(\alpha^{-1}\left(\dfrac{\gamma(\gamma^{-1}((\lambda^{(\gamma,\beta)}[f] - \varepsilon)\beta(R)))}{(\lambda^{(\gamma,\alpha)}[g] + \varepsilon)}\right)\right)$

$$= \dfrac{(\lambda^{(\gamma,\beta)}[f] - \varepsilon)}{(\lambda^{(\gamma,\alpha)}[g] + \varepsilon)}\beta(R)$$

i.e., $\dfrac{\alpha(M_{g,D}^{-1}(M_{f,D}(R)))}{\beta(R)} \geq \dfrac{(\lambda^{(\gamma,\beta)}[f] - \varepsilon)}{(\lambda^{(\gamma,\alpha)}[g] + \varepsilon)}.$

Since $\varepsilon\,(> 0)$ is arbitrary, we get from above that

$$\rho^{(\alpha,\beta)}[f]_g \geq \frac{\lambda^{(\gamma,\beta)}[f]}{\lambda^{(\gamma,\alpha)}[g]}. \tag{58}$$

Again in view of (54), we have from (49) for all sufficiently large values of $R$ that

$$\alpha(M_{g,D}^{-1}(M_{f,D}(R))) \leq \alpha(M_{g,D}^{-1}(\gamma^{-1}((\rho^{(\gamma,\beta)}[f] + \varepsilon)\beta(R))))$$

i.e., $\alpha(M_{g,D}^{-1}(M_{f,D}(R))) \leq \alpha\left(\alpha^{-1}\left(\dfrac{\gamma(\gamma^{-1}((\rho^{(\gamma,\beta)}[f] + \varepsilon)\beta(R)))}{(\lambda^{(\gamma,\alpha)}[g] - \varepsilon)}\right)\right)$

$$= \dfrac{(\rho^{(\gamma,\beta)}[f] + \varepsilon)}{(\lambda^{(\gamma,\alpha)}[g] - \varepsilon)}\beta(R)$$

$$i.e., \quad \frac{\alpha(M_{g,D}^{-1}(M_{f,D}(R)))}{\beta(R)} \leq \frac{\left(\rho^{(\gamma,\beta)}[f] + \varepsilon\right)}{\left(\lambda^{(\gamma,\alpha)}[g] - \varepsilon\right)}.$$

Since $\varepsilon \, (> 0)$ is arbitrary, we obtain that

$$\rho_{(\alpha,\beta)}[f] \leq \frac{\rho^{(\gamma,\beta)}[f]}{\lambda^{(\gamma,\alpha)}[g]}. \tag{59}$$

Again from (50) and in view of (53) with the same reasoning, we get that

$$\lambda^{(\alpha,\beta)}[f]_g \geq \frac{\lambda^{(\gamma,\beta)}[f]}{\rho^{(\gamma,\alpha)}[g]}. \tag{60}$$

Also in view of $(55)$, we get from (49) for a sequence of values of $R$ tending to infinity that

$$\alpha(M_{g,D}^{-1}(M_{f,D}(R))) \leq \alpha(M_{g,D}^{-1}\left(\gamma^{-1}((\rho^{(\gamma,\beta)}[f] + \varepsilon)\,\beta(R))\right))$$

$$i.e., \quad \alpha(M_{g,D}^{-1}(M_{f,D}(R))) \leq \alpha\left(\alpha^{-1}\left(\frac{\gamma(\gamma^{-1}((\rho^{(\gamma,\beta)}[f] + \varepsilon)\beta(R)))}{(\rho^{(\gamma,\alpha)}[g] - \varepsilon)}\right)\right)$$

$$= \frac{\left(\rho^{(\gamma,\beta)}[f] + \varepsilon\right)}{\left(\rho^{(\gamma,\alpha)}[g] - \varepsilon\right)}\beta(R)$$

$$i.e., \quad \frac{\alpha(M_{g,D}^{-1}(M_{f,D}(R)))}{\beta(R)} \leq \frac{\left(\rho^{(\gamma,\beta)}[f] + \varepsilon\right)}{\left(\rho^{(\gamma,\alpha)}[g] - \varepsilon\right)}.$$

Since $\varepsilon \, (> 0)$ is arbitrary, we get from above that

$$\lambda^{(\alpha,\beta)}[f]_g \leq \frac{\rho^{(\gamma,\beta)}[f]}{\rho^{(\gamma,\alpha)}[g]}. \tag{61}$$

Similarly, from (52) and in view of $(54)$, it follows for a sequence of values of $R$ tending to infinity that

$$\alpha(M_{g,D}^{-1}(M_{f,D}(R))) \leq \alpha(M_{g,D}^{-1}\left(\gamma^{-1}((\lambda^{(\gamma,\beta)}[f] + \varepsilon)\,\beta(R))\right))$$

$$i.e., \quad \alpha(M_{g,D}^{-1}(M_{f,D}(R))) \leq \alpha\left(\alpha^{-1}\left(\frac{\gamma((\gamma^{-1}(\lambda^{(\gamma,\beta)}[f] + \varepsilon)\beta(R)))}{(\lambda^{(\gamma,\alpha)}[g] - \varepsilon)}\right)\right)$$

$$= \frac{\left(\lambda^{(\gamma,\beta)}[f] + \varepsilon\right)}{\left(\lambda^{(\gamma,\alpha)}[g] - \varepsilon\right)}\beta(R)$$

$$i.e., \quad \frac{\alpha(M_{g,D}^{-1}(M_{f,D}(R)))}{\beta(R)} \leq \frac{\left(\lambda^{(\gamma,\beta)}[f] + \varepsilon\right)}{\left(\lambda^{(\gamma,\alpha)}[g] - \varepsilon\right)}.$$

As $\varepsilon \, (> 0)$ is arbitrary, we obtain from above that

$$\lambda^{(\alpha,\beta)}[f]_g \leq \frac{\lambda^{(\gamma,\beta)}[f]}{\lambda^{(\gamma,\alpha)}[g]}. \tag{62}$$

The theorem follows from $(57)$, $(58)$, $(59)$, $(60)$, $(61)$ and $(62)$. ∎

In view of Theorem 4.2.1, one can easily verify the following corollaries :

**Corollary 4.2.1** *Let $f(z)$ and $g(z)$ be two entire functions of $n$ complex variables such that $0 < \rho^{(\gamma,\beta)}[f] < \infty$ and $0 < \lambda^{(\gamma,\alpha)}[g] = \rho^{(\gamma,\alpha)}[g] < \infty$. Then*

$$\lambda^{(\alpha,\beta)}[f]_g = \frac{\lambda^{(\gamma,\beta)}[f]}{\rho^{(\gamma,\alpha)}[g]} \quad and \quad \rho^{(\alpha,\beta)}[f]_g = \frac{\rho^{(\gamma,\beta)}[f]}{\rho^{(\gamma,\alpha)}[g]}.$$

*In addition, if $\rho^{(\gamma,\beta)}[f] = \rho^{(\gamma,\alpha)}[g]$, then*

$$\rho^{(\alpha,\beta)}[f]_g = \lambda^{(\alpha,\beta)}[f]_g = 1.$$

**Corollary 4.2.2** *Let $f(z)$ and $g(z)$ be two entire functions of $n$ complex variables such that $0 < \lambda^{(\gamma,\beta)}[f] = \rho^{(\gamma,\beta)}[f] < \infty$ and $0 < \lambda^{(\gamma,\alpha)}[g] = \rho^{(\gamma,\alpha)}[g] < \infty$. Then*

$$\lambda^{(\alpha,\beta)}[f]_g = \rho^{(\alpha,\beta)}[f]_g = \frac{\rho^{(\gamma,\beta)}[f]}{\rho^{(\gamma,\alpha)}[g]}.$$

**Corollary 4.2.3** *Let $f(z)$ and $g(z)$ be two entire functions of $n$ complex variables such that $0 < \lambda^{(\gamma,\beta)}[f] = \rho^{(\gamma,\beta)}[f] < \infty$ and $0 < \lambda^{(\gamma,\alpha)}[g] = \rho^{(\gamma,\alpha)}[g] < \infty$. Also suppose that $\rho^{(\gamma,\beta)}[f] = \rho^{(\gamma,\alpha)}[g]$. Then*

$$\lambda^{(\alpha,\beta)}[f]_g = \rho^{(\alpha,\beta)}[f]_g = \lambda^{(\beta,\alpha)}[g]_f = \rho^{(\beta,\alpha)}[g]_f = 1.$$

**Corollary 4.2.4** *Let $f(z)$ and $g(z)$ be two entire functions of $n$ complex variables such that $0 < \lambda^{(\gamma,\beta)}[f] = \rho^{(\gamma,\beta)}[f] < \infty$ and $0 < \lambda^{(\gamma,\alpha)}[g] = \rho^{(\gamma,\alpha)}[g] < \infty$. Then*

$$\rho^{(\alpha,\beta)}[f]_g \cdot \rho^{(\beta,\alpha)}[g]_f = \lambda^{(\alpha,\beta)}[f]_g \cdot \lambda^{(\beta,\alpha)}[g]_f = 1.$$

**Corollary 4.2.5** *Let $f(z)$ and $g(z)$ be two entire functions of $n$ complex variables such that $0 < \rho^{(\gamma,\beta)}[f] < \infty$ and $0 < \rho^{(\gamma,\alpha)}[g] < \infty$. Also let either $f(z)$ is not of regular generalized relative Gol'dberg $(\gamma, \beta)$ growth or $g(z)$ is not of regular generalized relative Gol'dberg $(\gamma, \alpha)$ growth, then*

$$\lambda^{(\alpha,\beta)}[f]_g \cdot \lambda^{(\beta,\alpha)}[g]_f < 1 \ < \rho^{(\alpha,\beta)}[f]_g \cdot \rho^{(\beta,\alpha)}[g]_f.$$

**Corollary 4.2.6** *Let $f(z)$ be an entire function of $n$ complex variables such that $0 < \rho^{(\gamma,\beta)}[f] < \infty$. Then for any entire function $g(z)$,*

$$(i) \quad \lambda^{(\alpha,\beta)}[f]_g = \infty \ when \ \rho^{(\gamma,\alpha)}[g] = 0,$$

$$(ii) \quad \rho^{(\alpha,\beta)}[f]_g = \infty \ when \ \lambda^{(\gamma,\alpha)}[g] = 0,$$

$$(iii) \quad \lambda^{(\alpha,\beta)}[f]_g = 0 \ when \ \rho^{(\gamma,\alpha)}[g] = \infty$$

*and*

$$(iv) \quad \rho^{(\alpha,\beta)}[f]_g = 0 \ when \ \lambda^{(\gamma,\alpha)}[g] = \infty.$$

**Corollary 4.2.7** *Let $g(z)$ be an entire function of $n$ complex variables such that $0 < \rho^{(\gamma,\alpha)}[g] < \infty$. Then for any entire function $f(z)$,*

$$(i) \quad \rho^{(\alpha,\beta)}[f]_g = 0 \ when \ \rho^{(\gamma,\beta)}[f] = 0,$$

$$(ii) \quad \lambda^{(\alpha,\beta)}[f]_g = 0 \ when \ \lambda^{(\gamma,\beta)}[f] = 0,$$

$$(iii) \quad \rho^{(\alpha,\beta)}[f]_g = \infty \ when \ \rho^{(\gamma,\beta)}[f] = \infty$$

*and*

$$(iv) \quad \lambda^{(\alpha,\beta)}[f]_g = \infty \ when \ \lambda^{(\gamma,\beta)}[f] = \infty.$$

# 4.3   Conclusion.

The main aim of this chapter is to find out the limiting value of generalized relative Gol'dberg order $(\alpha, \beta)$ and generalized relative Gol'dberg lower order $(\alpha, \beta)$ under some different conditions. However in the case of generalized relative Gol'dberg order $(\alpha, \beta)$, it therefore seems reasonable to define suitably the generalized relative Gol'dberg type $(\alpha, \beta)$ and generalized relative Gol'dberg weak type $(\alpha, \beta)$ of entire functions of $n$ complex variables. In the next section we wish to introduce the definition of the generalized relative Gol'dberg type $(\alpha, \beta)$ and generalized relative Gol'dberg weak type $(\alpha, \beta)$ of entire functions of $n$ complex variables with respect to another entire functions of $n$ complex variables and establish their integral representations.

# Chapter 5

# Generalized relative Gol'dberg type $(\alpha, \beta)$ and generalized relative Gol'dberg weak type $(\alpha, \beta)$ of entire functions of several complex variables

**Abstract:** In this chapter, we develop some growth properties of entire functions of $n$ complex variables relating to generalized relative Gol'dberg order $(\alpha, \beta)$, generalized relative Gol'dberg type $(\alpha, \beta)$ and generalized relative Gol'dberg weak type $(\alpha, \beta)$.We also establish integral representations of generalized relative Gol'dberg type and weak type $(\alpha, \beta)$ of entire function of several complex variables and derive some interesting results, where $\alpha, \beta$ are continuous non-negative functions defined on $(-\infty, +\infty)$.

**Keywords:** Generalized relative Gol'dberg order $(\alpha, \beta)$, generalized relative Gol'dberg lower order $(\alpha, \beta)$, generalized relative Gol'dberg type $(\alpha, \beta)$, generalized relative Gol'dberg weak type $(\alpha, \beta)$, increasing function.

**Mathematics Subject Classification (2010) :** 32A15.

## 5.1　Introduction.

Mondal et al. [1] defined the concept of relative Gol'dberg order between two entire functions $f(z)$ and $g(z)$ for any bounded complete $n$-circular domain $D$ with center at all the origin $\mathbb{C}^n$. Extending this notion, we have already introduced the definitions of generalized relative Gol'dberg order $(\alpha, \beta)$ and generalized relative Gol'dberg lower order $(\alpha, \beta)$ between two entire functions of several complex variables  Now to compare the growth of entire functions of several complex variables having the same generalized relative Gol'dberg order $(\alpha, \beta)$ or generalized relative Gol'dberg lower order $(\alpha, \beta)$, we wish to introduce the definition of generalized relative Gol'dberg type $(\alpha, \beta)$ and  generalized relative Gol'dberg weak type $(\alpha, \beta)$ of an entire function of several complex variables with respect to another entire function of several complex variables and establish their integral representations. We also investigate their equivalence relations under certain conditions.

## 5.2   Preliminary remarks and definitions.

The definitions of generalized relative Gol'dberg order $(\alpha, \beta)$ and generalized relative Gol'dberg lower order $(\alpha, \beta)$ of $f(z)$ with respect to $g(z)$ where $f(z)$ and $g(z)$ be any two entire functions of $n$ complex variables are as follows:

**Definition 5.2.1** *Let $f(z)$ and $g(z)$ be any two entire functions of $n$ complex variables. The generalized relative Gol'dberg order $(\alpha, \beta)$ of $f(z)$ with respect to $g(z)$ is defined by:*

$$\rho^{(\alpha,\beta)}[f]_g = \limsup_{R \to \infty} \frac{\alpha(M_{g,D}^{-1}(M_{f,D}(R)))}{\beta(R)}.$$

*The generalized relative Gol'dberg lower order $(\alpha, \beta)$ of $f(z)$ with respect to $g(z)$ is defined as:*

$$\lambda^{(\alpha,\beta)}[f]_g = \liminf_{R \to \infty} \frac{\alpha(M_{g,D}^{-1}(M_{f,D}(R)))}{\beta(R)}.$$

In order to define the above growth scale, now we intend to introduce the definition of an another growth indicator, called generalized relative Gol'dberg type $(\alpha, \beta)$ of an entire function of $n$ complex variables with respect to another entire function of $n$ complex variables as follows:

**Definition 5.2.2** *Let $f(z)$ and $g(z)$ be any two entire functions of $n$ complex variables. The generalized relative Gol'dberg type $(\alpha, \beta)$ of entire function $f(z)$ with respect to the entire function $g(z)$ having finite positive generalized relative Gol'dberg order $(\alpha, \beta)$ denoted by $\rho^{(\alpha,\beta)}[f]_g \left(0 < \rho^{(\alpha,\beta)}[f]_g < \infty\right)$ is defined as :*

$$\sigma_D^{(\alpha,\beta)}[f]_g = \inf \left\{ \begin{array}{c} \phi > 0 : M_{f,D}(R) < M_{g,D}\left[\alpha^{-1} \log\left(\phi(\exp(\beta(R)))^{\rho^{(\alpha,\beta)}[f]_g}\right)\right] \\ \text{for all } R > R_0(\phi) > 0 \end{array} \right\}$$

$$= \limsup_{R \to \infty} \frac{\exp(\alpha(M_{g,D}^{-1}(M_{f,D}(R))))}{(\exp(\beta(R)))^{\rho^{(\alpha,\beta)}[f]_g}}.$$

The above definition can alternatively defined in the following manner:

**Definition 5.2.3** *Let $f(z)$ and $g(z)$ be any two entire functions of $n$ complex variables having finite positive generalized relative Gol'dberg order $(\alpha, \beta)$ denoted by $\rho^{(\alpha,\beta)}[f]_g$ $\left(0 < \rho^{(\alpha,\beta)}[f]_g < \infty\right)$, then the generalized relative Gol'dberg type $(\alpha, \beta)$ denoted by $\sigma_D^{(\alpha,\beta)}[f]_g$ of entire function $f(z)$ with respect to the entire function $g(z)$ is define as: The integral* $\int_{R_0}^{\infty} \frac{\exp^{[2]}(\alpha(M_{g,D}^{-1}(M_{f,D}(R))))}{\left[\exp\left((\exp(\beta(R)))^{\rho^{(\alpha,\beta)}[f]_g}\right)\right]^{k+1}} dR$ $(R_0 > 0)$ *converges for* $k > \sigma_D^{(\alpha,\beta)}[f]_g$ *and diverges for* $k < \sigma_D^{(\alpha,\beta)}[f]_g$.

Analogously, one can introduced the definition of generalized relative Gol'dberg weak type $(\alpha, \beta)$ denoted by $\tau_D^{(\alpha,\beta)} [f]_g$ of an entire function $f(z)$ with respect to another entire function $g(z)$ with finite positive generalized relative Gol'dberg lower order $(\alpha, \beta)$ denoted by $\lambda^{(\alpha,\beta)} [f]_g$ in the following way:

**Definition 5.2.4** *Let $f(z)$ and $g(z)$ be any two entire functions of $n$ complex variables. The generalized relative Gol'dberg weak type $(\alpha, \beta)$ of entire function $f(z)$ with respect to the entire function $g(z)$ having finite positive generalized relative Gol'dberg lower order $(\alpha, \beta)$ as $\lambda^{(\alpha,\beta)} [f]_g \left( 0 < \lambda^{(\alpha,\beta)} [f]_g < \infty \right)$ is defined as :*

$$\tau_D^{(\alpha,\beta)} [f]_g = \sup \left\{ \begin{array}{c} \phi > 0 : M_{f,D}(R) < M_{g,D} \left[ \alpha^{-1} \log \left( \phi \left( \exp(\beta(R)) \right)^{\lambda^{(\alpha,\beta)}[f]_g} \right) \right] \\ \text{for all } R > R_0(\phi) > 0 \end{array} \right\}$$

$$= \liminf_{R \to \infty} \frac{\exp(\alpha(M_{g,D}^{-1}(M_{f,D}(R))))}{(\exp(\beta(R)))^{\lambda^{(\alpha,\beta)}[f]_g}}.$$

The above definition can also alternatively defined as:

**Definition 5.2.5** *Let $f(z)$ and $g(z)$ be any two entire functions of $n$ complex variables having finite positive generalized relative Gol'dberg lower order $(\alpha, \beta)$ as $\lambda^{(\alpha,\beta)} [f]_g$ $\left( 0 < \lambda^{(\alpha,\beta)} [f]_g < \infty \right)$, then the generalized relative Gol'dberg weak type $(\alpha, \beta)$ denoted by $\tau_D^{(\alpha,\beta)} [f]_g$ of entire function $f(z)$ with respect to the entire function $g(z)$ is defined as:*

$$\text{The integral} \int_{R_0}^{\infty} \frac{\exp^{[2]}(\alpha(M_{g,D}^{-1}(M_{f,D}(R))))}{\left[ \exp \left( (\exp(\beta(R)))^{\lambda^{(\alpha,\beta)}[f]_g} \right) \right]^{k+1}} dR \, (R_0 > 0)$$

*converges for $k > \tau_D^{(\alpha,\beta)} [f]_g$ and diverges for $k < \tau_D^{(\alpha,\beta)} [f]_g$.*

**Remark 5.2.1** *As Gol'dberg has shown that (see [2]) Gol'dberg type depends on the domain $D$, so in general all the growth indicators defined in Definition 5.2.2 and Definition 5.2.4 also depend on $D$.*

Now a question may arise about the equivalence of the definitions of generalized relative Gol'dberg type $(\alpha, \beta)$ and generalized relative Gol'dberg weak type $(\alpha, \beta)$ with their integral representations. In the next section we would like to establish such equivalence of Definition 5.2.2 and Definition 5.2.3, and Definition 5.2.4 and Definition 5.2.5 and also investigate some growth properties related to generalized relative Gol'dberg type $(\alpha, \beta)$ and generalized relative Gol'dberg weak type $(\alpha, \beta)$ of entire function of $n$ complex variables with respect to another entire function of $n$ complex variables.

## 5.3  Lemma.

In this section we present a lemma which will be needed in the sequel.

**Lemma 5.3.1** *Let the integral* $\int_{R_0}^{\infty} \frac{\exp^{[2]}(\alpha(M_{g,D}^{-1}(M_{f,D}(R))))}{\left[\exp\left((\exp(\beta(R)))^A\right)\right]^{k+1}} dR$ $(R_0 > 0)$ *converges where* $0 < A < \infty$. *Then*

$$\lim_{R \to \infty} \frac{\exp^{[2]}(\alpha(M_{g,D}^{-1}(M_{f,D}(R))))}{\left[\exp\left((\exp(\beta(R)))^A\right)\right]^{k}} = 0 \ .$$

**Proof.** Since the integral $\int_{R_0}^{\infty} \frac{\exp^{[2]}(\alpha(M_{g,D}^{-1}(M_{f,D}(R))))}{\left[\exp\left((\exp(\beta(R)))^A\right)\right]^{k+1}} dR$ $(R_0 > 0)$ converges, then

$$\int_{R_0}^{\infty} \frac{\exp^{[2]}(\alpha(M_{g,D}^{-1}(M_{f,D}(R))))}{\left[\exp\left((\exp(\beta(R)))^A\right)\right]^{k+1}} dR < \varepsilon, \text{ if } R_0 > \Re(\varepsilon) \ .$$

Therefore,

$$\int_{R_0}^{\exp\left((\exp(\beta(R_0)))^A\right)+R_0} \frac{\exp^{[2]}(\alpha(M_{g,D}^{-1}(M_{f,D}(R))))}{\left[\exp\left((\exp(\beta(R)))^A\right)\right]^{k+1}} dR < \varepsilon.$$

Since here $\exp^{[2]}(\alpha(M_{g,D}^{-1}(M_{f,D}(R))))$ increases with $R$, so

$$\int_{R_0}^{\exp\left((\exp(\beta(R_0)))^A\right)+R_0} \frac{\exp^{[2]}(\alpha(M_{g,D}^{-1}(M_{f,D}(R))))}{\left[\exp\left((\exp(\beta(R)))^A\right)\right]^{k+1}} dR \geq$$

$$\frac{\exp^{[2]}(\alpha(M_{g,D}^{-1}(M_{f,D}(R_0))))}{\left[\exp\left((\exp(\beta(R_0)))^A\right)\right]^{k+1}} \cdot \left[\exp\left((\exp(\beta(R_0)))^A\right)\right].$$

i.e., for all sufficiently large values of $R$,

$$\int_{R_0}^{\exp\left((\exp(\beta(R_0)))^A\right)+R_0} \frac{\exp^{[2]}(\alpha(M_{g,D}^{-1}(M_{f,D}(R))))}{\left[\exp\left((\exp(\beta(R)))^A\right)\right]^{k+1}} dR \geq$$

$$\frac{\exp^{[2]}(\alpha(M_{g,D}^{-1}(M_{f,D}(R_0))))}{\left[\exp\left((\exp(\beta(R_0)))^A\right)\right]^{k}},$$

so that

$$\frac{\exp^{[2]}(\alpha(M_{g,D}^{-1}(M_{f,D}(R_0))))}{\left[\exp\left((\exp(\beta(R_0)))^A\right)\right]^{k}} < \varepsilon \text{ if } R_0 > \Re(\varepsilon) \ .$$

$$i.e., \lim_{R \to \infty} \frac{\exp^{[2]}(\alpha(M_{g,D}^{-1}(M_{f,D}(R_0))))}{\left[\exp\left((\exp(\beta(R_0)))^A\right)\right]^k} = 0.$$

This proves the lemma. ∎

## 5.4　Main Results.

In this section we state the main results of this chapter.

**Theorem 5.4.1** *Let $f(z)$ and $g(z)$ be any two entire functions of $n$ complex variables and $f(z)$ has finite positive generalized relative Gol'dberg order $(\alpha, \beta)$ as $\rho^{(\alpha,\beta)}[f]_g$ ( $0 < \rho^{(\alpha,\beta)}[f]_g < \infty$ ) and generalized relative Gol'dberg type $(\alpha, \beta)$ as $\sigma_D^{(\alpha,\beta)}[f]_g$ with respect to $g(z)$. Then Definition 5.2.2 and Definition 5.2.3 are equivalent.*

**Proof.** Let us consider $f(z)$ and $g(z)$ to be any two entire functions of $n$ complex variables such that $\rho^{(\alpha,\beta)}[f]_g \left(0 < \rho^{(\alpha,\beta)}[f]_g < \infty\right)$ exists.

**Case I.** Let $\sigma_D^{(\alpha,\beta)}[f]_g = \infty$.

**Definition 5.2.2 $\Rightarrow$ Definition 5.2.3.**

As $\sigma_D^{(\alpha,\beta)}[f]_g = \infty$, from Definition 5.2.2 we have for arbitrary positive $G$ and for a sequence of values of $R$ tending to infinity that

$$\exp(\alpha(M_{g,D}^{-1}(M_{f,D}(R)))) > G \cdot (\exp(\beta(R)))^{\rho^{(\alpha,\beta)}[f]_g}$$

$$i.e., \ \exp^{[2]}(\alpha(M_{g,D}^{-1}(M_{f,D}(R)))) > \left[\exp\left((\exp(\beta(R)))^{\rho^{(\alpha,\beta)}[f]_g}\right)\right]^G. \tag{63}$$

If possible let the integral $\int\limits_{R_0}^{\infty} \frac{\exp^{[2]}(\alpha(M_{g,D}^{-1}(M_{f,D}(R))))}{\left[\exp\left((\exp(\beta(R)))^{\rho^{(\alpha,\beta)}[f]_g}\right)\right]^{G+1}} dR \ (R_0 > 0)$ be converge.

Then by Lemma 5.3.1,

$$\limsup_{R \to \infty} \frac{\exp^{[2]}(\alpha(M_{g,D}^{-1}(M_{f,D}(R))))}{\left[\exp\left((\exp(\beta(R)))^{\rho^{(\alpha,\beta)}[f]_g}\right)\right]^G} = 0.$$

So for all sufficiently large values of $R$,

$$\exp^{[2]}(\alpha(M_{g,D}^{-1}(M_{f,D}(R)))) < \left[\exp\left((\exp(\beta(R)))^{\rho^{(\alpha,\beta)}[f]_g}\right)\right]^G. \tag{64}$$

Therefore from (63) and (64) we arrive at a contradiction.

Hence $\int\limits_{R_0}^{\infty} \frac{\exp^{[2]}(\alpha(M_{g,D}^{-1}(M_{f,D}(R))))}{\left[\exp\left((\exp(\beta(R)))^{\rho^{(\alpha,\beta)}[f]_g}\right)\right]^{G+1}} dR \ (R_0 > 0)$ diverges whenever $G$ is finite, which is the Definition 5.2.3.

**Definition 5.2.3 $\Rightarrow$ Definition 5.2.2.**

Let $G$ be any positive number. Since $\sigma_D^{(\alpha,\beta)}[f]_g = \infty$, from Definition 5.2.3, the divergence of the integral $\int\limits_{R_0}^{\infty} \dfrac{\exp^{[2]}(\alpha(M_{g,D}^{-1}(M_{f,D}(R))))}{\left[\exp\left((\exp(\beta(R)))^{\rho^{(\alpha,\beta)}[f]_g}\right)\right]^{G+1}} dR \ (R_0 > 0)$ gives for arbitrary positive $\varepsilon$ and for a sequence of values of $R$ tending to infinity

$$\exp^{[2]}(\alpha(M_{g,D}^{-1}(M_{f,D}(R)))) > \left[\exp\left((\exp(\beta(R)))^{\rho^{(\alpha,\beta)}[f]_g}\right)\right]^{G-\varepsilon}$$

$$i.e., \ \exp(\alpha(M_{g,D}^{-1}(M_{f,D}(R)))) > (G-\varepsilon)(\exp(\beta(R)))^{\rho^{(\alpha,\beta)}[f]_g},$$

which implies that

$$\limsup_{R\to\infty} \frac{\exp(\alpha(M_{g,D}^{-1}(M_{f,D}(R))))}{(\exp(\beta(R)))^{\rho^{(\alpha,\beta)}[f]_g}} \geq G - \varepsilon.$$

Since $G > 0$ is arbitrary, it follows that

$$\limsup_{R\to\infty} \frac{\exp(\alpha(M_{g,D}^{-1}(M_{f,D}(R))))}{(\exp(\beta(R)))^{\rho^{(\alpha,\beta)}[f]_g}} = \infty.$$

Thus Definition 5.2.2 follows.
**Case II.** $0 \leq \sigma_D^{(\alpha,\beta)}[f]_g < \infty$.
**Definition 5.2.2 $\Rightarrow$ Definition 5.2.3.**
**Subcase (A).** $0 < \sigma_D^{(\alpha,\beta)}[f]_g < \infty$.

Let $f(z)$ and $g(z)$ be any two entire functions of $n$ complex variables such that $0 < \sigma_D^{(\alpha,\beta)}[f]_g < \infty$ exists. Then according to the Definition 5.2.2, for arbitrary positive $\varepsilon$ and for all sufficiently large values of $R$, we obtain that

$$\exp(\alpha(M_{g,D}^{-1}(M_{f,D}(R)))) < \left(\sigma_D^{(\alpha,\beta)}[f]_g + \varepsilon\right)(\exp(\beta(R)))^{\rho^{(\alpha,\beta)}[f]_g}$$

$$i.e., \ \exp^{[2]}(\alpha(M_{g,D}^{-1}(M_{f,D}(R)))) < \left[\exp\left((\exp(\beta(R)))^{\rho^{(\alpha,\beta)}[f]_g}\right)\right]^{\sigma_D^{(\alpha,\beta)}[f]_g+\varepsilon}$$

$$i.e., \ \frac{\exp^{[2]}(\alpha(M_{g,D}^{-1}(M_{f,D}(R))))}{\left[\exp\left((\exp(\beta(R)))^{\rho^{(\alpha,\beta)}[f]_g}\right)\right]^k} < \frac{\left[\exp\left((\exp(\beta(R)))^{\rho^{(\alpha,\beta)}[f]_g}\right)\right]^{\sigma_D^{(\alpha,\beta)}[f]_g+\varepsilon}}{\left[\exp\left((\exp(\beta(R)))^{\rho^{(\alpha,\beta)}[f]_g}\right)\right]^k}$$

$$i.e., \ \frac{\exp^{[2]}(\alpha(M_{g,D}^{-1}(M_{f,D}(R))))}{\left[\exp\left((\exp(\beta(R)))^{\rho^{(\alpha,\beta)}[f]_g}\right)\right]^k} <$$

$$\frac{1}{\left[\exp\left((\exp(\beta(R)))^{\rho^{(\alpha,\beta)}[f]_g}\right)\right]^{k-\left(\sigma_D^{(\alpha,\beta)}[f]_g+\varepsilon\right)}}.$$

Therefore $\int\limits_{R_0}^{\infty} \dfrac{\exp^{[2]}(\alpha(M_{g,D}^{-1}(M_{f,D}(R))))}{\left[\exp\left((\exp(\beta(R)))^{\rho^{(\alpha,\beta)}[f]_g}\right)\right]^{k+1}} dR \ (R_0 > 0)$ converges for $k > \sigma_D^{(\alpha,\beta)}[f]_g$.

Again by Definition 5.2.2, we obtain for a sequence values of $R$ tending to infinity that

$$\exp(\alpha(M_{g,D}^{-1}(M_{f,D}(R)))) > \left(\sigma_D^{(\alpha,\beta)}[f]_g - \varepsilon\right)(\exp(\beta(R)))^{\rho^{(\alpha,\beta)}[f]_g}$$

$$i.e., \ \ \exp^{[2]}(\alpha(M_{g,D}^{-1}(M_{f,D}(R)))) > \left[\exp\left((\exp(\beta(R)))^{\rho^{(\alpha,\beta)}[f]_g}\right)\right]^{\sigma_D^{(\alpha,\beta)}[f]_g - \varepsilon}. \tag{65}$$

So for $k < \sigma_D^{(\alpha,\beta)}[f]_g$, we get from (65) that

$$\frac{\exp^{[2]}(\alpha(M_{g,D}^{-1}(M_{f,D}(R))))}{\left[\exp\left((\exp(\beta(R)))^{\rho^{(\alpha,\beta)}[f]_g}\right)\right]^k} > \frac{1}{\left[\exp\left((\exp(\beta(R)))^{\rho^{(\alpha,\beta)}[f]_g}\right)\right]^{k - \left(\sigma_D^{(\alpha,\beta)}[f]_g - \varepsilon\right)}}.$$

Therefore $\int_{R_0}^{\infty} \frac{\exp^{[2]}(\alpha(M_{g,D}^{-1}(M_{f,D}(R))))}{\left[\exp\left((\exp(\beta(R)))^{\rho^{(\alpha,\beta)}[f]_g}\right)\right]^{k+1}} dR$ $(R_0 > 0)$ diverges for $k < \sigma_D^{(\alpha,\beta)}[f]_g$.

Hence $\int_{R_0}^{\infty} \frac{\exp^{[2]}(\alpha(M_{g,D}^{-1}(M_{f,D}(R))))}{\left[\exp\left((\exp(\beta(R)))^{\rho^{(\alpha,\beta)}[f]_g}\right)\right]^{k+1}} dR$ $(R_0 > 0)$ converges for $k > \sigma_D^{(\alpha,\beta)}[f]_g$ and diverges

for $k < \sigma_D^{(\alpha,\beta)}[f]_g$.

**Subcase (B).** $\sigma_D^{(\alpha,\beta)}[f]_g = 0$.

When $\sigma_D^{(\alpha,\beta)}[f]_g = 0$, Definition 5.2.2 gives for all sufficiently large values of $R$ that

$$\frac{\exp(\alpha(M_{g,D}^{-1}(M_{f,D}(R))))}{(\exp(\beta(R)))^{\rho^{(\alpha,\beta)}[f]_g}} < \varepsilon.$$

Then as before we obtain that $\int_{R_0}^{\infty} \frac{\exp^{[2]}(\alpha(M_{g,D}^{-1}(M_{f,D}(R))))}{\left[\exp\left((\exp(\beta(R)))^{\rho^{(\alpha,\beta)}[f]_g}\right)\right]^{k+1}} dR$ $(R_0 > 0)$ converges for $k > 0$

and diverges for $k < 0$. Thus combining Subcase $(A)$ and Subcase $(B)$, Definition 5.2.3 follows.

**Definition 5.2.3 $\Rightarrow$ Definition 5.2.2.**

From Definition 5.2.3 and for arbitrary positive $\varepsilon$ the integral

$$\int_{R_0}^{\infty} \frac{\exp^{[2]}(\alpha(M_{g,D}^{-1}(M_{f,D}(R))))}{\left[\exp\left((\exp(\beta(R)))^{\rho^{(\alpha,\beta)}[f]_g}\right)\right]^{\sigma_D^{(\alpha,\beta)}[f]_g + \varepsilon + 1}} dR \ (R_0 > 0)$$

converges. Then by Lemma 5.3.1, we get that

$$\limsup_{R \to \infty} \frac{\exp^{[2]}(\alpha(M_{g,D}^{-1}(M_{f,D}(R))))}{\left[\exp\left((\exp(\beta(R)))^{\rho^{(\alpha,\beta)}[f]_g}\right)\right]^{\sigma_D^{(\alpha,\beta)}[f]_g + \varepsilon}} = 0.$$

So we obtain all sufficiently large values of $R$ that

$$\frac{\exp^{[2]}(\alpha(M_{g,D}^{-1}(M_{f,D}(R))))}{\left[\exp\left((\exp(\beta(R)))^{\rho^{(\alpha,\beta)}[f]_g}\right)\right]^{\sigma_D^{(\alpha,\beta)}[f]_g+\varepsilon}} < \varepsilon$$

$$i.e., \ \exp^{[2]}(\alpha(M_{g,D}^{-1}(M_{f,D}(R)))) < \varepsilon \cdot \left[\exp\left((\exp(\beta(R)))^{\rho^{(\alpha,\beta)}[f]_g}\right)\right]^{\sigma_D^{(\alpha,\beta)}[f]_g+\varepsilon}$$

$$i.e., \ \exp(\alpha(M_{g,D}^{-1}(M_{f,D}(R)))) < \log\varepsilon + \left(\sigma_D^{(\alpha,\beta)}[f]_g+\varepsilon\right)(\exp(\beta(R)))^{\rho^{(\alpha,\beta)}[f]_g}$$

$$i.e., \ \limsup_{R\to\infty}\frac{\exp(\alpha(M_{g,D}^{-1}(M_{f,D}(R))))}{(\exp(\beta(R)))^{\rho^{(\alpha,\beta)}[f]_g}} \leq \sigma_D^{(\alpha,\beta)}[f]_g+\varepsilon.$$

Since $\varepsilon\,(>0)$ is arbitrary, it follows from above that

$$\limsup_{R\to\infty}\frac{\exp(\alpha(M_{g,D}^{-1}(M_{f,D}(R))))}{(\exp(\beta(R)))^{\rho^{(\alpha,\beta)}[f]_g}} \leq \sigma_D^{(\alpha,\beta)}[f]_g. \tag{66}$$

On the other hand the divergence of the integral

$$\int_{R_0}^{\infty}\frac{\exp^{[2]}(\alpha(M_{g,D}^{-1}(M_{f,D}(R))))}{\left[\exp\left((\exp(\beta(R)))^{\rho^{(\alpha,\beta)}[f]_g}\right)\right]^{\sigma_D^{(\alpha,\beta)}[f]_g-\varepsilon+1}}dR\ (R_0 > 0)$$

implies that there exists a sequence of values of $R$ tending to infinity such that

$$\frac{\exp^{[2]}(\alpha(M_{g,D}^{-1}(M_{f,D}(R))))}{\left[\exp\left((\exp(\beta(R)))^{\rho^{(\alpha,\beta)}[f]_g}\right)\right]^{\sigma_D^{(\alpha,\beta)}[f]_g-\varepsilon+1}} > \frac{1}{\left[\exp\left((\exp(\beta(R)))^{\rho^{(\alpha,\beta)}[f]_g}\right)\right]^{1+\varepsilon}}$$

$$i.e., \ \exp^{[2]}(\alpha(M_{g,D}^{-1}(M_{f,D}(R)))) > \left[\exp\left((\exp(\beta(R)))^{\rho^{(\alpha,\beta)}[f]_g}\right)\right]^{\sigma_D^{(\alpha,\beta)}[f]_g-2\varepsilon}$$

$$i.e., \ \exp(\alpha(M_{g,D}^{-1}(M_{f,D}(R)))) > \left(\sigma_D^{(\alpha,\beta)}[f]_g-2\varepsilon\right)\left((\exp(\beta(R)))^{\rho^{(\alpha,\beta)}[f]_g}\right)$$

$$i.e., \ \frac{\exp(\alpha(M_{g,D}^{-1}(M_{f,D}(R))))}{(\exp(\beta(R)))^{\rho^{(\alpha,\beta)}[f]_g}} > \left(\sigma_D^{(\alpha,\beta)}[f]_g-2\varepsilon\right).$$

As $\varepsilon\,(>0)$ is arbitrary, it follows from above that

$$\limsup_{R\to\infty}\frac{\exp(\alpha(M_{g,D}^{-1}(M_{f,D}(R))))}{(\exp(\beta(R)))^{\rho^{(\alpha,\beta)}[f]_g}} \geq \sigma_D^{(\alpha,\beta)}[f]_g. \tag{67}$$

So from (66) and (67) , we obtain that

$$\limsup_{R\to\infty}\frac{\exp(\alpha(M_{g,D}^{-1}(M_{f,D}(R))))}{(\exp(\beta(R)))^{\rho^{(\alpha,\beta)}[f]_g}} = \sigma_D^{(\alpha,\beta)}[f]_g.$$

This proves the theorem. ∎

**Theorem 5.4.2** *Let $f(z)$ and $g((z))$ be any two entire functions of $n$ complex variables with $f(z)$ has finite positive generalized relative Gol'dberg lower order $(\alpha, \beta)$ i.e. $\lambda^{(\alpha,\beta)}[f]_g$ $\left(0 < \lambda^{(\alpha,\beta)}[f]_g < \infty\right)$ and generalized relative Gol'dberg weak type $(\alpha, \beta)$ i.e. $\tau_D^{(\alpha,\beta)}[f]_g$ with respect to the entire function $g(z)$. Then Definition 5.2.4 and Definition 5.2.5 are equivalent.*

**Proof.** Let us consider $f(z)$ and $g(z)$ be any two entire functions of $n$ complex variables such that $\lambda^{(\alpha,\beta)}[f]_g$ $\left(0 < \lambda^{(\alpha,\beta)}[f]_g < \infty\right)$ exists.

    **Case I.** $\tau_D^{(\alpha,\beta)}[f]_g = \infty$.

**Definition 5.2.4 $\Rightarrow$ Definition 5.2.5.**

As $\tau_D^{(\alpha,\beta)}[f]_g = \infty$, from Definition 5.2.4 we obtain for arbitrary positive $G$ and for all sufficiently large values of $R$ that

$$\exp(\alpha(M_{g,D}^{-1}(M_{f,D}(R)))) > G \cdot (\exp(\beta(R)))^{\lambda^{(\alpha,\beta)}[f]_g}$$

$$i.e., \quad \exp^{[2]}(\alpha(M_{g,D}^{-1}(M_{f,D}(R)))) > \left[\exp\left((\exp(\beta(R)))^{\lambda^{(\alpha,\beta)}[f]_g}\right)\right]^G. \tag{68}$$

Now if let the integral $\int\limits_{R_0}^{\infty} \dfrac{\exp^{[2]}(\alpha(M_{g,D}^{-1}(M_{f,D}(R))))}{\left[\exp\left((\exp(\beta(R)))^{\lambda^{(\alpha,\beta)}[f]_g}\right)\right]^{G+1}} dR \; (R_0 > 0)$ be converge.

Then by Lemma 5.3.1,

$$\liminf_{R\to\infty} \frac{\exp^{[2]}(\alpha(M_{g,D}^{-1}(M_{f,D}(R))))}{\left[\exp\left((\exp(\beta(R)))^{\lambda^{(\alpha,\beta)}[f]_g}\right)\right]^G} = 0.$$

So for a sequence of values of $R$ tending to infinity we get that

$$\exp^{[2]}(\alpha(M_{g,D}^{-1}(M_{f,D}(R)))) < \left[\exp\left((\exp(\beta(R)))^{\lambda^{(\alpha,\beta)}[f]_g}\right)\right]^G. \tag{69}$$

Therefore from (68) and (69), we arrive at a contradiction.

    Hence $\int\limits_{R_0}^{\infty} \dfrac{\exp^{[2]}(\alpha(M_{g,D}^{-1}(M_{f,D}(R))))}{\left[\exp\left((\exp(\beta(R)))^{\lambda^{(\alpha,\beta)}[f]_g}\right)\right]^{G+1}} dR \; (R_0 > 0)$ diverges whenever $G$ is finite, which is the Definition 5.2.5.

**Definition 5.2.5 $\Rightarrow$ Definition 5.2.4.**

Let $G$ be any positive number. Since $\tau_D^{(\alpha,\beta)}[f]_g = \infty$, from Definition 5.2.5, the divergence of the integral $\int\limits_{R_0}^{\infty} \dfrac{\exp^{[2]}(\alpha(M_{g,D}^{-1}(M_{f,D}(R))))}{\left[\exp\left((\exp(\beta(R)))^{\lambda^{(\alpha,\beta)}[f]_g}\right)\right]^{G+1}} dR \; (R_0 > 0)$ gives for arbitrary positive $\varepsilon$ and for all sufficiently large values of $R$ that

$$\exp^{[2]}(\alpha(M_{g,D}^{-1}(M_{f,D}(R)))) > \left[\exp\left((\exp(\beta(R)))^{\lambda^{(\alpha,\beta)}[f]_g}\right)\right]^{G-\varepsilon}$$

$$i.e., \quad \exp(\alpha(M_{g,D}^{-1}(M_{f,D}(R)))) > (G - \varepsilon)(\exp(\beta(R)))^{\lambda^{(\alpha,\beta)}[f]_g},$$

which implies that

$$\liminf_{R \to \infty} \frac{\exp(\alpha(M_{g,D}^{-1}(M_{f,D}(R))))}{(\exp(\beta(R)))^{\lambda^{(\alpha,\beta)}[f]_g}} \geq G - \varepsilon.$$

Since $G > 0$ is arbitrary, it follows that

$$\liminf_{R \to \infty} \frac{\exp(\alpha(M_{g,D}^{-1}(M_{f,D}(R))))}{(\exp(\beta(R)))^{\lambda^{(\alpha,\beta)}[f]_g}} = \infty.$$

Thus Definition 5.2.4 follows.
**Case II.** $0 \leq \tau_D^{(\alpha,\beta)}[f]_g < \infty.$
**Definition 5.2.4 $\Rightarrow$ Definition 5.2.5.**
**Subcase (C).** $0 < \tau_D^{(\alpha,\beta)}[f]_g < \infty.$

Let $f(z)$ and $g(z)$ be any two entire functions of $n$ complex variables such that $0 < \tau_D^{(\alpha,\beta)}[f]_g < \infty$ exists. Then according to the Definition 5.2.4, for a sequence of values of $R$ tending to infinity we get that

$$\exp(\alpha(M_{g,D}^{-1}(M_{f,D}(R)))) < \left(\tau_D^{(\alpha,\beta)}[f]_g + \varepsilon\right)(\exp(\beta(R)))^{\lambda^{(\alpha,\beta)}[f]_g}$$

$$i.e., \ \exp^{[2]}(\alpha(M_{g,D}^{-1}(M_{f,D}(R)))) < \left[\exp\left((\exp(\beta(R)))^{\lambda^{(\alpha,\beta)}[f]_g}\right)\right]^{\tau_D^{(\alpha,\beta)}[f]_g + \varepsilon}$$

$$i.e., \ \frac{\exp^{[2]}(\alpha(M_{g,D}^{-1}(M_{f,D}(R))))}{\left[\exp\left((\exp(\beta(R)))^{\lambda^{(\alpha,\beta)}[f]_g}\right)\right]^k} < \frac{\left[\exp\left((\exp(\beta(R)))^{\lambda^{(\alpha,\beta)}[f]_g}\right)\right]^{\tau_D^{(\alpha,\beta)}[f]_g + \varepsilon}}{\left[\exp\left((\exp(\beta(R)))^{\lambda^{(\alpha,\beta)}[f]_g}\right)\right]^k}$$

$$i.e., \ \frac{\exp^{[2]}(\alpha(M_{g,D}^{-1}(M_{f,D}(R))))}{\left[\exp\left((\exp(\beta(R)))^{\lambda^{(\alpha,\beta)}[f]_g}\right)\right]^k} < \frac{1}{\left[\exp\left((\exp(\beta(R)))^{\lambda^{(\alpha,\beta)}[f]_g}\right)\right]^{k - \left(\tau_D^{(\alpha,\beta)}[f]_g + \varepsilon\right)}}.$$

Therefore $\int_{R_0}^{\infty} \frac{\exp^{[2]}(\alpha(M_{g,D}^{-1}(M_{f,D}(R))))}{\left[\exp\left((\exp(\beta(R)))^{\lambda^{(\alpha,\beta)}[f]_g}\right)\right]^{k+1}} dR \ (R_0 > 0)$ converges for $k > \tau_D^{(\alpha,\beta)}[f]_g$.

Again by Definition 5.2.4, we obtain for all sufficiently large values of $R$ that

$$\exp(\alpha(M_{g,D}^{-1}(M_{f,D}(R)))) > \left(\tau_D^{(\alpha,\beta)}[f]_g - \varepsilon\right)\left(\log^{[q-1]} R\right)^{\lambda^{(\alpha,\beta)}[f]_g}$$

$$i.e., \ \exp^{[2]}(\alpha(M_{g,D}^{-1}(M_{f,D}(R)))) > \left[\exp\left((\exp(\beta(R)))^{\lambda^{(\alpha,\beta)}[f]_g}\right)\right]^{\tau_D^{(\alpha,\beta)}[f]_g - \varepsilon}. \quad (70)$$

So for $k < \tau_D^{(\alpha,\beta)}[f]_g$, we get from (70) that

$$\frac{\exp^{[2]}(\alpha(M_{g,D}^{-1}(M_{f,D}(R))))}{\left[\exp\left((\exp(\beta(R)))^{\lambda^{(\alpha,\beta)}[f]_g}\right)\right]^k} > \frac{1}{\left[\exp\left((\exp(\beta(R)))^{\lambda^{(\alpha,\beta)}[f]_g}\right)\right]^{k - \left(\tau_D^{(\alpha,\beta)}[f]_g - \varepsilon\right)}}.$$

Therefore $\int_{R_0}^{\infty} \dfrac{\exp^{[2]}(\alpha(M_{g,D}^{-1}(M_{f,D}(R))))}{\left[\exp\left((\exp(\beta(R)))^{\lambda^{(\alpha,\beta)}[f]_g}\right)\right]^{k+1}} dR \, (R_0 > 0)$ diverges for $k < \tau_D^{(\alpha,\beta)}[f]_g$.

Hence $\int_{R_0}^{\infty} \dfrac{\exp^{[2]}(\alpha(M_{g,D}^{-1}(M_{f,D}(R))))}{\left[\exp\left((\exp(\beta(R)))^{\lambda^{(\alpha,\beta)}[f]_g}\right)\right]^{k+1}} dR \, (R_0 > 0)$ converges for $k > \tau_D^{(\alpha,\beta)}[f]_g$ and diverges

for $k < \tau_D^{(\alpha,\beta)}[f]_g$.

**Subcase (D).** $\tau_D^{(\alpha,\beta)}[f]_g = 0$.

When $\tau_D^{(\alpha,\beta)}[f]_g = 0$, Definition 5.2.4 gives for a sequence of values of $R$ tending to infinity that

$$\frac{\exp(\alpha(M_{g,D}^{-1}(M_{f,D}(R))))}{(\exp(\beta(R)))^{\lambda^{(\alpha,\beta)}[f]_g}} < \varepsilon.$$

Then as before we obtain that $\int_{R_0}^{\infty} \dfrac{\exp^{[2]}(\alpha(M_{g,D}^{-1}(M_{f,D}(R))))}{\left[\exp\left((\exp(\beta(R)))^{\lambda^{(\alpha,\beta)}[f]_g}\right)\right]^{k+1}} dR \, (R_0 > 0)$ converges for $k >$

$0$ and diverges for $k < 0$.

Thus combining Subcase $(C)$ and Subcase $(D)$, Definition 5.2.5 follows.

**Definition 5.2.5 $\Rightarrow$ Definition 5.2.4.**

From Definition **5.2.5** and for arbitrary positive $\varepsilon$ the integral

$$\int_{R_0}^{\infty} \frac{\exp^{[2]}(\alpha(M_{g,D}^{-1}(M_{f,D}(R))))}{\left[\exp\left((\exp(\beta(R)))^{\lambda^{(\alpha,\beta)}[f]_g}\right)\right]^{\tau_D^{(\alpha,\beta)}[f]_g+\varepsilon+1}} dR \, (R_0 > 0)$$

converges.

Then by Lemma 5.3.1, we get that

$$\liminf_{R\to\infty} \frac{\exp^{[2]}(\alpha(M_{g,D}^{-1}(M_{f,D}(R))))}{\left[\exp\left((\exp(\beta(R)))^{\lambda^{(\alpha,\beta)}[f]_g}\right)\right]^{\tau_D^{(\alpha,\beta)}[f]_g+\varepsilon}} = 0.$$

So we get for a sequence of values of $R$ tending to infinity that

$$\frac{\exp^{[2]}(\alpha(M_{g,D}^{-1}(M_{f,D}(R))))}{\left[\exp\left((\exp(\beta(R)))^{\lambda^{(\alpha,\beta)}[f]_g}\right)\right]^{\tau_D^{(\alpha,\beta)}[f]_g+\varepsilon}} < \varepsilon$$

*i.e.,* $\exp^{[2]}(\alpha(M_{g,D}^{-1}(M_{f,D}(R)))) < \varepsilon \cdot \left[\exp\left((\exp(\beta(R)))^{\lambda^{(\alpha,\beta)}[f]_g}\right)\right]^{\tau_D^{(\alpha,\beta)}[f]_g+\varepsilon}$

*i.e.,* $\exp(\alpha(M_{g,D}^{-1}(M_{f,D}(R)))) < \log\varepsilon + \left(\tau_D^{(\alpha,\beta)}[f]_g + \varepsilon\right)(\exp(\beta(R)))^{\lambda^{(\alpha,\beta)}[f]_g}$

*i.e.,* $\displaystyle\liminf_{R\to\infty} \frac{\exp(\alpha(M_{g,D}^{-1}(M_{f,D}(R))))}{(\exp(\beta(R)))^{\lambda^{(\alpha,\beta)}[f]_g}} \le \tau_D^{(\alpha,\beta)}[f]_g + \varepsilon.$

Since $\varepsilon \, (> 0)$ is arbitrary, it follows from above that

$$\liminf_{R\to\infty} \frac{\exp(\alpha(M_{g,D}^{-1}(M_{f,D}(R))))}{(\exp(\beta(R)))^{\lambda^{(\alpha,\beta)}[f]_g}} \le \tau_D^{(\alpha,\beta)}[f]_g. \tag{71}$$

On the other hand the divergence of the integral

$$\int_{R_0}^{\infty} \frac{\exp^{[2]}(\alpha(M_{g,D}^{-1}(M_{f,D}(R))))}{\left[\exp\left((\exp(\beta(R)))^{\lambda^{(\alpha,\beta)}[f]_g}\right)\right]^{\tau_D^{(\alpha,\beta)}[f]_g-\varepsilon+1}} dR \, (R_0 > 0)$$

implies for all sufficiently large values of $R$ that

$$\frac{\exp^{[2]}(\alpha(M_{g,D}^{-1}(M_{f,D}(R))))}{\left[\exp\left((\exp(\beta(R)))^{\lambda^{(\alpha,\beta)}[f]_g}\right)\right]^{\tau_D^{(\alpha,\beta)}[f]_g-\varepsilon+1}} > \frac{1}{\exp\left((\exp(\beta(R)))^{\lambda^{(\alpha,\beta)}[f]_g}\right)^{1+\varepsilon}}$$

$$i.e., \ \exp^{[2]}(\alpha(M_{g,D}^{-1}(M_{f,D}(R)))) > \left[\exp\left((\exp(\beta(R)))^{\lambda^{(\alpha,\beta)}[f]_g}\right)\right]^{\tau_D^{(\alpha,\beta)}[f]_g-2\varepsilon}$$

$$i.e., \ \exp(\alpha(M_{g,D}^{-1}(M_{f,D}(R)))) > \left(\tau_D^{(\alpha,\beta)}[f]_g - 2\varepsilon\right)\left((\exp(\beta(R)))^{\lambda^{(\alpha,\beta)}[f]_g}\right)$$

$$i.e., \ \frac{\exp(\alpha(M_{g,D}^{-1}(M_{f,D}(R))))}{(\exp(\beta(R)))^{\lambda^{(\alpha,\beta)}[f]_g}} > \left(\tau_D^{(\alpha,\beta)}[f]_g - 2\varepsilon\right).$$

As $\varepsilon \, (> 0)$ is arbitrary, it follows from above that

$$\liminf_{R \to \infty} \frac{\exp(\alpha(M_{g,D}^{-1}(M_{f,D}(R))))}{(\exp(\beta(R)))^{\lambda^{(\alpha,\beta)}[f]_g}} \geq \tau_D^{(\alpha,\beta)}[f]_g. \tag{72}$$

So from (71) and (72) we obtain that

$$\liminf_{R \to \infty} \frac{\exp(\alpha(M_{g,D}^{-1}(M_{f,D}(R))))}{(\exp(\beta(R)))^{\lambda^{(\alpha,\beta)}[f]_g}} = \tau_D^{(\alpha,\beta)}[f]_g.$$

This proves the theorem. ∎

Next we introduce the following two Relative growth indicators which will also enable help our subsequent study.

**Definition 5.4.1** *Let $f(z)$ and $g(z)$ be any two entire functions of $n$ complex variables and $f(z)$ has finite positive generalized relative Gol'dberg order $(\alpha, \beta)$ as $\rho^{(\alpha,\beta)}[f]_g$ $\left(0 < \rho^{(\alpha,\beta)}[f]_g < \infty\right)$ with respect to $g(z)$. Then generalized relative Gol'dberg lower type $(\alpha, \beta)$ denoted by $\overline{\sigma}_D^{(\alpha,\beta)}[f]_g$ of $f(z)$ with respect to $g(z)$ is defined as :*

$$\overline{\sigma}_D^{(\alpha,\beta)}[f]_g = \liminf_{R \to \infty} \frac{\exp(\alpha(M_{g,D}^{-1}(M_{f,D}(R))))}{(\exp(\beta(R)))^{\rho^{(\alpha,\beta)}[f]_g}}.$$

The above definition can alternatively be defined in the following manner:

**Definition 5.4.2** *Let $f(z)$ and $g(z)$ be any two entire functions of $n$ complex variables and $f(z)$ has finite positive generalized relative Gol'dberg order $(\alpha, \beta)$ as $\rho^{(\alpha,\beta)}[f]_g$ $\left(0 < \rho^{(\alpha,\beta)}[f]_g < \infty\right)$ with respect to $g(z)$. Then generalized relative Gol'dberg lower type $(\alpha, \beta)$ denoted by $\overline{\sigma}_D^{(\alpha,\beta)}[f]_g$ of $f(z)$ with respect to $g(z)$ is defined as : The integral $\int_{R_0}^{\infty} \dfrac{\exp^{[2]}(\alpha(M_{g,D}^{-1}(M_{f,D}(R))))}{\left[\exp\left((\exp(\beta(R)))^{\rho^{(\alpha,\beta)}[f]_g}\right)\right]^{k+1}} dR \, (R_0 > 0)$ converges for $k > \overline{\sigma}_D^{(\alpha,\beta)}[f]_g$ and diverges for $k < \overline{\sigma}_D^{(\alpha,\beta)}[f]_g$.*

**Definition 5.4.3** *Let $f(z)$ and $g(z)$ be any two entire functions of $n$ complex variables with $f(z)$ has finite positive generalized relative Gol'dberg lower order $(\alpha, \beta)$ i.e. $\lambda^{(\alpha,\beta)}[f]_g$ $\left(0 < \lambda^{(\alpha,\beta)}[f]_g < \infty\right)$ with respect to $g(z)$. Then the generalized relative Gol'dberg upper weak type $(\alpha, \beta)$ denoted by $\overline{\tau}_D^{(\alpha,\beta)}[f]_g$ of $f(z)$ with respect to $g(z)$ is defined as :*

$$\overline{\tau}_D^{(\alpha,\beta)}[f]_g = \limsup_{R \to \infty} \frac{\exp(\alpha(M_{g,D}^{-1}(M_{f,D}(R))))}{(\exp(\beta(R)))^{\lambda^{(\alpha,\beta)}[f]_g}}.$$

The above definition can also alternatively defined as:

**Definition 5.4.4** *Let $f(z)$ and $g(z)$ be any two entire functions of $n$ complex variables with $f(z)$ has finite positive generalized relative Gol'dberg lower order $(\alpha, \beta)$ i.e. $\lambda^{(\alpha,\beta)}[f]_g$ $\left(0 < \lambda^{(\alpha,\beta)}[f]_g < \infty\right)$ with respect to $g(z)$. Then the generalized relative Gol'dberg upper weak type $(\alpha, \beta)$ denoted by $\overline{\tau}_D^{(\alpha,\beta)}[f]_g$ of $f(z)$ with respect to $g(z)$ is defined as : The integral $\int_{R_0}^{\infty} \dfrac{\exp^{[2]}(\alpha(M_{g,D}^{-1}(M_{f,D}(R))))}{\left[\exp\left((\exp(\beta(R)))^{\lambda^{(\alpha,\beta)}[f]_g}\right)\right]^{k+1}} dR \, (R_0 > 0)$ converges for $k > \overline{\tau}_D^{(\alpha,\beta)}[f]_g$ and diverges for $k < \overline{\tau}_D^{(\alpha,\beta)}[f]_g$.*

**Remark 5.4.1** *As Gol'dberg has shown that (see [2]) Gol'dberg type depends on the domain $D$, therefore in general all the growth indicators defined in Definition 5.4.1 and Definition 5.4.3 also depend on $D$.*

Now we state the following two theorems without their proofs as those can easily be carried out with help of Lemma 5.3.1 and in the line of Theorem 5.4.1 and Theorem 5.4.2 respectively.

**Theorem 5.4.3** *Let $f(z)$ and $g(z)$ be any two entire functions of $n$ complex variables and $f(z)$ has finite positive generalized relative Gol'dberg order $(\alpha, \beta)$ i.e., $\rho^{(\alpha,\beta)}[f]_g$ $\left(0 < \rho^{(\alpha,\beta)}[f]_g < \infty\right)$ and generalized relative Gol'dberg lower type $(\alpha, \beta)$ denoted by $\overline{\sigma}_D^{(\alpha,\beta)}[f]_g$ with respect to $g(z)$ Then Definition 5.4.1 and Definition 5.4.2 are equivalent.*

**Theorem 5.4.4** *Let $f(z)$ and $g(z)$ be any two entire functions of $n$ complex variables with $f(z)$ has finite positive generalized relative Gol'dberg lower order $(\alpha, \beta)$ denoted by $\lambda^{(\alpha,\beta)}[f]_g \left(0 < \lambda^{(\alpha,\beta)}[f]_g < \infty\right)$ and the generalized relative Gol'dberg upper weak type $(\alpha, \beta)$ denoted by with respect to $g(z)$. Then Definition 5.4.3 and Definition 5.4.4 are equivalent.*

**Theorem 5.4.5** *Let $f(z)$ and $g(z)$ be any two entire functions of $n$ complex variables with $0 < \lambda^{(\alpha,\beta)}[f]_g \leq \rho^{(\alpha,\beta)}[f]_g < \infty$, then*

$$(i) \quad \sigma_D^{(\alpha,\beta)}[f]_g = \limsup_{R \to \infty} \frac{\exp(\alpha(M_{g,D}^{-1}(R)))}{\left[\exp(\beta(M_{f,D}^{-1}(R)))\right]^{\rho^{(\alpha,\beta)}[f]_g}},$$

$$(ii) \quad \overline{\sigma}_D^{(\alpha,\beta)}[f]_g = \liminf_{R \to \infty} \frac{\exp(\alpha(M_{g,D}^{-1}(R)))}{\left[\exp(\beta(M_{f,D}^{-1}(R)))\right]^{\rho^{(\alpha,\beta)}[f]_g}},$$

$$(iii) \quad \tau_D^{(\alpha,\beta)}[f]_g = \liminf_{R \to \infty} \frac{\exp(\alpha(M_{g,D}^{-1}(R)))}{\left[\exp(\beta(M_{f,D}^{-1}(R)))\right]^{\lambda^{(\alpha,\beta)}[f]_g}}$$

*and*

$$(iv) \quad \overline{\tau}_D^{(\alpha,\beta)}[f]_g = \limsup_{R \to \infty} \frac{\exp(\alpha(M_{g,D}^{-1}(R)))}{\left[\exp(\beta(M_{f,D}^{-1}(R)))\right]^{\lambda^{(\alpha,\beta)}[f]_g}}.$$

**Proof.** Taking $M_{f,D}(R) = R$, theorem follows from the definitions of $\sigma_D^{(\alpha,\beta)}[f]_g$, $\overline{\sigma}_D^{(\alpha,\beta)}[f]_g$, $\tau_D^{(\alpha,\beta)}[f]_g$ and $\overline{\tau}_D^{(\alpha,\beta)}[f]_g$ respectively. ■

In the following theorem we obtain a relationship between $\sigma_D^{(\alpha,\beta)}[f]_g$, $\overline{\sigma}_D^{(\alpha,\beta)}[f]_g$, $\tau_D^{(\alpha,\beta)}[f]_g$ and $\overline{\tau}_D^{(\alpha,\beta)}[f]_g$.

**Theorem 5.4.6** *Let $f(z)$ and $g(z)$ be any two entire functions of $n$ complex variables such that $f((z))$ is of regular generalized relative Gol'dberg $(\alpha, \beta)$ growth with respect to $g(z)$ i.e., $\lambda^{(\alpha,\beta)}[f]_g = \rho^{(\alpha,\beta)}[f]_g \left(0 < \lambda^{(\alpha,\beta)}[f]_g = \rho^{(\alpha,\beta)}[f]_g < \infty\right)$, then the following quantities*

$$(i) \ \sigma_D^{(\alpha,\beta)}[f]_g, \quad (ii) \ \tau_D^{(\alpha,\beta)}[f]_g, \quad (iii) \ \overline{\sigma}_D^{(\alpha,\beta)}[f]_g \ and \quad (iv) \ \overline{\tau}_D^{(\alpha,\beta)}[f]_g$$

*are all equivalent.*

**Proof.**     From     Definition     5.2.5,     it     follows     that     the     integral

$$\int_{R_0}^{\infty} \frac{\exp^{[2]}(\alpha(M_{g,D}^{-1}(M_{f,D}(R))))}{\left[\exp\left((\exp(\beta(R)))^{\lambda^{(\alpha,\beta)}[f]_g}\right)\right]^{k+1}} dR \ (R_0 > 0)$$

converges for $k > \tau_D^{(\alpha,\beta)}[f]_g$ and diverges for $k < \tau_D^{(\alpha,\beta)}[f]_g$. On the other hand, Definition 5.2.3     implies     that     the     integral

$$\int_{R_0}^{\infty} \frac{\exp^{[2]}(\alpha(M_{g,D}^{-1}(M_{f,D}(R))))}{\left[\exp\left((\exp(\beta(R)))^{\rho^{(\alpha,\beta)}[f]_g}\right)\right]^{k+1}} dR \, (R_0 > 0)$$ converges for $k > \sigma_D^{(\alpha,\beta)}[f]_g$ and diverges for

$k < \sigma_D^{(\alpha,\beta)}[f]_g$.

$(i) \Rightarrow (ii)$.

Now it is obvious that all the quantities in the expression

$$\left[ \frac{\exp^{[2]}(\alpha(M_{g,D}^{-1}(M_{f,D}(R))))}{\left[\exp\left((\exp(\beta(R)))^{\lambda^{(\alpha,\beta)}[f]_g}\right)\right]^{k+1}} - \frac{\exp^{[2]}(\alpha(M_{g,D}^{-1}(M_{f,D}(R))))}{\left[\exp\left((\exp(\beta(R)))^{\rho^{(\alpha,\beta)}[f]_g}\right)\right]^{k+1}} \right]$$

are of non negative type. So

$$\int_{R_0}^{\infty} \left[ \frac{\exp^{[2]}(\alpha(M_{g,D}^{-1}(M_{f,D}(R))))}{\left[\exp\left((\exp(\beta(R)))^{\lambda^{(\alpha,\beta)}[f]_g}\right)\right]^{k+1}} \right.$$

$$\left. - \frac{\exp^{[2]}(\alpha(M_{g,D}^{-1}(M_{f,D}(R))))}{\left[\exp\left((\exp(\beta(R)))^{\rho^{(\alpha,\beta)}[f]_g}\right)\right]^{k+1}} \right] dR \, (R_0 > 0) \geq 0$$

i.e., $$\int_{R_0}^{\infty} \frac{\exp^{[2]}(\alpha(M_{g,D}^{-1}(M_{f,D}(R))))}{\left[\exp\left((\exp(\beta(R)))^{\lambda^{(\alpha,\beta)}[f]_g}\right)\right]^{k+1}} dR \geq$$

$$\int_{R_0}^{\infty} \frac{\exp^{[2]}(\alpha(M_{g,D}^{-1}(M_{f,D}(R))))}{\left[\exp\left((\exp(\beta(R)))^{\rho^{(\alpha,\beta)}[f]_g}\right)\right]^{k+1}} dR \text{ for } R_0 > 0 \ .$$

i.e., $\tau_D^{(\alpha,\beta)}[f]_g \geq \sigma_D^{(\alpha,\beta)}[f]_g$. $\qquad (73)$

Further $f(z)$ is of regular generalized relative Gol'dberg $(\alpha, \beta)$ growth with respect to $g(z)$. Therefore we get that

$$\sigma_D^{(\alpha,\beta)}[f]_g = \limsup_{R\to\infty} \frac{\exp(\alpha(M_{g,D}^{-1}(M_{f,D}(R))))}{(\exp(\beta(R)))^{\rho^{(\alpha,\beta)}[f]_g}}$$

$$\geq \liminf_{R\to\infty} \frac{\exp(\alpha(M_{g,D}^{-1}(M_{f,D}(R))))}{(\exp(\beta(R)))^{\rho^{(\alpha,\beta)}[f]_g}}$$

$$= \liminf_{R\to\infty} \frac{\exp(\alpha(M_{g,D}^{-1}(M_{f,D}(R))))}{(\exp(\beta(R)))^{\lambda^{(\alpha,\beta)}[f]_g}} = \tau_D^{(\alpha,\beta)}[f]_g. \qquad (74)$$

Hence from (73) and (74) we obtain that

$$\sigma_D^{(\alpha,\beta)}[f]_g = \tau_D^{(\alpha,\beta)}[f]_g. \qquad (75)$$

$(ii) \Rightarrow (iii)$.

Since $f(z)$ is of regular generalized relative Gol'dberg $(\alpha, \beta)$ growth with respect to $g(z)$ i.e., $\rho^{(\alpha,\beta)}[f]_g = \lambda^{(\alpha,\beta)}[f]_g$, we get that

$$\tau_D^{(\alpha,\beta)}[f]_g = \liminf_{R \to \infty} \frac{\exp(\alpha(M_{g,D}^{-1}(M_{f,D}(R))))}{(\exp(\beta(R)))^{\lambda^{(\alpha,\beta)}[f]_g}}$$

$$= \liminf_{R \to \infty} \frac{\exp(\alpha(M_{g,D}^{-1}(M_{f,D}(R))))}{(\exp(\beta(R)))^{\rho^{(\alpha,\beta)}[f]_g}} = \overline{\sigma}_D^{(\alpha,\beta)}[f]_g.$$

$(iii) \Rightarrow (iv)$.

In view of (75) and the condition $\rho^{(\alpha,\beta)}[f]_g = \lambda^{(\alpha,\beta)}[f]_g$, it follows that

$$\overline{\sigma}_D^{(\alpha,\beta)}[f]_g = \liminf_{R \to \infty} \frac{\exp(\alpha(M_{g,D}^{-1}(M_{f,D}(R))))}{(\exp(\beta(R)))^{\rho^{(\alpha,\beta)}[f]_g}}$$

$$i.e., \ \overline{\sigma}_D^{(\alpha,\beta)}[f]_g = \liminf_{R \to \infty} \frac{\exp(\alpha(M_{g,D}^{-1}(M_{f,D}(R))))}{(\exp(\beta(R)))^{\lambda^{(\alpha,\beta)}[f]_g}}$$

$$i.e., \ \overline{\sigma}_D^{(\alpha,\beta)}[f]_g = \tau_D^{(\alpha,\beta)}[f]_g$$

$$i.e., \ \overline{\sigma}_D^{(\alpha,\beta)}[f]_g = \sigma_D^{(\alpha,\beta)}[f]_g$$

$$i.e., \ \overline{\sigma}_D^{(\alpha,\beta)}[f]_g = \limsup_{R \to \infty} \frac{\exp(\alpha(M_{g,D}^{-1}(M_{f,D}(R))))}{(\exp(\beta(R)))^{\rho^{(\alpha,\beta)}[f]_g}}$$

$$i.e., \ \overline{\sigma}_D^{(\alpha,\beta)}[f]_g = \limsup_{R \to \infty} \frac{\exp(\alpha(M_{g,D}^{-1}(M_{f,D}(R))))}{(\exp(\beta(R)))^{\lambda^{(\alpha,\beta)}[f]_g}}$$

$$i.e., \ \overline{\sigma}_D^{(\alpha,\beta)}[f]_g = \overline{\tau}_D^{(\alpha,\beta)}[f]_g.$$

$(iv) \Rightarrow (i)$.

As $f(z)$ is of regular generalized relative Gol'dberg $(\alpha, \beta)$ growth with respect to $g(z)$ i.e., $\rho^{(\alpha,\beta)}[f]_g = \lambda^{(\alpha,\beta)}[f]_g$, we obtain that

$$\overline{\tau}_D^{(\alpha,\beta)}[f]_g = \limsup_{R \to \infty} \frac{\exp(\alpha(M_{g,D}^{-1}(M_{f,D}(R))))}{(\exp(\beta(R)))^{\lambda^{(\alpha,\beta)}[f]_g}}$$

$$= \limsup_{R \to \infty} \frac{\exp(\alpha(M_{g,D}^{-1}(M_{f,D}(R))))}{(\exp(\beta(R)))^{\rho^{(\alpha,\beta)}[f]_g}} = \sigma_D^{(\alpha,\beta)}[f]_g.$$

Thus the theorem follows. ∎

**Theorem 5.4.7** *Let* $f(z)$, $g(z)$, $h(z)$ *and* $k(z)$ *be four entire functions of* $n$ *complex variables such that* $0 < \overline{\sigma}_D^{(\alpha_1,\beta_1)} [f]_h \leq \sigma_D^{(\alpha_1,\beta_1)} [f]_h < \infty$, $0 < \overline{\sigma}_D^{(\alpha_2,\beta_2)} [g]_k \leq \sigma_D^{(\alpha_2,\beta_2)} [g]_k < \infty$ *and* $\rho^{(\alpha_1,\beta_1)} [f]_h = \rho^{(\alpha_2,\beta_2)} [g]_k$. *Then*

$$\frac{\overline{\sigma}_D^{(\alpha_1,\beta_1)} [f]_h}{\sigma_D^{(\alpha_2,\beta_2)} [g]_k} \leq \liminf_{R \to \infty} \frac{\exp(\alpha_1(M_{h,D}^{-1}(M_{f,D}(R))))}{\exp(\alpha_2(M_{k,D}^{-1}(M_{g,D}(\beta_2^{-1}(\beta_1(R))))))} \leq \frac{\overline{\sigma}_D^{(\alpha_1,\beta_1)} [f]_h}{\sigma_D^{(\alpha_2,\beta_2)} [g]_k}$$

$$\leq \limsup_{R \to \infty} \frac{\exp(\alpha_1(M_{h,D}^{-1}(M_{f,D}(R))))}{\exp(\alpha_2(M_{k,D}^{-1}(M_{g,D}(\beta_2^{-1}(\beta_1(R))))))} \leq \frac{\sigma_D^{(\alpha_1,\beta_1)} [f]_h}{\overline{\sigma}_D^{(\alpha_2,\beta_2)} [g]_k}.$$

**Proof.** From the definition of $\sigma_D^{(\alpha_2,\beta_2)} [g]_k$ and $\overline{\sigma}_D^{(\alpha_1,\beta_1)} [f]_h$, we have for arbitrary positive $\varepsilon$ and for all large values of $R$ that

$$\exp(\alpha_1(M_{h,D}^{-1}(M_{f,D}(R)))) \geqslant \left(\overline{\sigma}_D^{(\alpha_1,\beta_1)} [f]_h - \varepsilon\right) (\exp \beta_1(R))^{\rho^{(\alpha_1,\beta_1)}[f]_h} \qquad (76)$$

and

$$\exp(\alpha_2(M_{k,D}^{-1}(M_{g,D}(\beta_2^{-1}(\beta_1(R)))))) \leq \left(\sigma_D^{(\alpha_2,\beta_2)} [g]_k + \varepsilon\right) (\exp \beta_1(R))^{\rho^{(\alpha_2,\beta_2)}[g]_k}. \qquad (77)$$

Now from (76), (77) and the condition $\rho^{(\alpha_1,\beta_1)} [f]_h = \rho^{(\alpha_2,\beta_2)} [g]_k$, it follows for all large values of $R$ that

$$\frac{\exp(\alpha_1(M_{h,D}^{-1}(M_{f,D}(R))))}{\exp(\alpha_2(M_{k,D}^{-1}(M_{g,D}(\beta_2^{-1}(\beta_1(R))))))} \geqslant \frac{\overline{\sigma}_D^{(\alpha_1,\beta_1)} [f]_h - \varepsilon}{\sigma_D^{(\alpha_2,\beta_2)} [g]_k + \varepsilon}.$$

As $\varepsilon \, (> 0)$ is arbitrary , we obtain that

$$\liminf_{R \to \infty} \frac{\exp(\alpha_1(M_{h,D}^{-1}(M_{f,D}(R))))}{\exp(\alpha_2(M_{k,D}^{-1}(M_{g,D}(\beta_2^{-1}(\beta_1(R))))))} \geqslant \frac{\overline{\sigma}_D^{(\alpha_1,\beta_1)} [f]_h}{\sigma_D^{(\alpha_2,\beta_2)} [g]_k}. \qquad (78)$$

Again for a sequence of values of $R$ tending to infinity,

$$\exp(\alpha_1(M_{h,D}^{-1}(M_{f,D}(R)))) \leq \left(\overline{\sigma}_D^{(\alpha_1,\beta_1)} [f]_h + \varepsilon\right) (\exp \beta_1(R))^{\rho^{(\alpha_1,\beta_1)}[f]_h} \qquad (79)$$

and for all sufficiently large values of $R$,

$$\exp(\alpha_2(M_{k,D}^{-1}(M_{g,D}(\beta_2^{-1}(\beta_1(R)))))) \geqslant \left(\overline{\sigma}_D^{(\alpha_2,\beta_2)} [g]_k - \varepsilon\right) (\exp \beta_1(R))^{\rho^{(\alpha_2,\beta_2)}[g]_k}. \qquad (80)$$

Combining the condition $\rho^{(\alpha_1,\beta_1)} [f]_h = \rho_D^{(\alpha_2,\beta_2)} [g]_k$, (79) and (80) we get for a sequence of values of $R$ tending to infinity that

$$\frac{\exp(\alpha_1(M_{h,D}^{-1}(M_{f,D}(R))))}{\exp(\alpha_2(M_{k,D}^{-1}(M_{g,D}(\beta_2^{-1}(\beta_1(R))))))} \leq \frac{\overline{\sigma}_D^{(\alpha_1,\beta_1)} [f]_h + \varepsilon}{\overline{\sigma}_D^{(\alpha_2,\beta_2)} [g]_k - \varepsilon}.$$

Since $\varepsilon \, (> 0)$ is arbitrary, it follows that

$$\liminf_{R \to \infty} \frac{\exp(\alpha_1(M_{h,D}^{-1}(M_{f,D}(R))))}{\exp(\alpha_2(M_{k,D}^{-1}(M_{g,D}(\beta_2^{-1}(\beta_1(R)))))))} \leq \frac{\overline{\sigma}_D^{(\alpha_1,\beta_1)}[f]_h}{\overline{\sigma}_D^{(\alpha_2,\beta_2)}[g]_k}. \tag{81}$$

Also for a sequence of values of $R$ tending to infinity that

$$\exp(\alpha_2(M_{k,D}^{-1}(M_{g,D}(\beta_2^{-1}(\beta_1(R)))))) \leq \left(\overline{\sigma}_D^{(\alpha_2,\beta_2)}[g]_k + \varepsilon\right)(\exp\beta_1(R))^{\rho^{(\alpha_2,\beta_2)}[g]_k}. \tag{82}$$

Now from (76), (82) and the condition $\rho^{(\alpha_1,\beta_1)}[f]_h = \rho^{(\alpha_2,\beta_2)}[g]_k$, we obtain for a sequence of values of $R$ tending to infinity that

$$\frac{\exp(\alpha_1(M_{h,D}^{-1}(M_{f,D}(R))))}{\exp(\alpha_2(M_{k,D}^{-1}(M_{g,D}(\beta_2^{-1}(\beta_1(R)))))))} \geq \frac{\overline{\sigma}_D^{(\alpha_1,\beta_1)}[f]_h - \varepsilon}{\overline{\sigma}_D^{(\alpha_2,\beta_2)}[g]_k + \varepsilon}.$$

As $\varepsilon \, (> 0)$ is arbitrary, we get from above that

$$\limsup_{R \to \infty} \frac{\exp(\alpha_1(M_{h,D}^{-1}(M_{f,D}(R))))}{\exp(\alpha_2(M_{k,D}^{-1}(M_{g,D}(\beta_2^{-1}(\beta_1(R)))))))} \geq \frac{\overline{\sigma}_D^{(\alpha_1,\beta_1)}[f]_h}{\overline{\sigma}_D^{(\alpha_2,\beta_2)}[g]_k}. \tag{83}$$

Also for all sufficiently large values of $R$,

$$\exp(\alpha_1(M_{h,D}^{-1}(M_{f,D}(R)))) \leq \left(\sigma_D^{(\alpha_1,\beta_1)}[f]_h + \varepsilon\right)(\exp\beta_1(R))^{\rho^{(\alpha_1,\beta_1)}[f]_h}. \tag{84}$$

As the condition $\rho^{(\alpha_1,\beta_1)}[f]_h = \rho^{(\alpha_2,\beta_2)}[g]_k$, it follows from (80) and (84) for all large values of $R$ that

$$\frac{\exp(\alpha_1(M_{h,D}^{-1}(M_{f,D}(R))))}{\exp(\alpha_2(M_{k,D}^{-1}(M_{g,D}(\beta_2^{-1}(\beta_1(R)))))))} \leq \frac{\left(\sigma_D^{(\alpha_1,\beta_1)}[f]_h + \varepsilon\right)}{\left(\overline{\sigma}_D^{(\alpha_2,\beta_2)}[g]_k - \varepsilon\right)}.$$

Since $\varepsilon \, (> 0)$ is arbitrary, we obtain that

$$\limsup_{R \to \infty} \frac{\exp(\alpha_1(M_{h,D}^{-1}(M_{f,D}(R))))}{\exp(\alpha_2(M_{k,D}^{-1}(M_{g,D}(\beta_2^{-1}(\beta_1(R)))))))} \leq \frac{\sigma_D^{(\alpha_1,\beta_1)}[f]_h}{\overline{\sigma}_D^{(\alpha_2,\beta_2)}[g]_k}. \tag{85}$$

Thus the theorem follows from $(78), (81), (83)$ and $(85)$. ∎

**Theorem 5.4.8** *Let $f(z)$, $g(z)$, $h(z)$ and $k(z)$ be four entire functions of $n$ complex variables such that $0 < \sigma_D^{(\alpha_1,\beta_1)}[f]_h < \infty$, $0 < \sigma_D^{(\alpha_2,\beta_2)}[g]_k < \infty$ and $\rho^{(\alpha_1,\beta_1)}[f]_h = \rho^{(\alpha_2,\beta_2)}[g]_k$. Then*

$$\liminf_{R \to \infty} \frac{\exp(\alpha_1(M_{h,D}^{-1}(M_{f,D}(R))))}{\exp(\alpha_2(M_{k,D}^{-1}(M_{g,D}(\beta_2^{-1}(\beta_1(R)))))))} \leq \frac{\sigma_D^{(\alpha_1,\beta_1)}[f]_h}{\sigma_D^{(\alpha_2,\beta_2)}[g]_k}$$

$$\leq \limsup_{R \to \infty} \frac{\exp(\alpha_1(M_{h,D}^{-1}(M_{f,D}(R))))}{\exp(\alpha_2(M_{k,D}^{-1}(M_{g,D}(\beta_2^{-1}(\beta_1(R)))))))}.$$

**Proof.** From the definition of $\sigma_D^{(\alpha_2,\beta_2)}[g]_k$, we get for a sequence of values of $R$ tending to infinity that

$$\exp(\alpha_2(M_{k,D}^{-1}(M_{g,D}\left(\beta_2^{-1}(\beta_1(R))\right)))) \geqslant \left(\sigma_D^{(\alpha_2,\beta_2)}[g]_k - \varepsilon\right)(\exp\beta_1(R))^{\rho^{(\alpha_2,\beta_2)}[g]_k}. \qquad (86)$$

Now from (84), (86) and the condition $\rho^{(\alpha_1,\beta_1)}[f]_h = \rho^{(\alpha_2,\beta_2)}[g]_k$, it follows for a sequence of values of $R$ tending to infinity that

$$\frac{\exp(\alpha_1(M_{h,D}^{-1}(M_{f,D}(R))))}{\exp(\alpha_2(M_{k,D}^{-1}(M_{g,D}\left(\beta_2^{-1}(\beta_1(R))\right))))} \leq \frac{\sigma_D^{(\alpha_1,\beta_1)}[f]_h + \varepsilon}{\sigma_D^{(\alpha_2,\beta_2)}[g]_k - \varepsilon}.$$

As $\varepsilon\,(> 0)$ is arbitrary, we obtain that

$$\liminf_{R\to\infty}\frac{\exp(\alpha_1(M_{h,D}^{-1}(M_{f,D}(R))))}{\exp(\alpha_2(M_{k,D}^{-1}(M_{g,D}\left(\beta_2^{-1}(\beta_1(R))\right))))} \leq \frac{\sigma_D^{(\alpha_1,\beta_1)}[f]_h}{\sigma_D^{(\alpha_2,\beta_2)}[g]_k}. \qquad (87)$$

Again for a sequence of values of $R$ tending to infinity that

$$\exp(\alpha_1(M_{h,D}^{-1}(M_{f,D}(R)))) \geqslant \left(\sigma_D^{(\alpha_1,\beta_1)}[f]_h - \varepsilon\right)(\exp\beta_1(R))^{\rho^{(\alpha_1,\beta_1)}[f]_h}. \qquad (88)$$

So combining the condition $\rho^{(\alpha_1,\beta_1)}[f]_h = \rho^{(\alpha_2,\beta_2)}[g]_k$, (77) and (88) we get for a sequence of values of $R$ tending to infinity ,

$$\frac{\exp(\alpha_1(M_{h,D}^{-1}(M_{f,D}(R))))}{\exp(\alpha_2(M_{k,D}^{-1}(M_{g,D}\left(\beta_2^{-1}(\beta_1(R))\right))))} \geqslant \frac{\sigma_D^{(\alpha_1,\beta_1)}[f]_h - \varepsilon}{\sigma_D^{(\alpha_2,\beta_2)}[g]_k + \varepsilon}.$$

Since $\varepsilon\,(> 0)$ is arbitrary, it follows that

$$\limsup_{R\to\infty}\frac{\exp(\alpha_1(M_{h,D}^{-1}(M_{f,D}(R))))}{\exp(\alpha_2(M_{k,D}^{-1}(M_{g,D}\left(\beta_2^{-1}(\beta_1(R))\right))))} \geqslant \frac{\sigma_D^{(\alpha_1,\beta_1)}[f]_h}{\sigma_D^{(\alpha_2,\beta_2)}[g]_k}. \qquad (89)$$

Thus the theorem follows from (87) and (89). ∎

The following theorem is a natural consequence of Theorem 5.4.7 and Theorem 5.4.8.

**Theorem 5.4.9** *Let $f(z)$, $g(z)$, $h(z)$ and $k(z)$ be four entire functions of $n$ complex variables such that $0 < \overline{\sigma}_D^{(\alpha_1,\beta_1)}[f]_h \leq \sigma_D^{(\alpha_1,\beta_1)}[f]_h < \infty$, $0 < \overline{\sigma}_D^{(\alpha_2,\beta_2)}[g]_k \leq \sigma_D^{(\alpha_2,\beta_2)}[g]_k < \infty$ and $\rho^{(\alpha_1,\beta_1)}[f]_h = \rho^{(\alpha_2,\beta_2)}[g]_k$. Then*

$$\liminf_{R\to\infty}\frac{\exp(\alpha_1(M_{h,D}^{-1}(M_{f,D}(R))))}{\exp(\alpha_2(M_{k,D}^{-1}(M_{g,D}\left(\beta_2^{-1}(\beta_1(R))\right))))} \leq \min\left\{\frac{\overline{\sigma}_D^{(\alpha_1,\beta_1)}[f]_h}{\overline{\sigma}_D^{(\alpha_2,\beta_2)}[g]_k}, \frac{\sigma_D^{(\alpha_1,\beta_1)}[f]_h}{\sigma_D^{(\alpha_2,\beta_2)}[g]_k}\right\}$$

$$\leq \max\left\{\frac{\overline{\sigma}_D^{(\alpha_1,\beta_1)}[f]_h}{\overline{\sigma}_D^{(\alpha_2,\beta_2)}[g]_k}, \frac{\sigma_D^{(\alpha_1,\beta_1)}[f]_h}{\sigma_D^{(\alpha_2,\beta_2)}[g]_k}\right\} \leq \limsup_{R\to\infty}\frac{\exp(\alpha_1(M_{h,D}^{-1}(M_{f,D}(R))))}{\exp(\alpha_2(M_{k,D}^{-1}(M_{g,D}\left(\beta_2^{-1}(\beta_1(R))\right))))}.$$

The proof is omitted.

Now in the line of Theorem 5.4.7, Theorem 5.4.8 and Theorem 5.4.9 respectively one can easily prove the following six theorems using the notion of generalized relative Gol'dberg weak type $(\alpha, \beta)$ and therefore their proofs are omitted.

**Theorem 5.4.10** *Let $f(z)$, $g(z)$, $h(z)$ and $k(z)$ be four entire functions of $n$ complex variables such that $0 < \tau_D^{(\alpha_1,\beta_1)}[f]_h \leq \overline{\tau}_D^{(\alpha_1,\beta_1)}[f]_h < \infty$, $0 < \tau_D^{(\alpha_2,\beta_2)}[g]_k \leq \overline{\tau}_D^{(\alpha_2,\beta_2)}[g]_k < \infty$ and $\lambda^{(\alpha_1,\beta_1)}[f]_h = \lambda^{(\alpha_2,\beta_2)}[g]_k$. Then*

$$\frac{\tau_D^{(\alpha_1,\beta_1)}[f]_h}{\overline{\tau}_D^{(\alpha_2,\beta_2)}[g]_k} \leq \liminf_{R \to \infty} \frac{\exp(\alpha_1(M_{h,D}^{-1}(M_{f,D}(R))))}{\exp(\alpha_2(M_{k,D}^{-1}(M_{g,D}(\beta_2^{-1}(\beta_1(R))))))} \leq \frac{\tau_D^{(\alpha_1,\beta_1)}[f]_h}{\tau_D^{(\alpha_2,\beta_2)}[g]_k}$$

$$\leq \limsup_{R \to \infty} \frac{\exp(\alpha_1(M_{h,D}^{-1}(M_{f,D}(R))))}{\exp(\alpha_2(M_{k,D}^{-1}(M_{g,D}(\beta_2^{-1}(\beta_1(R))))))} \leq \frac{\overline{\tau}_D^{(\alpha_1,\beta_1)}[f]_h}{\tau_D^{(\alpha_2,\beta_2)}[g]_k}.$$

**Theorem 5.4.11** *Let $f(z)$, $g(z)$, $h(z)$ and $k(z)$ be four entire functions of $n$ complex variables such that $0 < \overline{\tau}_D^{(\alpha_1,\beta_1)}[f]_h < \infty$, $0 < \overline{\tau}_D^{(\alpha_2,\beta_2)}[g]_k < \infty$ and $\lambda^{(\alpha_1,\beta_1)}[f]_h = \lambda^{(\alpha_2,\beta_2)}[g]_k$. Then*

$$\liminf_{R \to \infty} \frac{\exp(\alpha_1(M_{h,D}^{-1}(M_{f,D}(R))))}{\exp(\alpha_2(M_{k,D}^{-1}(M_{g,D}(\beta_2^{-1}(\beta_1(R))))))} \leq \frac{\overline{\tau}_D^{(\alpha_1,\beta_1)}[f]_h}{\overline{\tau}_D^{(\alpha_2,\beta_2)}[g]_k}$$

$$\leq \limsup_{R \to \infty} \frac{\exp(\alpha_1(M_{h,D}^{-1}(M_{f,D}(R))))}{\exp(\alpha_2(M_{k,D}^{-1}(M_{g,D}(\beta_2^{-1}(\beta_1(R))))))}.$$

**Theorem 5.4.12** *Let $f(z)$, $g(z)$, $h(z)$ and $k(z)$ be four entire functions of $n$ complex variables such that $0 < \tau_D^{(\alpha_1,\beta_1)}[f]_h \leq \overline{\tau}_D^{(\alpha_1,\beta_1)}[f]_h < \infty$, $0 < \tau_D^{(\alpha_2,\beta_2)}[g]_k \leq \overline{\tau}_D^{(\alpha_2,\beta_2)}[g]_k < \infty$ and $\lambda^{(\alpha_1,\beta_1)}[f]_h = \lambda^{(\alpha_2,\beta_2)}[g]_k$. Then*

$$\liminf_{R \to \infty} \frac{\exp(\alpha_1(M_{h,D}^{-1}(M_{f,D}(R))))}{\exp(\alpha_2(M_{k,D}^{-1}(M_{g,D}(\beta_2^{-1}(\beta_1(R))))))} \leq \min\left\{\frac{\tau_D^{(\alpha_1,\beta_1)}[f]_h}{\tau_D^{(\alpha_2,\beta_2)}[g]_k}, \frac{\overline{\tau}_D^{(\alpha_1,\beta_1)}[f]_h}{\overline{\tau}_D^{(\alpha_2,\beta_2)}[g]_k}\right\}$$

$$\leq \max\left\{\frac{\tau_D^{(\alpha_1,\beta_1)}[f]_h}{\tau_D^{(\alpha_2,\beta_2)}[g]_k}, \frac{\overline{\tau}_D^{(\alpha_1,\beta_1)}[f]_h}{\overline{\tau}_D^{(\alpha_2,\beta_2)}[g]_k}\right\} \leq \limsup_{R \to \infty} \frac{\exp(\alpha_1(M_{h,D}^{-1}(M_{f,D}(R))))}{\exp(\alpha_2(M_{k,D}^{-1}(M_{g,D}(\beta_2^{-1}(\beta_1(R))))))}.$$

We may now state the following theorems without their proofs based on generalized relative Gol'dberg type $(\alpha, \beta)$ and generalized relative Gol'dberg weak type $(\alpha, \beta)$.

**Theorem 5.4.13** *Let $f(z)$, $g(z)$, $h(z)$ and $k(z)$ be four entire functions of $n$ complex variables such that $0 < \overline{\sigma}_D^{(\alpha_1,\beta_1)}[f]_h \leq \sigma_D^{(\alpha_1,\beta_1)}[f]_h < \infty$, $0 < \tau_D^{(\alpha_2,\beta_2)}[g]_k \leq \overline{\tau}_D^{(\alpha_2,\beta_2)}[g]_k < \infty$ and $\rho^{(\alpha_1,\beta_1)}[f]_h = \lambda^{(\alpha_2,\beta_2)}[g]_k$. Then*

$$\frac{\overline{\sigma}_D^{(\alpha_1,\beta_1)}[f]_h}{\overline{\tau}_D^{(\alpha_2,\beta_2)}[g]_k} \leq \liminf_{R \to \infty} \frac{\exp(\alpha_1(M_{h,D}^{-1}(M_{f,D}(R))))}{\exp(\alpha_2(M_{k,D}^{-1}(M_{g,D}(\beta_2^{-1}(\beta_1(R))))))} \leq \frac{\overline{\sigma}_D^{(\alpha_1,\beta_1)}[f]_h}{\tau_D^{(\alpha_2,\beta_2)}[g]_k}$$

$$\leq \limsup_{R \to \infty} \frac{\exp(\alpha_1(M_{h,D}^{-1}(M_{f,D}(R))))}{\exp(\alpha_2(M_{k,D}^{-1}(M_{g,D}(\beta_2^{-1}(\beta_1(R))))))} \leq \frac{\sigma_D^{(\alpha_1,\beta_1)}[f]_h}{\tau_D^{(\alpha_2,\beta_2)}[g]_k}.$$

**Theorem 5.4.14** *Let* $f(z)$, $g(z)$, $h(z)$ *and* $k(z)$ *be four entire functions of* $n$ *complex variables such that* $0 < \sigma_D^{(\alpha_1,\beta_1)}[f]_h < \infty$, $0 < \overline{\tau}_D^{(\alpha_2,\beta_2)}[g]_k < \infty$ *and* $\rho^{(\alpha_1,\beta_1)}[f]_h = \lambda^{(\alpha_2,\beta_2)}[g]_k$. *Then*

$$\liminf_{R\to\infty} \frac{\exp(\alpha_1(M_{h,D}^{-1}(M_{f,D}(R))))}{\exp(\alpha_2(M_{k,D}^{-1}(M_{g,D}(\beta_2^{-1}(\beta_1(R))))))} \leq \frac{\sigma_D^{(\alpha_1,\beta_1)}[f]_h}{\overline{\tau}_D^{(\alpha_2,\beta_2)}[g]_k}$$

$$\leq \limsup_{R\to\infty} \frac{\exp(\alpha_1(M_{h,D}^{-1}(M_{f,D}(R))))}{\exp(\alpha_2(M_{k,D}^{-1}(M_{g,D}(\beta_2^{-1}(\beta_1(R))))))}.$$

**Theorem 5.4.15** *Let* $f(z)$, $g(z)$, $h(z)$ *and* $k(z)$ *be four entire functions of* $n$ *complex variables such that* $0 < \overline{\sigma}_D^{(\alpha_1,\beta_1)}[f]_h \leq \sigma_D^{(\alpha_1,\beta_1)}[f]_h < \infty$, $0 < \tau_D^{(\alpha_2,\beta_2)}[g]_k \leq \overline{\tau}_D^{(\alpha_2,\beta_2)}[g]_k < \infty$ *and* $\rho^{(\alpha_1,\beta_1)}[f]_h = \lambda^{(\alpha_2,\beta_2)}[g]_k$. *Then*

$$\liminf_{R\to\infty} \frac{\exp(\alpha_1(M_{h,D}^{-1}(M_{f,D}(R))))}{\exp(\alpha_2(M_{k,D}^{-1}(M_{g,D}(\beta_2^{-1}(\beta_1(R))))))} \leq \min\left\{\frac{\overline{\sigma}_D^{(\alpha_1,\beta_1)}[f]_h}{\tau_D^{(\alpha_2,\beta_2)}[g]_k}, \frac{\sigma_D^{(\alpha_1,\beta_1)}[f]_h}{\overline{\tau}_D^{(\alpha_2,\beta_2)}[g]_k}\right\}$$

$$\leq \max\left\{\frac{\overline{\sigma}_D^{(\alpha_1,\beta_1)}[f]_h}{\tau_D^{(\alpha_2,\beta_2)}[g]_k}, \frac{\sigma_D^{(\alpha_1,\beta_1)}[f]_h}{\overline{\tau}_D^{(\alpha_2,\beta_2)}[g]_k}\right\} \leq \limsup_{R\to\infty} \frac{\exp(\alpha_1(M_{h,D}^{-1}(M_{f,D}(R))))}{\exp(\alpha_2(M_{k,D}^{-1}(M_{g,D}(\beta_2^{-1}(\beta_1(R))))))}.$$

**Theorem 5.4.16** *Let* $f(z)$, $g(z)$, $h(z)$ *and* $k(z)$ *be four entire functions of* $n$ *complex variables such that* $0 < \tau_D^{(\alpha_1,\beta_1)}[f]_h \leq \overline{\tau}_D^{(\alpha_1,\beta_1)}[f]_h < \infty$, $0 < \overline{\sigma}_D^{(\alpha_2,\beta_2)}[g]_k \leq \sigma_D^{(\alpha_2,\beta_2)}[g]_k < \infty$ *and* $\lambda^{(\alpha_1,\beta_1)}[f]_h = \rho^{(\alpha_2,\beta_2)}[g]_k$. *Then*

$$\frac{\tau_D^{(\alpha_1,\beta_1)}[f]_h}{\sigma_D^{(\alpha_2,\beta_2)}[g]_k} \leq \liminf_{R\to\infty} \frac{\exp(\alpha_1(M_{h,D}^{-1}(M_{f,D}(R))))}{\exp(\alpha_2(M_{k,D}^{-1}(M_{g,D}(\beta_2^{-1}(\beta_1(R))))))} \leq \frac{\tau_D^{(\alpha_1,\beta_1)}[f]_h}{\overline{\sigma}_D^{(\alpha_2,\beta_2)}[g]_k}$$

$$\leq \limsup_{R\to\infty} \frac{\exp(\alpha_1(M_{h,D}^{-1}(M_{f,D}(R))))}{\exp(\alpha_2(M_{k,D}^{-1}(M_{g,D}(\beta_2^{-1}(\beta_1(R))))))} \leq \frac{\overline{\tau}_D^{(\alpha_1,\beta_1)}[f]_h}{\overline{\sigma}_D^{(\alpha_2,\beta_2)}[g]_k}.$$

**Theorem 5.4.17** *Let* $f(z)$, $g(z)$, $h(z)$ *and* $k(z)$ *be four entire functions of* $n$ *complex variables such that* $0 < \overline{\tau}_D^{(\alpha_1,\beta_1)}[f]_h < \infty$, $0 < \sigma_D^{(\alpha_2,\beta_2)}[g]_k < \infty$ *and* $\lambda^{(\alpha_1,\beta_1)}[f]_h = \rho^{(\alpha_2,\beta_2)}[g]_k$. *Then*

$$\liminf_{R\to\infty} \frac{\exp(\alpha_1(M_{h,D}^{-1}(M_{f,D}(R))))}{\exp(\alpha_2(M_{k,D}^{-1}(M_{g,D}(\beta_2^{-1}(\beta_1(R))))))} \leq \frac{\overline{\tau}_D^{(\alpha_1,\beta_1)}[f]_h}{\sigma_D^{(\alpha_2,\beta_2)}[g]_k}$$

$$\leq \limsup_{R\to\infty} \frac{\exp(\alpha_1(M_{h,D}^{-1}(M_{f,D}(R))))}{\exp(\alpha_2(M_{k,D}^{-1}(M_{g,D}(\beta_2^{-1}(\beta_1(R))))))}.$$

**Theorem 5.4.18** *Let* $f(z)$, $g(z)$, $h(z)$ *and* $k(z)$ *be four entire functions of* $n$ *complex variables such that* $0 < \tau_D^{(\alpha_1,\beta_1)}[f]_h \leq \overline{\tau}_D^{(\alpha_1,\beta_1)}[f]_h < \infty$, $0 < \overline{\sigma}_D^{(\alpha_2,\beta_2)}[g]_k \leq \sigma_D^{(\alpha_2,\beta_2)}[g]_k$

$< \infty$ *and* $\lambda^{(\alpha_1,\beta_1)} [f]_h = \rho^{(\alpha_2,\beta_2)} [g]_k$. *Then*

$$\liminf_{R \to \infty} \frac{\exp(\alpha_1(M_{h,D}^{-1}(M_{f,D}(R))))}{\exp(\alpha_2(M_{k,D}^{-1}(M_{g,D}(\beta_2^{-1}(\beta_1(R))))))} \leq \min \left\{ \frac{\tau_D^{(\alpha_1,\beta_1)} [f]_h}{\overline{\sigma}_D^{(\alpha_2,\beta_2)} [g]_k}, \frac{\overline{\tau}_D^{(\alpha_1,\beta_1)} [f]_h}{\sigma_D^{(\alpha_2,\beta_2)} [g]_k} \right\}$$

$$\leq \max \left\{ \frac{\tau_D^{(\alpha_1,\beta_1)} [f]_h}{\overline{\sigma}_D^{(\alpha_2,\beta_2)} [g]_k}, \frac{\overline{\tau}_D^{(\alpha_1,\beta_1)} [f]_h}{\sigma_D^{(\alpha_2,\beta_2)} [g]_k} \right\} \leq \limsup_{R \to \infty} \frac{\exp(\alpha_1(M_{h,D}^{-1}(M_{f,D}(R))))}{\exp(\alpha_2(M_{k,D}^{-1}(M_{g,D}(\beta_2^{-1}(\beta_1(R))))))}.$$

## 5.5   Conclusion.

The main aim of this chapter is actually to extend and to modify the notion of generalized Gol'dberg type $(\alpha, \beta)$ (generalized Gol'dberg weak type $(\alpha, \beta)$) to generalized relative Gol'dberg type $(\alpha, \beta)$ (generalized relative Gol'dberg weak type $(\alpha, \beta)$) of higher dimensions in case of entire functions of several complex variables and establish its integral representation. Here we see that the previous definitions are easily generated as particular cases of the present definitions, e.g. if $\alpha(R) = \log^{[p]} R$ and $\beta(R) = \log^{[q]} R$ respectively, then Definition 5.2.2 reduces to Definition 1.1.22. Further, from the conclusion of Theorem 5.4.6, one can easily state that an entire function $f(z)$ of several complex variables for which generalized relative Gol'dberg order $(\alpha, \beta)$ and generalized relative Gol'dberg lower order $(\alpha, \beta)$ with respect to another entire function $g(z)$ of several complex variables are the same may be called a function of regular generalized relative Gol'dberg $(\alpha, \beta)$ growth as well as perfectly regular generalized relative Gol'dberg $(\alpha, \beta)$ growth with respect to $g(z)$.

## References

[1] B. C. Mondal and C. Roy: Relative gol'dberg order of an entire function of several variables, Bull Cal. Math. Soc., 102(4) (2010), 371-380.

[2] A. A. Gol'dberg: Elementary remarks on the formulas defining order and type of functions of several variables, Dokl. Akad. Nauk Arm. SSR, 29 (1959), 145-151 (Russian).

# Chapter 6

# Derivation of some inequalities using generalized relative Gol'dberg type $(\alpha, \beta)$ and generalized relative Gol'dberg weak type $(\alpha, \beta)$ of entire functions of several complex variables

**Abstract:** In this chapter,we establish some important relations relating to generalized relative Gol'dberg type and weak type $(\alpha, \beta)$ with generalized Gol'dberg type and weak type $(\alpha, \beta)$ of entire functions of $n$ complex variables, where $\alpha, \beta$ are continuous non-negative functions defined on $(-\infty, +\infty)$.

**Keywords:** Generalized Gol'dberg order $(\alpha, \beta)$, generalized Gol'dberg lower order $(\alpha, \beta)$, generalized Gol'dberg type $(\alpha, \beta)$, generalized Gol'dberg weak type $(\alpha, \beta)$, generalized relative Gol'dberg order $(\alpha, \beta)$, generalized relative Gol'dberg lower order $(\alpha, \beta)$, generalized relative Gol'dberg type $(\alpha, \beta)$, generalized relative Gol'dberg weak type $(\alpha, \beta)$, increasing function.

**Mathematics Subject Classification (2010) :** 32A15.

## 6.1 Introduction.

The relative growth indicators gives a quantitative assessment of how different functions scale each other and until what extent they are self-similar in growth. The concepts of generalized relative Gol'dberg type $(\alpha, \beta)$ and generalized relative Gol'dberg weak type $(\alpha, \beta)$ of entire functions of $n$ complex variables are not at all known to the researchers of this area. Therefore the studies of the growths of entire functions of $n$ complex variables in the light of their generalized relative Gol'dberg type $(\alpha, \beta)$ and generalized relative

Gol'dberg weak type $(\alpha, \beta)$ are the prime concern of this chapter. Actually in this chapter we study some relative growth rates of entire functions of $n$ complex variables with respect to another entire function of $n$ complex variables on the basis of their generalized relative Gol'dberg type $(\alpha, \beta)$ and generalized relative Gol'dberg weak type $(\alpha, \beta)$. In this present chapter $\alpha, \beta$ and $\gamma$ always denote the functions belonging to $L^0$.

## 6.2 Lemmas.

From the conclusion of Theorem 4.2.1, we present the following two lemmas which will be needed in the sequel.

**Lemma 6.2.1** *Let $f(z)$ and $g(z)$ be two entire functions of $n$ complex variables such that $0 < \rho^{(\gamma,\beta)}[f] < \infty$ and $0 < \lambda^{(\gamma,\alpha)}[g] = \rho^{(\gamma,\alpha)}[g] < \infty$. Then*

$$\rho^{(\alpha,\beta)}[f]_g = \frac{\rho^{(\gamma,\beta)}[f]}{\rho^{(\gamma,\alpha)}[g]} \quad and \quad \lambda^{(\alpha,\beta)}[f]_g = \frac{\lambda^{(\gamma,\beta)}[f]}{\lambda^{(\gamma,\alpha)}[g]}.$$

**Lemma 6.2.2** *Let $f(z)$ and $g(z)$ be two entire functions of $n$ complex variables such that $0 < \lambda^{(\gamma,\beta)}[f] = \rho^{(\gamma,\beta)}[f] < \infty$ and $0 < \rho^{(\gamma,\alpha)}[g] < \infty$. Then*

$$\rho^{(\alpha,\beta)}[f]_g = \frac{\lambda^{(\gamma,\beta)}[f]}{\lambda^{(\gamma,\alpha)}[g]} \quad and \quad \lambda^{(\alpha,\beta)}[f]_g = \frac{\rho^{(\gamma,\beta)}[f]}{\rho^{(\gamma,\alpha)}[g]}.$$

## 6.3 Main Results.

In this section we state the main results of the chapter.

**Theorem 6.3.1** *Let $f(z)$ and $g(z)$ be two entire functions of $n$ complex variables such that $0 < \rho^{(\gamma,\beta)}[f] < \infty$ and $0 < \lambda^{(\gamma,\alpha)}[g] = \rho^{(\gamma,\alpha)}[g] < \infty$. Then*

$$\left[\frac{\overline{\sigma}_D^{(\gamma,\beta)}[f]}{\sigma_D^{(\gamma,\alpha)}[g]}\right]^{\frac{1}{\rho^{(\gamma,\alpha)}[g]}} \leq \overline{\sigma}_D^{(\alpha,\beta)}[f]_g \leq \min\left\{\left[\frac{\overline{\sigma}_D^{(\gamma,\beta)}[f]}{\overline{\sigma}_D^{(\gamma,\alpha)}[g]}\right]^{\frac{1}{\rho^{(\gamma,\alpha)}[g]}}, \left[\frac{\sigma_D^{(\gamma,\beta)}[f]}{\sigma_D^{(\gamma,\alpha)}[g]}\right]^{\frac{1}{\rho^{(\gamma,\alpha)}[g]}}\right\}$$

$$\leq \max\left\{\left[\frac{\overline{\sigma}_D^{(\gamma,\beta)}[f]}{\overline{\sigma}_D^{(\gamma,\alpha)}[g]}\right]^{\frac{1}{\rho^{(\gamma,\alpha)}[g]}}, \left[\frac{\sigma_D^{(\gamma,\beta)}[f]}{\sigma_D^{(\gamma,\alpha)}[g]}\right]^{\frac{1}{\rho^{(\gamma,\alpha)}[g]}}\right\} \leq \sigma_D^{(\alpha,\beta)}[f]_g \leq \left[\frac{\sigma_D^{(\gamma,\beta)}[f]}{\overline{\sigma}_D^{(\gamma,\alpha)}[g]}\right]^{\frac{1}{\rho^{(\gamma,\alpha)}[g]}}.$$

**Proof.** From the definitions of $\sigma_D^{(\gamma,\beta)}[f]$ and $\overline{\sigma}_D^{(\gamma,\beta)}[f]$, we have for all sufficiently large values of $R$ that

$$M_{f,D}(R) \leq \gamma^{-1}\left(\log\left(\left(\sigma_D^{(\gamma,\beta)}[f] + \varepsilon\right)(\exp\beta(R))^{\rho^{(\gamma,\beta)}[f]}\right)\right), \tag{90}$$

$$M_{f,D}(R) \geq \gamma^{-1}\left(\log\left(\left(\overline{\sigma}_D^{(\gamma,\beta)}[f] - \varepsilon\right)(\exp\beta(R))^{\rho^{(\gamma,\beta)}[f]}\right)\right) \tag{91}$$

and also for a sequence of values of $R$ tending to infinity we get that

$$M_{f,D}(R) \geq \gamma^{-1}\left(\log\left(\left(\sigma_D^{(\gamma,\beta)}[f] - \varepsilon\right)(\exp\beta(R))^{\rho^{(\gamma,\beta)}[f]}\right)\right), \tag{92}$$

$$M_{f,D}(R) \leq \gamma^{-1}\left(\log\left(\left(\overline{\sigma}_D^{(\gamma,\beta)}[f] + \varepsilon\right)(\exp\beta(R))^{\rho^{(\gamma,\beta)}[f]}\right)\right). \tag{93}$$

Similarly from the definitions of $\sigma_D^{(\gamma,\alpha)}[g]$ and $\overline{\sigma}_D^{(\gamma,\alpha)}[g]$ it follows for all sufficiently large values of $R$ that

$$M_{g,D}^{-1}(R) \leq \gamma^{-1}\left(\log\left(\left(\sigma_D^{(\gamma,\alpha)}[g] + \varepsilon\right)(\exp(\alpha(R)))^{\rho^{(\gamma,\alpha)}[g]}\right)\right)$$

$$i.e., \ R \leq M_{g,D}^{-1}\left(\gamma^{-1}\left(\log\left(\left(\sigma_D^{(\gamma,\alpha)}[g] + \varepsilon\right)(\exp(\alpha(R)))^{\rho^{(\gamma,\alpha)}[g]}\right)\right)\right)$$

$$i.e., \ M_{g,D}^{-1}(R) \geq \alpha^{-1}\left(\log\left(\frac{\exp(\gamma(R))}{\left(\sigma_D^{(\gamma,\alpha)}[g] + \varepsilon\right)}\right)^{\frac{1}{\rho^{(\gamma,\alpha)}[g]}}\right), \tag{94}$$

$$M_{g,D}^{-1}(R) \geq \gamma^{-1}\left(\log\left(\left(\overline{\sigma}_D^{(\gamma,\alpha)}[g] - \varepsilon\right)(\exp\alpha(R))^{\rho^{(\gamma,\alpha)}[g]}\right)\right)$$

$$i.e., \ R \geq M_{g,D}^{-1}\left(\gamma^{-1}\left(\log\left(\left(\overline{\sigma}_D^{(\gamma,\alpha)}[g] - \varepsilon\right)(\exp\alpha(R))^{\rho^{(\gamma,\alpha)}[g]}\right)\right)\right)$$

$$i.e., \ M_{g,D}^{-1}(R) \leq \alpha^{-1}\left(\log\left(\frac{\exp(\gamma(R))}{\left(\overline{\sigma}_D^{(\gamma,\alpha)}[g] - \varepsilon\right)}\right)^{\frac{1}{\rho^{(\gamma,\alpha)}[g]}}\right) \tag{95}$$

and for a sequence of values of $R$ tending to infinity we obtain that

$$M_{g,D}^{-1}(R) \geq \gamma^{-1}\left(\log\left(\left(\sigma_D^{(\gamma,\alpha)}[g] - \varepsilon\right)(\exp\alpha(R))^{\rho^{(\gamma,\alpha)}[g]}\right)\right)$$

$$i.e., \ R \geq M_{g,D}^{-1}\left(\gamma^{-1}\left(\log\left(\left(\sigma_D^{(\gamma,\alpha)}[g] - \varepsilon\right)(\exp\alpha(R))^{\rho^{(\gamma,\alpha)}[g]}\right)\right)\right)$$

$$i.e., \ M_{g,D}^{-1}(R) \leq \alpha^{-1}\left(\log\left(\frac{\exp(\gamma(R))}{\left(\sigma_D^{(\gamma,\alpha)}[g] - \varepsilon\right)}\right)^{\frac{1}{\rho^{(\gamma,\alpha)}[g]}}\right), \tag{96}$$

$$M_{g,D}^{-1}(R) \leq \gamma^{-1}\left(\log\left(\left(\overline{\sigma}_D^{(\gamma,\alpha)}[g] + \varepsilon\right)(\exp\alpha(R))^{\rho^{(\gamma,\alpha)}[g]}\right)\right)$$

$$i.e., \ R \leq M_{g,D}^{-1}\left(\gamma^{-1}\left(\log\left(\left(\overline{\sigma}_D^{(\gamma,\alpha)}[g] + \varepsilon\right)(\exp\alpha(R))^{\rho^{(\gamma,\alpha)}[g]}\right)\right)\right)$$

$$i.e., \ M_{g,D}^{-1}(R) \geq \alpha^{-1}\left(\log\left(\left(\frac{\exp(\gamma(R))}{\left(\overline{\sigma}_D^{(\gamma,\alpha)}[g] - \varepsilon\right)}\right)^{\frac{1}{\rho^{(\gamma,\alpha)}[g]}}\right)\right). \tag{97}$$

Now from (92) and in view of (94), we get for a sequence of values of $R$ tending to infinity that

$$\exp(\alpha(M_{g,D}^{-1}(M_{f,D}(R)))) \geq$$
$$\exp\left(\alpha\left(M_{g,D}^{-1}\left(\gamma^{-1}\left(\log\left(\left(\sigma_D^{(\gamma,\beta)}[f] - \varepsilon\right)(\exp\beta(R))^{\rho^{(\gamma,\beta)}[f]}\right)\right)\right)\right)\right)$$

$$i.e., \ \exp(\alpha(M_{g,D}^{-1}(M_{f,D}(R)))) \geq \left(\frac{\left(\sigma_D^{(\gamma,\beta)}[f] - \varepsilon\right)(\exp\beta(R))^{\rho^{(\gamma,\beta)}[f]}}{\left(\sigma_D^{(\gamma,\alpha)}[g] + \varepsilon\right)}\right)^{\frac{1}{\rho^{(\gamma,\alpha)}[g]}}$$

$$i.e., \ \exp(\alpha(M_{g,D}^{-1}(M_{f,D}(R)))) \geq \left[\frac{\left(\sigma_D^{(\gamma,\beta)}[f] - \varepsilon\right)}{\left(\sigma_D^{(\gamma,\alpha)}[g] + \varepsilon\right)}\right]^{\frac{1}{\rho^{(\gamma,\alpha)}[g]}} (\exp\beta(R))^{\frac{\rho^{(\gamma,\beta)}[f]}{\rho^{(\gamma,\alpha)}[g]}}$$

$$i.e., \ \frac{\exp(\alpha(M_{g,D}^{-1}(M_{f,D}(R))))}{(\exp\beta(R))^{\frac{\rho^{(\gamma,\beta)}[f]}{\rho^{(\gamma,\alpha)}[g]}}} \geq \left[\frac{\left(\sigma_D^{(\gamma,\beta)}[f] - \varepsilon\right)}{\left(\sigma_D^{(\gamma,\alpha)}[g] + \varepsilon\right)}\right]^{\frac{1}{\rho^{(\gamma,\alpha)}[g]}}. \tag{98}$$

As $\varepsilon \, (> 0)$ is arbitrary, in view of Lemma 6.2.1 it follows that

$$\limsup_{R\to\infty} \frac{\exp(\alpha(M_{g,D}^{-1}(M_{f,D}(R))))}{(\exp\beta(R))^{\rho^{(\alpha,\beta)}[f]g}} \geq \left[\frac{\sigma_D^{(\gamma,\beta)}[f]}{\sigma_D^{(\gamma,\alpha)}[g]}\right]^{\frac{1}{\rho^{(\gamma,\alpha)}[g]}}$$

$$i.e., \ \sigma_D^{(\alpha,\beta)}[f]_g \geq \left[\frac{\sigma_D^{(\gamma,\beta)}[f]}{\sigma_D^{(\gamma,\alpha)}[g]}\right]^{\frac{1}{\rho^{(\gamma,\alpha)}[g]}}. \tag{99}$$

Analogously from (91) and in view of (97), it follows for a sequence of values of $R$ tending to infinity that

$$\exp(\alpha(M_{g,D}^{-1}(M_{f,D}(R)))) \geq$$
$$\exp\left(\alpha\left(M_{g,D}^{-1}\left(\gamma^{-1}\left(\log\left(\left(\overline{\sigma}_D^{(\gamma,\beta)}[f] - \varepsilon\right)(\exp\beta(R))^{\rho^{(\gamma,\beta)}[f]}\right)\right)\right)\right)\right)$$

$$i.e., \ \exp(\alpha(M_{g,D}^{-1}(M_{f,D}(R)))) \geq \left(\frac{\left(\overline{\sigma}_D^{(\gamma,\beta)}[f] - \varepsilon\right)(\exp\beta(R))^{\rho^{(\gamma,\beta)}[f]}}{\left(\overline{\sigma}_D^{(\gamma,\alpha)}[g] + \varepsilon\right)}\right)^{\frac{1}{\rho^{(\gamma,\alpha)}[g]}}$$

$$i.e., \ \exp(\alpha(M_{g,D}^{-1}(M_{f,D}(R)))) \geq \left[\frac{\left(\overline{\sigma}_D^{(\gamma,\beta)}[f] - \varepsilon\right)}{\left(\overline{\sigma}_D^{(\gamma,\alpha)}[g] + \varepsilon\right)}\right]^{\frac{1}{\rho^{(\gamma,\alpha)}[g]}} (\exp\beta(R))^{\frac{\rho^{(\gamma,\beta)}[f]}{\rho^{(\gamma,\alpha)}[g]}}$$

$$i.e., \ \frac{\exp(\alpha(M_{g,D}^{-1}(M_{f,D}(R))))}{(\exp\beta(R))^{\frac{\rho^{(\gamma,\beta)}[f]}{\rho^{(\gamma,\alpha)}[g]}}} \geq \left[\frac{\left(\overline{\sigma}_D^{(\gamma,\beta)}[f] - \varepsilon\right)}{\left(\overline{\sigma}_D^{(\gamma,\alpha)}[g] + \varepsilon\right)}\right]^{\frac{1}{\rho^{(\gamma,\alpha)}[g]}}. \tag{100}$$

Since $\varepsilon \, (> 0)$ is arbitrary, we get from above and Lemma 6.2.1 that

$$\limsup_{R \to \infty} \frac{\exp(\alpha(M_{g,D}^{-1}(M_{f,D}(R))))}{(\exp \beta(R))^{\rho^{(\alpha,\beta)}[f]_g}} \geq \left[ \frac{\overline{\sigma}_D^{(\gamma,\beta)}[f]}{\overline{\sigma}_D^{(\gamma,\alpha)}[g]} \right]^{\frac{1}{\rho^{(\gamma,\alpha)}[g]}}$$

$$i.e., \ \sigma_D^{(\alpha,\beta)}[f]_g \geq \left[ \frac{\overline{\sigma}_D^{(\gamma,\beta)}[f]}{\overline{\sigma}_D^{(\gamma,\alpha)}[g]} \right]^{\frac{1}{\rho^{(\gamma,\alpha)}[g]}} . \tag{101}$$

Again in view of (95), we have from (90) for all sufficiently large values of $R$ that

$$\exp(\alpha(M_{g,D}^{-1}(M_{f,D}(R)))) \leq$$

$$\exp \left( \alpha \left( M_{g,D}^{-1} \left( \gamma^{-1} \left( \log \left( \left( \sigma_D^{(\gamma,\beta)}[f] + \varepsilon \right) (\exp \beta(R))^{\rho^{(\gamma,\beta)}[f]} \right) \right) \right) \right) \right)$$

$$i.e., \ \exp(\alpha(M_{g,D}^{-1}(M_{f,D}(R)))) \leq \left( \frac{\left( \sigma_D^{(\gamma,\beta)}[f] + \varepsilon \right) (\exp \beta(R))^{\rho^{(\gamma,\beta)}[f]}}{\left( \overline{\sigma}_D^{(\gamma,\alpha)}[g] - \varepsilon \right)} \right)^{\frac{1}{\rho^{(\gamma,\alpha)}[g]}}$$

$$i.e., \ \exp(\alpha(M_{g,D}^{-1}(M_{f,D}(R)))) \leq \left[ \frac{\left( \sigma_D^{(\gamma,\beta)}[f] + \varepsilon \right)}{\left( \overline{\sigma}_D^{(\gamma,\alpha)}[g] - \varepsilon \right)} \right]^{\frac{1}{\rho^{(\gamma,\alpha)}[g]}} (\exp \beta(R))^{\frac{\rho^{(\gamma,\beta)}[f]}{\rho^{(\gamma,\alpha)}[g]}}$$

$$i.e., \ \frac{\exp(\alpha(M_{g,D}^{-1}(M_{f,D}(R))))}{(\exp \beta(R))^{\frac{\rho^{(\gamma,\beta)}[f]}{\rho^{(\gamma,\alpha)}[g]}}} \leq \left[ \frac{\left( \sigma_D^{(\gamma,\beta)}[f] + \varepsilon \right)}{\left( \overline{\sigma}_D^{(\gamma,\alpha)}[g] - \varepsilon \right)} \right]^{\frac{1}{\rho^{(\gamma,\alpha)}[g]}} . \tag{102}$$

Since $\varepsilon \, (> 0)$ is arbitrary, we obtain in view of Lemma 6.2.1 that

$$\limsup_{R \to \infty} \frac{\exp(\alpha(M_{g,D}^{-1}(M_{f,D}(R))))}{(\exp \beta(R))^{\rho^{(\alpha,\beta)}[f]_g}} \leq \left[ \frac{\sigma_D^{(\gamma,\beta)}[f]}{\overline{\sigma}_D^{(\gamma,\alpha)}[g]} \right]^{\frac{1}{\rho^{(\gamma,\alpha)}[g]}}$$

$$i.e., \ \sigma_D^{(\alpha,\beta)}[f]_g \leq \left[ \frac{\sigma_D^{(\gamma,\beta)}[f]}{\overline{\sigma}_D^{(\gamma,\alpha)}[g]} \right]^{\frac{1}{\rho^{(\gamma,\alpha)}[g]}} . \tag{103}$$

Again from (91) and in view of (94), we get for all sufficiently large values of $R$ that

$$\exp(\alpha(M_{g,D}^{-1}(M_{f,D}(R)))) \geq$$

$$\exp \left( \alpha \left( M_{g,D}^{-1} \left( \gamma^{-1} \left( \log \left( \left( \overline{\sigma}_D^{(\gamma,\beta)}[f] - \varepsilon \right) (\exp \beta(R))^{\rho^{(\gamma,\beta)}[f]} \right) \right) \right) \right) \right)$$

$$i.e., \ \exp(\alpha(M_{g,D}^{-1}(M_{f,D}(R)))) \geq \left( \frac{\left( \overline{\sigma}_D^{(\gamma,\beta)}[f] - \varepsilon \right) (\exp \beta(R))^{\rho^{(\gamma,\beta)}[f]}}{\left( \sigma_D^{(\gamma,\alpha)}[g] + \varepsilon \right)} \right)^{\frac{1}{\rho^{(\gamma,\alpha)}[g]}}$$

$$i.e., \ \exp(\alpha(M_{g,D}^{-1}(M_{f,D}(R)))) \geq \left[\frac{\left(\overline{\sigma}_D^{(\gamma,\beta)}[f] - \varepsilon\right)}{\sigma_D^{(\gamma,\alpha)}[g] + \varepsilon}\right]^{\frac{1}{\rho^{(\gamma,\alpha)}[g]}} (\exp \beta(R))^{\frac{\rho^{(\gamma,\beta)}[f]}{\rho^{(\gamma,\alpha)}[g]}}$$

$$i.e., \ \frac{\exp(\alpha(M_{g,D}^{-1}(M_{f,D}(R))))}{(\exp \beta(R))^{\frac{\rho^{(\gamma,\beta)}[f]}{\rho^{(\gamma,\alpha)}[g]}}} \geq \left[\frac{\left(\overline{\sigma}_D^{(\gamma,\beta)}[f] - \varepsilon\right)}{\sigma_D^{(\gamma,\alpha)}[g] + \varepsilon}\right]^{\frac{1}{\rho^{(\gamma,\alpha)}[g]}}. \tag{104}$$

As $\varepsilon \,(> 0)$ is arbitrary, it follows from above and Lemma 6.2.1 that

$$\liminf_{R \to \infty} \frac{\exp(\alpha(M_{g,D}^{-1}(M_{f,D}(R))))}{(\exp \beta(R))^{\rho^{(\alpha,\beta)}[f]g}} \geq \left[\frac{\overline{\sigma}_D^{(\gamma,\beta)}[f]}{\sigma_D^{(\gamma,\alpha)}[g]}\right]^{\frac{1}{\rho^{(\gamma,\alpha)}[g]}}$$

$$i.e., \ \overline{\sigma}_D^{(\alpha,\beta)}[f]_g \geq \left[\frac{\overline{\sigma}_D^{(\gamma,\beta)}[f]}{\sigma_D^{(\gamma,\alpha)}[g]}\right]^{\frac{1}{\rho^{(\gamma,\alpha)}[g]}}. \tag{105}$$

Also in view of (96), we get from (90) for a sequence of values of $R$ tending to infinity that

$$\exp(\alpha(M_{g,D}^{-1}(M_{f,D}(R)))) \leq$$
$$\exp\left(\alpha\left(M_{g,D}^{-1}\left(\gamma^{-1}\left(\log\left(\left(\sigma_D^{(\gamma,\beta)}[f] + \varepsilon\right)(\exp \beta(R))^{\rho^{(\gamma,\beta)}[f]}\right)\right)\right)\right)\right)$$

$$i.e., \ \exp(\alpha(M_{g,D}^{-1}(M_{f,D}(R)))) \leq \left(\frac{\left(\sigma_D^{(\gamma,\beta)}[f] + \varepsilon\right)(\exp \beta(R))^{\rho^{(\gamma,\beta)}[f]}}{\left(\sigma_D^{(\gamma,\alpha)}[g] - \varepsilon\right)}\right)^{\frac{1}{\rho^{(\gamma,\alpha)}[g]}}$$

$$i.e., \ \exp(\alpha(M_{g,D}^{-1}(M_{f,D}(R)))) \leq \left[\frac{\left(\sigma_D^{(\gamma,\beta)}[f] + \varepsilon\right)}{\left(\sigma_D^{(\gamma,\alpha)}[g] - \varepsilon\right)}\right]^{\frac{1}{\rho^{(\gamma,\alpha)}[g]}} (\exp \beta(R))^{\frac{\rho^{(\gamma,\beta)}[f]}{\rho^{(\gamma,\alpha)}[g]}}$$

$$i.e., \ \frac{\exp(\alpha(M_{g,D}^{-1}(M_{f,D}(R))))}{(\exp \beta(R))^{\frac{\rho^{(\gamma,\beta)}[f]}{\rho^{(\gamma,\alpha)}[g]}}} \leq \left[\frac{\left(\sigma_D^{(\gamma,\beta)}[f] + \varepsilon\right)}{\left(\sigma_D^{(\gamma,\alpha)}[g] - \varepsilon\right)}\right]^{\frac{1}{\rho^{(\gamma,\alpha)}[g]}}. \tag{106}$$

Since $\varepsilon \,(> 0)$ is arbitrary, we get from Lemma 6.2.1 and above that

$$\liminf_{R \to \infty} \frac{\exp(\alpha(M_{g,D}^{-1}(M_{f,D}(R))))}{(\exp \beta(R))^{\rho^{(\alpha,\beta)}[f]g}} \leq \left[\frac{\sigma_D^{(\gamma,\beta)}[f]}{\sigma_D^{(\gamma,\alpha)}[g]}\right]^{\frac{1}{\rho^{(\gamma,\alpha)}[g]}}$$

$$i.e., \ \overline{\sigma}_D^{(\alpha,\beta)}[f]_g \leq \left[\frac{\sigma_D^{(\gamma,\beta)}[f]}{\sigma_D^{(\gamma,\alpha)}[g]}\right]^{\frac{1}{\rho^{(\gamma,\alpha)}[g]}}. \tag{107}$$

Similarly from (93) and in view of (95), it follows for a sequence of values of $R$ tending to infinity that

$$\exp(\alpha(M_{g,D}^{-1}(M_{f,D}(R)))) \leq$$

$$\exp\left(\alpha\left(M_{g,D}^{-1}\left(\gamma^{-1}\left(\log\left(\left(\overline{\sigma}_D^{(\gamma,\beta)}[f]+\varepsilon\right)(\exp\beta(R))^{\rho^{(\gamma,\beta)}[f]}\right)\right)\right)\right)\right)$$

$$i.e.,\ \ \exp(\alpha(M_{g,D}^{-1}(M_{f,D}(R)))) \leq \left(\frac{\left(\overline{\sigma}_D^{(\gamma,\beta)}[f]+\varepsilon\right)(\exp\beta(R))^{\rho^{(\gamma,\beta)}[f]}}{\left(\overline{\sigma}_D^{(\gamma,\alpha)}[g]-\varepsilon\right)}\right)^{\frac{1}{\rho^{(\gamma,\alpha)}[g]}}$$

$$i.e.,\ \ \exp(\alpha(M_{g,D}^{-1}(M_{f,D}(R)))) \leq \left[\frac{\left(\overline{\sigma}_D^{(\gamma,\beta)}[f]+\varepsilon\right)}{\left(\overline{\sigma}_D^{(\gamma,\alpha)}[g]-\varepsilon\right)}\right]^{\frac{1}{\rho^{(\gamma,\alpha)}[g]}}(\exp\beta(R))^{\frac{\rho^{(\gamma,\beta)}[f]}{\rho^{(\gamma,\alpha)}[g]}}$$

$$i.e.,\ \ \frac{\exp(\alpha(M_{g,D}^{-1}(M_{f,D}(R))))}{(\exp\beta(R))^{\frac{\rho^{(\gamma,\beta)}[f]}{\rho^{(\gamma,\alpha)}[g]}}} \leq \left[\frac{\left(\overline{\sigma}_D^{(\gamma,\beta)}[f]+\varepsilon\right)}{\left(\overline{\sigma}_D^{(\gamma,\alpha)}[g]-\varepsilon\right)}\right]^{\frac{1}{\rho^{(\gamma,\alpha)}[g]}}. \tag{108}$$

As $\varepsilon\,(>0)$ is arbitrary, we obtain from Lemma 6.2.1 and above that

$$\liminf_{R\to\infty}\frac{\exp(\alpha(M_{g,D}^{-1}(M_{f,D}(R))))}{(\exp\beta(R))^{\rho^{(\alpha,\beta)}[f]g}} \leq \left[\frac{\overline{\sigma}_D^{(\gamma,\beta)}[f]}{\overline{\sigma}_D^{(\gamma,\alpha)}[g]}\right]^{\frac{1}{\rho^{(\gamma,\alpha)}[g]}}$$

$$i.e.,\ \ \overline{\sigma}_D^{(\alpha,\beta)}[f]_g \leq \left[\frac{\overline{\sigma}_D^{(\gamma,\beta)}[f]}{\overline{\sigma}_D^{(\gamma,\alpha)}[g]}\right]^{\frac{1}{\rho^{(\gamma,\alpha)}[g]}}. \tag{109}$$

Thus the theorem follows from $(99),(101),(103),(105),(107)$ and $(109)$ . ∎

In view of Theorem 6.3.1, one can easily verify the following corollaries :

**Corollary 6.3.1** *Let $f(z)$ and $g(z)$ be two entire functions of $n$ complex variables such that $0<\rho^{(\gamma,\beta)}[f]<\infty$ and $0<\lambda^{(\gamma,\alpha)}[g]=\rho^{(\gamma,\alpha)}[g]<\infty$. Then*

$$\sigma_D^{(\alpha,\beta)}[f]_g = \left[\frac{\sigma_D^{(\gamma,\beta)}[f]}{\sigma_D^{(\gamma,\alpha)}[g]}\right]^{\frac{1}{\rho^{(\gamma,\alpha)}[g]}} \ \ and \ \ \overline{\sigma}_D^{(\alpha,\beta)}[f]_g = \left[\frac{\overline{\sigma}_D^{(\gamma,\beta)}[f]}{\sigma_D^{(\gamma,\alpha)}[g]}\right]^{\frac{1}{\rho^{(\gamma,\alpha)}[g]}}.$$

*In addition, if $\sigma_D^{(\gamma,\beta)}[f]=\sigma_D^{(\gamma,\alpha)}[g]$ and $0<\lambda^{(\gamma,\beta)}[f]=\rho^{(\gamma,\beta)}[f]<\infty$, then*

$$\sigma_D^{(\alpha,\beta)}[f]_g = \overline{\sigma}_D^{(\beta,\alpha)}[g]_f = 1.$$

**Corollary 6.3.2** *Let $f(z)$ and $g(z)$ be two entire functions of $n$ complex variables such that $\sigma_D^{(\gamma,\beta)}[f]=\overline{\sigma}_D^{(\gamma,\beta)}[f]$ and $0<\lambda^{(\gamma,\alpha)}[g]=\rho^{(\gamma,\alpha)}[g]<\infty$ . Then*

$$\sigma_D^{(\alpha,\beta)}[f]_g = \overline{\sigma}_D^{(\alpha,\beta)}[f]_g = \left[\frac{\sigma_D^{(\gamma,\beta)}[f]}{\sigma_D^{(\gamma,\alpha)}[g]}\right]^{\frac{1}{\rho^{(\gamma,\alpha)}[g]}}.$$

*In addition, if $\sigma_D^{(\gamma,\beta)}[f]=\sigma_D^{(\gamma,\alpha)}[g]$ and $0<\lambda^{(\gamma,\beta)}[f]=\rho^{(\gamma,\beta)}[f]<\infty$, then*

$$\sigma_D^{(\alpha,\beta)}[f]_g = \overline{\sigma}_D^{(\alpha,\beta)}[f]_g = \sigma_D^{(\beta,\alpha)}[g]_f = \overline{\sigma}_D^{(\beta,\alpha)}[g]_f = 1.$$

**Corollary 6.3.3** *Let $f(z)$ and $g(z)$ be two entire functions of $n$ complex variables such that $0 < \rho^{(\gamma,\beta)}[f] < \infty$ and $0 < \lambda^{(\gamma,\alpha)}[g] = \rho^{(\gamma,\alpha)}[g] < \infty$. Then*

$$(i) \quad \sigma_D^{(\alpha,\beta)}[f]_g = \overline{\sigma}_D^{(\alpha,\beta)}[f]_g = \infty \text{ when } \sigma_D^{(\gamma,\alpha)}[g] = 0$$

*and*

$$(ii) \quad \sigma_D^{(\alpha,\beta)}[f]_g = \overline{\sigma}_D^{(\alpha,\beta)}[f]_g = 0 \text{ when } \sigma_D^{(\gamma,\alpha)}[g] = \infty.$$

**Corollary 6.3.4** *Let $g(z)$ be an entire function of $n$ complex variables such that $0 < \lambda^{(\gamma,\alpha)}[g] = \rho^{(\gamma,\alpha)}[g] < \infty$. Then for any entire function $f(z)$ of $n$ complex variables*

$$(i) \quad \sigma_D^{(\alpha,\beta)}[f]_g = 0 \text{ when } \sigma_D^{(\gamma,\beta)}[f] = 0,$$

$$(ii) \quad \overline{\sigma}_D^{(\alpha,\beta)}[f]_g = 0 \text{ when } \overline{\sigma}_D^{(\gamma,\beta)}[f] = 0,$$

$$(iii) \quad \sigma_D^{(\alpha,\beta)}[f]_g = \infty \text{ when } \sigma_D^{(\gamma,\beta)}[f] = \infty$$

*and*

$$(iv) \quad \overline{\sigma}_D^{(\alpha,\beta)}[f]_g = \infty \text{ when } \overline{\sigma}_D^{(\gamma,\beta)}[f] = \infty.$$

**Theorem 6.3.2** *Let $f(z)$ and $g(z)$ be two entire functions of $n$ complex variables such that $0 < \lambda^{(\gamma,\beta)}[f] = \rho^{(\gamma,\beta)}[f] < \infty$ and $0 < \lambda^{(\gamma,\alpha)}[g] \leq \rho^{(\gamma,\alpha)}[g] < \infty$. Then*

$$\left[\frac{\overline{\sigma}_D^{(\gamma,\beta)}[f]}{\sigma_D^{(\gamma,\alpha)}[g]}\right]^{\frac{1}{\rho^{(\gamma,\alpha)}[g]}} \leq \tau_D^{(\alpha,\beta)}[f]_g \leq \min\left\{\left[\frac{\overline{\sigma}_D^{(\gamma,\beta)}[f]}{\overline{\sigma}_D^{(\gamma,\alpha)}[g]}\right]^{\frac{1}{\rho^{(\gamma,\alpha)}[g]}}, \left[\frac{\sigma_D^{(\gamma,\beta)}[f]}{\sigma_D^{(\gamma,\alpha)}[g]}\right]^{\frac{1}{\rho^{(\gamma,\alpha)}[g]}}\right\}$$

$$\leq \max\left\{\left[\frac{\overline{\sigma}_D^{(\gamma,\beta)}[f]}{\overline{\sigma}_D^{(\gamma,\alpha)}[g]}\right]^{\frac{1}{\rho^{(\gamma,\alpha)}[g]}}, \left[\frac{\sigma_D^{(\gamma,\beta)}[f]}{\sigma_D^{(\gamma,\alpha)}[g]}\right]^{\frac{1}{\rho^{(\gamma,\alpha)}[g]}}\right\} \leq \overline{\tau}_D^{(\alpha,\beta)}[f]_g \leq \left[\frac{\sigma_D^{(\gamma,\beta)}[f]}{\overline{\sigma}_D^{(\gamma,\alpha)}[g]}\right]^{\frac{1}{\rho^{(\gamma,\alpha)}[g]}}.$$

**Proof.** From the definitions of $\overline{\tau}_D^{(\gamma,\beta)}[f]$ and $\tau_D^{(\gamma,\beta)}[f]$, we have for all sufficiently large values of $R$ that

$$M_{f,D}(R) \leq \gamma^{-1}\left(\log\left(\left(\overline{\tau}_D^{(\gamma,\beta)}[f] + \varepsilon\right)(\exp\beta(R))^{\lambda^{(\gamma,\beta)}[f]}\right)\right), \tag{110}$$

$$M_{f,D}(R) \geq \gamma^{-1}\left(\log\left(\left(\tau_D^{(\gamma,\beta)}[f] - \varepsilon\right)(\exp\beta(R))^{\lambda^{(\gamma,\beta)}[f]}\right)\right) \tag{111}$$

and also for a sequence of values of $R$ tending to infinity we get that

$$M_{f,D}(R) \geq \gamma^{-1}\left(\log\left(\left(\overline{\tau}_D^{(\gamma,\beta)}[f] - \varepsilon\right)(\exp\beta(R))^{\lambda^{(\gamma,\beta)}[f]}\right)\right), \tag{112}$$

$$M_{f,D}(R) \leq \gamma^{-1}\left(\log\left(\left(\tau_D^{(\gamma,\beta)}[f] + \varepsilon\right)(\exp\beta(R))^{\lambda^{(\gamma,\beta)}[f]}\right)\right). \tag{113}$$

Similarly from the definitions of $\overline{\tau}_D^{(\gamma,\alpha)}[g]$ and $\tau_D^{(\gamma,\beta)}[f]$ it follows for all sufficiently large values of $R$ that

$$M_{g,D}^{-1}(R) \leq \gamma^{-1}\left(\log\left(\left(\overline{\tau}_D^{(\gamma,\alpha)}[g]+\varepsilon\right)(\exp\alpha(R))^{\lambda^{(\gamma,\alpha)}[g]}\right)\right)$$

$$\text{i.e., } R \leq M_{g,D}^{-1}\left(\gamma^{-1}\log\left(\left(\overline{\tau}_D^{(\gamma,\alpha)}[g]+\varepsilon\right)(\exp\alpha(R))^{\lambda^{(\gamma,\alpha)}[g]}\right)\right)$$

$$\text{i.e., } M_{g,D}^{-1}(R) \geq \alpha^{-1}\left(\log\left(\frac{\exp\gamma(R)}{\left(\overline{\tau}_D^{(\gamma,\alpha)}[g]+\varepsilon\right)}\right)^{\frac{1}{\lambda^{(\gamma,\alpha)}[g]}}\right), \tag{114}$$

$$M_{g,D}^{-1}(R) \geq \gamma^{-1}\left(\log\left(\left(\tau_D^{(\gamma,\alpha)}[g]-\varepsilon\right)(\exp\alpha(R))^{\lambda^{(\gamma,\alpha)}[g]}\right)\right)$$

$$\text{i.e., } R \geq M_{g,D}^{-1}\left(\gamma^{-1}\left(\log\left(\left(\tau_D^{(\gamma,\alpha)}[g]-\varepsilon\right)(\exp\alpha(R))^{\lambda^{(\gamma,\alpha)}[g]}\right)\right)\right)$$

$$\text{i.e., } M_{g,D}^{-1}(R) \leq \alpha^{-1}\left(\log\left(\frac{\exp\gamma(R)}{\left(\tau_D^{(\gamma,\alpha)}[g]-\varepsilon\right)}\right)^{\frac{1}{\lambda^{(\gamma,\alpha)}[g]}}\right) \tag{115}$$

and for a sequence of values of $R$ tending to infinity we obtain that

$$M_{g,D}^{-1}(R) \geq \gamma^{-1}\left(\log\left(\left(\overline{\tau}_D^{(\gamma,\alpha)}[g]-\varepsilon\right)(\exp\alpha(R))^{\lambda^{(\gamma,\alpha)}[g]}\right)\right)$$

$$\text{i.e., } R \geq M_{g,D}^{-1}\left(\gamma^{-1}\left(\log\left(\left(\overline{\tau}_D^{(\gamma,\alpha)}[g]-\varepsilon\right)(\exp\alpha(R))^{\lambda^{(\gamma,\alpha)}[g]}\right)\right)\right)$$

$$\text{i.e., } M_{g,D}^{-1}(R) \leq \alpha^{-1}\left(\log\left(\frac{\exp\gamma(R)}{\left(\overline{\tau}_D^{(\gamma,\alpha)}[g]-\varepsilon\right)}\right)^{\frac{1}{\lambda^{(\gamma,\alpha)}[g]}}\right), \tag{116}$$

$$M_{g,D}^{-1}(R) \leq \gamma^{-1}\left(\log\left(\left(\tau_D^{(\gamma,\alpha)}[g]+\varepsilon\right)(\exp\alpha R)^{\lambda^{(\gamma,\alpha)}[g]}\right)\right)$$

$$\text{i.e., } R \leq M_{g,D}^{-1}\left(\gamma^{-1}\left(\log\left(\left(\tau_D^{(\gamma,\alpha)}[g]+\varepsilon\right)(\exp\alpha R)^{\lambda^{(\gamma,\alpha)}[g]}\right)\right)\right)$$

$$\text{i.e., } M_{g,D}^{-1}(R) \geq \alpha^{-1}\left(\log\left(\frac{\exp\gamma(R)}{\left(\tau_D^{(\gamma,\alpha)}[g]-\varepsilon\right)}\right)^{\frac{1}{\lambda^{(\gamma,\alpha)}[g]}}\right). \tag{117}$$

■

Now using the same technique of Theorem 6.3.1, one can easily prove the conclusion of the present theorem by the help of Lemma 6.2.2. Therefore the remaining part of the proof of the present theorem is omitted.

In view of Theorem 6.3.2, one can easily derive the following corollaries:

**Corollary 6.3.5** *Let $f(z)$ and $g(z)$ be two entire functions of $n$ complex variables such that $0 < \lambda^{(\gamma,\beta)}[f] = \rho^{(\gamma,\beta)}[f] < \infty$ and $0 < \rho^{(\gamma,\alpha)}[g] < \infty$ . Then*

$$\overline{\tau}_D^{(\alpha,\beta)}[f]_g = \left[\frac{\sigma_D^{(\gamma,\beta)}[f]}{\overline{\sigma}_D^{(\gamma,\alpha)}[g]}\right]^{\frac{1}{\rho^{(\gamma,\alpha)}[g]}} \quad and \quad \tau_D^{(\alpha,\beta)}[f]_g = \left[\frac{\sigma_D^{(\gamma,\beta)}[f]}{\sigma_D^{(\gamma,\alpha)}[g]}\right]^{\frac{1}{\rho^{(\gamma,\alpha)}[g]}} .$$

*In addition, if $\sigma_D^{(\gamma,\beta)}[f] = \overline{\sigma}_D^{(\gamma,\alpha)}[g]$ and $0 < \lambda^{(\gamma,\alpha)}[g] = \rho^{(\gamma,\alpha)}[g] < \infty$, then*

$$\overline{\tau}_D^{(\alpha,\beta)}[f]_g = \tau_D^{(\beta,\alpha)}[g]_f = 1.$$

**Corollary 6.3.6** *Let $f(z)$ and $g(z)$ be two entire functions of $n$ complex variables such that $\sigma_D^{(\gamma,\alpha)}[g] = \overline{\sigma}_D^{(\gamma,\alpha)}[g$ and $0 < \lambda^{(\gamma,\beta)}[f] = \rho^{(\gamma,\beta)}[f] < \infty$ . Then*

$$\overline{\tau}_D^{(\alpha,\beta)}[f]_g = \tau_D^{(\alpha,\beta)}[f]_g = \left[\frac{\sigma_D^{(\gamma,\beta)}[f]}{\sigma_D^{(\gamma,\alpha)}[g]}\right]^{\frac{1}{\rho^{(\gamma,\alpha)}[g]}} .$$

*In addition, if $\sigma_D^{(\gamma,\beta)}[f] = \overline{\sigma}_D^{(\gamma,\alpha)}[g]$ and $0 < \lambda^{(\gamma,\alpha)}[g] = \rho^{(\gamma,\alpha)}[g] < \infty$, then*

$$\overline{\tau}_D^{(\alpha,\beta)}[f]_g = \tau_D^{(\alpha,\beta)}[f]_g = \overline{\tau}_D^{(\beta,\alpha)}[g]_f = \tau_D^{(\beta,\alpha)}[g]_f = 1.$$

**Corollary 6.3.7** *Let $f(z)$ and $g(z)$ be two entire functions of $n$ complex variables such that $0 < \lambda^{(\gamma,\beta)}[f] = \rho^{(\gamma,\beta)}[f] < \infty$ and $0 < \rho^{(\gamma,\alpha)}[g] < \infty$ . Then*

$$(i) \ \ \overline{\tau}_D^{(\alpha,\beta)}[f]_g = \tau_D^{(\alpha,\beta)}[f]_g = \infty \ \ when \ \sigma_D^{(\gamma,\alpha)}[g] = 0$$

*and*

$$(ii) \ \ \overline{\tau}_D^{(\alpha,\beta)}[f]_g = \tau_D^{(\alpha,\beta)}[f]_g = 0 \ \ when \ \sigma_D^{(\gamma,\alpha)}[g] = \infty.$$

**Corollary 6.3.8** *Let $f(z)$ be an entire function of $n$ complex variables such that $0 < \lambda^{(\gamma,\beta)}[f] = \rho^{(\gamma,\beta)}[f] < \infty$. Then for any entire function $g(z)$ of $n$ complex variables*

$$(i) \quad \overline{\tau}_D^{(\alpha,\beta)}[f]_g = 0 \ \ when \ \sigma_D^{(\gamma,\beta)}[f] = 0,$$
$$(ii) \quad \tau_D^{(\alpha,\beta)}[f]_g = 0 \ \ when \ \overline{\sigma}_D^{(\gamma,\beta)}[f] = 0,$$
$$(iii) \quad \overline{\tau}_D^{(\alpha,\beta)}[f]_g = \infty \ \ when \ \sigma_D^{(\gamma,\beta)}[f] = \infty$$

*and*

$$(iv) \ \ \tau_D^{(\alpha,\beta)}[f]_g = \infty \ \ when \ \overline{\sigma}_D^{(\gamma,\beta)}[f] = \infty.$$

Similarly in the line of Theorem 6.3.1 and Theorem 6.3.2 and with the help of Lemma 6.2.1 and Lemma 6.2.2, one may easily prove the following two theorems and therefore their proofs are omitted:

**Theorem 6.3.3** *Let $f(z)$ and $g(z)$ be two entire functions of $n$ complex variables such that $0 < \rho^{(\gamma,\beta)}[f] < \infty$ and $0 < \lambda^{(\gamma,\alpha)}[g] = \rho^{(\gamma,\alpha)}[g] < \infty$. Then*

$$\left[\frac{\tau_D^{(\gamma,\beta)}[f]}{\overline{\tau}_D^{(\gamma,\alpha)}[g]}\right]^{\frac{1}{\lambda^{(\gamma,\alpha)}[g]}} \leq \tau_D^{(\alpha,\beta)}[f]_g \leq \min\left\{\left[\frac{\tau_D^{(\gamma,\beta)}[f]}{\tau_D^{(\gamma,\alpha)}[g]}\right]^{\frac{1}{\lambda^{(\gamma,\alpha)}[g]}}, \left[\frac{\overline{\tau}_D^{(\gamma,\beta)}[f]}{\overline{\tau}_D^{(\gamma,\alpha)}[g]}\right]^{\frac{1}{\lambda^{(\gamma,\alpha)}[g]}}\right\}$$

$$\leq \max\left\{\left[\frac{\tau_D^{(\gamma,\beta)}[f]}{\tau_D^{(\gamma,\alpha)}[g]}\right]^{\frac{1}{\lambda^{(\gamma,\alpha)}[g]}}, \left[\frac{\overline{\tau}_D^{(\gamma,\beta)}[f]}{\overline{\tau}_D^{(\gamma,\alpha)}[g]}\right]^{\frac{1}{\lambda^{(\gamma,\alpha)}[g]}}\right\} \leq \overline{\tau}_D^{(\alpha,\beta)}[f]_g \leq \left[\frac{\overline{\tau}_D^{(\gamma,\beta)}[f]}{\tau_D^{(\gamma,\alpha)}[g]}\right]^{\frac{1}{\lambda^{(\gamma,\alpha)}[g]}}.$$

In view of Theorem 6.3.3, the following corollaries may also be obtained:

**Corollary 6.3.9** *Let $f(z)$ and $g(z)$ be two entire functions of $n$ complex variables such that $0 < \rho^{(\gamma,\beta)}[f] < \infty$ and $0 < \lambda^{(\gamma,\alpha)}[g] = \rho^{(\gamma,\alpha)}[g] < \infty$. Then*

$$\overline{\tau}_D^{(\alpha,\beta)}[f]_g = \left[\frac{\overline{\tau}_D^{(\gamma,\beta)}[f]}{\overline{\tau}_D^{(\gamma,\alpha)}[g]}\right]^{\frac{1}{\lambda^{(\gamma,\alpha)}[g]}} \quad \text{and} \quad \tau_D^{(\alpha,\beta)}[f]_g = \left[\frac{\tau_D^{(\gamma,\beta)}[f]}{\overline{\tau}_D^{(\gamma,\alpha)}[g]}\right]^{\frac{1}{\lambda^{(\gamma,\alpha)}[g]}}.$$

*In addition, if $\overline{\tau}_D^{(\gamma,\beta)}[f] = \overline{\tau}_D^{(\gamma,\alpha)}[g]$ and $0 < \lambda^{(\gamma,\beta)}[f] = \rho^{(\gamma,\beta)}[f] < \infty$, then*

$$\overline{\tau}_D^{(\alpha,\beta)}[f]_g = \tau_D^{(\beta,\alpha)}[g]_f = 1.$$

**Corollary 6.3.10** *Let $f(z)$ and $g(z)$ be two entire functions of $n$ complex variables such that $\overline{\tau}_D^{(\gamma,\beta)}[f] = \tau_D^{(\gamma,\beta)}[f]$ and $0 < \lambda^{(\gamma,\alpha)}[g] = \rho^{(\gamma,\alpha)}[g] < \infty$. Then*

$$\overline{\tau}_D^{(\alpha,\beta)}[f]_g = \tau_D^{(\alpha,\beta)}[f]_g = \left[\frac{\overline{\tau}_D^{(\gamma,\beta)}[f]}{\overline{\tau}_D^{(\gamma,\alpha)}[g]}\right]^{\frac{1}{\lambda^{(\gamma,\alpha)}[g]}}.$$

*In addition, if $\overline{\tau}_D^{(\gamma,\beta)}[f] = \overline{\tau}_D^{(\gamma,\alpha)}[g]$ and $0 < \lambda^{(\gamma,\beta)}[f] = \rho^{(\gamma,\beta)}[f] < \infty$, then*

$$\overline{\tau}_D^{(\alpha,\beta)}[f]_g = \tau_D^{(\alpha,\beta)}[f]_g = \overline{\tau}_D^{(\beta,\alpha)}[g]_f = \tau_D^{(\beta,\alpha)}[g]_f = 1.$$

**Corollary 6.3.11** *Let $f(z)$ and $g(z)$ be two entire functions of $n$ complex variables such that $0 < \rho^{(\gamma,\beta)}[f] < \infty$ and $0 < \lambda^{(\gamma,\alpha)}[g] = \rho^{(\gamma,\alpha)}[g] < \infty$. Then*

$$(i) \quad \overline{\tau}_D^{(\alpha,\beta)}[f]_g = \tau_D^{(\alpha,\beta)}[f]_g = \infty \text{ when } \overline{\tau}_D^{(\gamma,\alpha)}[g] = 0$$

*and*

$$(ii) \quad \overline{\tau}_D^{(\alpha,\beta)}[f]_g = \tau_D^{(\alpha,\beta)}[f]_g = 0 \text{ when } \overline{\tau}_D^{(\gamma,\alpha)}[g] = \infty.$$

**Corollary 6.3.12** *Let $g(z)$ be an entire function of $n$ complex variables such that $0 < \lambda^{(\gamma,\alpha)}[g] = \rho^{(\gamma,\alpha)}[g] < \infty$. Then for any entire function $f(z)$ of $n$ complex variables*

$$(i) \quad \overline{\tau}_D^{(\alpha,\beta)}[f]_g = 0 \text{ when } \overline{\tau}_D^{(\gamma,\beta)}[f] = 0 ,$$

$$(ii) \quad \tau_D^{(\alpha,\beta)}[f]_g = 0 \text{ when } \tau_D^{(\gamma,\beta)}[f] = 0 ,$$

$$(iii) \quad \overline{\tau}_D^{(\alpha,\beta)}[f]_g = \infty \text{ when } \overline{\tau}_D^{(\gamma,\beta)}[f] = \infty$$

*and*

$$(iv) \quad \tau_D^{(\alpha,\beta)}[f]_g = \infty \text{ when } \tau_D^{(\gamma,\beta)}[f] = \infty.$$

**Theorem 6.3.4** *Let $f(z)$ and $g(z)$ be two entire functions of $n$ complex variables such that $0 < \lambda^{(\gamma,\beta)}[f] = \rho^{(\gamma,\beta)}[f] < \infty$ and $0 < \rho^{(\gamma,\alpha)}[g] < \infty$ . Then*

$$\left[\frac{\tau_D^{(\gamma,\beta)}[f]}{\overline{\tau}_D^{(\gamma,\alpha)}[g]}\right]^{\frac{1}{\lambda^{(\gamma,\alpha)}[g]}} \leq \overline{\sigma}_D^{(\alpha,\beta)}[f]_g \leq \min\left\{\left[\frac{\tau_D^{(\gamma,\beta)}[f]}{\tau_D^{(\gamma,\alpha)}[g]}\right]^{\frac{1}{\lambda^{(\gamma,\alpha)}[g]}}, \left[\frac{\overline{\tau}_D^{(\gamma,\beta)}[f]}{\overline{\tau}_D^{(\gamma,\alpha)}[g]}\right]^{\frac{1}{\lambda^{(\gamma,\alpha)}[g]}}\right\}$$

$$\leq \max\left\{\left[\frac{\tau_D^{(\gamma,\beta)}[f]}{\tau_D^{(\gamma,\alpha)}[g]}\right]^{\frac{1}{\lambda^{(\gamma,\alpha)}[g]}}, \left[\frac{\overline{\tau}_D^{(\gamma,\beta)}[f]}{\overline{\tau}_D^{(\gamma,\alpha)}[g]}\right]^{\frac{1}{\lambda^{(\gamma,\alpha)}[g]}}\right\} \leq \sigma_D^{(\alpha,\beta)}[f]_g \leq \left[\frac{\overline{\tau}_D^{(\gamma,\beta)}[f]}{\tau_D^{(\gamma,\alpha)}[g]}\right]^{\frac{1}{\lambda^{(\gamma,\alpha)}[g]}} .$$

From Theorem 6.3.4 the following corollaries are immediate:

**Corollary 6.3.13** *Let $f(z)$ and $g(z)$ be two entire functions of $n$ complex variables such that $0 < \lambda^{(\gamma,\beta)}[f] = \rho^{(\gamma,\beta)}[f] < \infty$ and $0 < \rho^{(\gamma,\alpha)}[g] < \infty$ . Then*

$$\sigma_D^{(\alpha,\beta)}[f]_g = \left[\frac{\overline{\tau}_D^{(\gamma,\beta)}[f]}{\tau_D^{(\gamma,\alpha)}[g]}\right]^{\frac{1}{\lambda^{(\gamma,\alpha)}[g]}} \quad and \quad \overline{\sigma}_D^{(\alpha,\beta)}[f]_g = \left[\frac{\overline{\tau}_D^{(\gamma,\beta)}[f]}{\overline{\tau}_D^{(\gamma,\alpha)}[g]}\right]^{\frac{1}{\lambda^{(\gamma,\alpha)}[g]}} .$$

*In addition, if $\overline{\tau}_D^{(\gamma,\beta)}[f] = \tau_D^{(\gamma,\alpha)}[g]$ and $0 < \lambda^{(\gamma,\alpha)}[g] = \rho^{(\gamma,\alpha)}[g] < \infty$, then*

$$\sigma_D^{(\alpha,\beta)}[f]_g = \overline{\sigma}_D^{(\beta,\alpha)}[g]_f = 1.$$

**Corollary 6.3.14** *Let $f(z)$ and $g(z)$ be two entire functions of $n$ complex variables such that $0 < \lambda^{(\gamma,\beta)}[f] = \rho^{(\gamma,\beta)}[f] < \infty$ and $\overline{\tau}_D^{(\gamma,\alpha)}[g] = \tau_D^{(\gamma,\alpha)}[g]$ . Then*

$$\sigma_D^{(\alpha,\beta)}[f]_g = \overline{\sigma}_D^{(\alpha,\beta)}[f]_g = \left[\frac{\overline{\tau}_D^{(\gamma,\beta)}[f]}{\overline{\tau}_D^{(\gamma,\alpha)}[g]}\right]^{\frac{1}{\rho^{(\gamma,\alpha)}[g]}} .$$

*In addition, if $\overline{\tau}_D^{(\gamma,\beta)}[f] = \tau_D^{(\gamma,\alpha)}[g]$ and $0 < \lambda^{(\gamma,\alpha)}[g] = \rho^{(\gamma,\alpha)}[g] < \infty$, then*

$$\sigma_D^{(\alpha,\beta)}[f]_g = \overline{\sigma}_D^{(\alpha,\beta)}[f]_g = \sigma_D^{(\beta,\alpha)}[g]_f = \overline{\sigma}_D^{(\beta,\alpha)}[g]_f = 1.$$

**Corollary 6.3.15** *Let $f(z)$ and $g(z)$ be two entire functions of $n$ complex variables such that $0 < \lambda^{(\gamma,\beta)}[f] = \rho^{(\gamma,\beta)}[f] < \infty$ and $0 < \rho^{(\gamma,\alpha)}[g] < \infty$. Then*

$$(i) \;\; \sigma_D^{(\alpha,\beta)}[f]_g = \overline{\sigma}_D^{(\alpha,\beta)}[f]_g = \infty \;\; when \; \overline{\tau}_D^{(\gamma,\alpha)}[g] = 0$$

*and*

$$(ii) \;\; \sigma_D^{(\alpha,\beta)}[f]_g = \overline{\sigma}_D^{(\alpha,\beta)}[f]_g = 0 \;\; when \; \overline{\tau}_D^{(\gamma,\alpha)}[g] = \infty.$$

**Corollary 6.3.16** *Let $f(z)$ be an entire functions of $n$ complex variables such that $0 < \lambda^{(\gamma,\beta)}[f] = \rho^{(\gamma,\beta)}[f] < \infty$. Then for any entire function $g(z)$ of $n$ complex variables*

$$(i) \qquad \sigma_D^{(\alpha,\beta)}[f]_g = 0 \;\; when \; \overline{\tau}_D^{(\gamma,\beta)}[f] = 0,$$

$$(ii) \qquad \overline{\sigma}_D^{(\alpha,\beta)}[f]_g = 0 \;\; when \; \tau_D^{(\gamma,\beta)}[f] = 0,$$

$$(iii) \quad \sigma_D^{(\alpha,\beta)}[f]_g = \infty \;\; when \; \overline{\tau}_D^{(\gamma,\beta)}[f] = \infty$$

*and*

$$(iv) \quad \overline{\sigma}_D^{(\alpha,\beta)}[f]_g = \infty \;\; when \; \tau_D^{(\gamma,\beta)}[f] = \infty.$$

**Theorem 6.3.5** *Let $f(z)$ and $g(z)$ be two entire functions of $n$ complex variables such that $0 < \rho^{(\gamma,\beta)}[f] < \infty$ and $0 < \rho^{(\gamma,\alpha)}[g] < \infty$. Then*

$$\max\left\{\left[\frac{\overline{\sigma}_D^{(\gamma,\beta)}[f]}{\tau_D^{(\gamma,\alpha)}[g]}\right]^{\frac{1}{\lambda^{(\gamma,\alpha)}[g]}}, \left[\frac{\sigma_D^{(\gamma,\beta)}[f]}{\overline{\tau}_D^{(\gamma,\alpha)}[g]}\right]^{\frac{1}{\lambda^{(\gamma,\alpha)}[g]}}\right\} \le \sigma_D^{(\alpha,\beta)}[f]_g$$

$$\le \min\left\{\left[\frac{\overline{\tau}_D^{(\gamma,\beta)}[f]}{\tau_D^{(\gamma,\alpha)}[g]}\right]^{\frac{1}{\lambda^{(\gamma,\alpha)}[g]}}, \left[\frac{\sigma_D^{(\gamma,\beta)}[f]}{\overline{\sigma}_D^{(\gamma,\alpha)}[g]}\right]^{\frac{1}{\rho^{(\gamma,\alpha)}[g]}}, \left[\frac{\overline{\tau}_D^{(\gamma,\beta)}[f]}{\overline{\sigma}_D^{(\gamma,\alpha)}[g]}\right]^{\frac{1}{\rho^{(\gamma,\alpha)}[g]}}\right\}$$

*and*

$$\left[\frac{\overline{\sigma}_D^{(\gamma,\beta)}[f]}{\overline{\tau}_D^{(\gamma,\alpha)}[g]}\right]^{\frac{1}{\lambda^{(\gamma,\alpha)}[g]}} \le \overline{\sigma}_D^{(\alpha,\beta)}[f]_g$$

$$\le \min\left\{\begin{array}{ccc}\left[\dfrac{\overline{\sigma}_D^{(\gamma,\beta)}[f]}{\overline{\sigma}_D^{(\gamma,\alpha)}[g]}\right]^{\frac{1}{\rho^{(\gamma,\alpha)}[g]}}, & \left[\dfrac{\sigma_D^{(\gamma,\beta)}[f]}{\sigma_D^{(\gamma,\alpha)}[g]}\right]^{\frac{1}{\rho^{(\gamma,\alpha)}[g]}}, & \left[\dfrac{\tau_D^{(\gamma,\beta)}[f]}{\tau_D^{(\gamma,\alpha)}[g]}\right]^{\frac{1}{\lambda^{(\gamma,\alpha)}[g]}}, \\[20pt] \left[\dfrac{\overline{\tau}_D^{(\gamma,\beta)}[f]}{\overline{\tau}_D^{(\gamma,\alpha)}[g]}\right]^{\frac{1}{\lambda^{(\gamma,\alpha)}[g]}}, & \left[\dfrac{\overline{\tau}_D^{(\gamma,\beta)}[f]}{\sigma_D^{(\gamma,\alpha)}[g]}\right]^{\frac{1}{\rho^{(\gamma,\alpha)}[g]}}, & \left[\dfrac{\tau_D^{(\gamma,\beta)}[f]}{\overline{\sigma}_D^{(\gamma,\alpha)}[g]}\right]^{\frac{1}{\rho^{(\gamma,\alpha)}[g]}}\end{array}\right\}.$$

**Proof.** Now from (92) and in view of (114), we get for a sequence of values of $R$ tending to infinity that

$$\exp(\alpha(M_{g,D}^{-1}(M_{f,D}(R)))) \ge$$

$$\exp\left(\alpha\left(M_{g,D}^{-1}\left(\gamma^{-1}\left(\log\left(\left(\sigma_D^{(\gamma,\beta)}[f] - \varepsilon\right)(\exp\beta(R))^{\rho^{(\gamma,\beta)}[f]}\right)\right)\right)\right)\right)$$

$$i.e., \quad \exp(\alpha(M_{g,D}^{-1}(M_{f,D}(R)))) \ge \left(\frac{\left(\sigma_D^{(\gamma,\beta)}[f] - \varepsilon\right)(\exp\beta(R))^{\rho^{(\gamma,\beta)}[f]}}{\left(\overline{\tau}_D^{(\gamma,\alpha)}[g] + \varepsilon\right)}\right)^{\frac{1}{\lambda^{(\gamma,\alpha)}[g]}}$$

$$i.e., \quad \exp(\alpha(M_{g,D}^{-1}(M_{f,D}(R)))) \ge \left[\frac{\left(\sigma_D^{(\gamma,\beta)}[f] - \varepsilon\right)}{\left(\overline{\tau}_D^{(\gamma,\alpha)}[g] + \varepsilon\right)}\right]^{\frac{1}{\lambda^{(\gamma,\alpha)}[g]}}(\exp\beta(R))^{\frac{\rho^{(\gamma,\beta)}[f]}{\lambda^{(\gamma,\alpha)}[g]}}$$

$$i.e., \quad \frac{\exp(\alpha(M_{g,D}^{-1}(M_{f,D}(R))))}{(\exp\beta(R))^{\frac{\rho^{(\gamma,\beta)}[f]}{\lambda^{(\gamma,\alpha)}[g]}}} \ge \left[\frac{\left(\sigma_D^{(\gamma,\beta)}[f] - \varepsilon\right)}{\left(\overline{\tau}_D^{(\gamma,\alpha)}[g] + \varepsilon\right)}\right]^{\frac{1}{\lambda^{(\gamma,\alpha)}[g]}}.$$

Since in view of Theorem 4.2.1, we get that $\frac{\rho^{(\gamma,\beta)}[f]}{\lambda^{(\gamma,\alpha)}[g]} \geq \rho^{(\alpha,\beta)}[f]_g$ and as $\varepsilon\,(>0)$ is arbitrary therefore it follows from above that

$$\limsup_{R\to\infty} \frac{\exp(\alpha(M_{g,D}^{-1}(M_{f,D}(R))))}{(\exp\beta(R))^{\rho^{(\alpha,\beta)}[f]_g}} \geq \left[\frac{\sigma_D^{(\gamma,\beta)}[f]}{\overline{\tau}_D^{(\gamma,\alpha)}[g]}\right]^{\frac{1}{\lambda^{(\gamma,\alpha)}[g]}}$$

$$i.e.,\ \sigma_D^{(\alpha,\beta)}[f]_g \geq \left[\frac{\sigma_D^{(\gamma,\beta)}[f]}{\overline{\tau}_D^{(\gamma,\alpha)}[g]}\right]^{\frac{1}{\lambda^{(\gamma,\alpha)}[g]}}. \tag{118}$$

Analogously from (91) and in view of (117), it follows for a sequence of values of $R$ tending to infinity that

$$\exp(\alpha(M_{g,D}^{-1}(M_{f,D}(R)))) \geq$$

$$\exp\left(\alpha\left(M_{g,D}^{-1}\left(\gamma^{-1}\left(\log\left(\left(\overline{\sigma}_D^{(\gamma,\beta)}[f]-\varepsilon\right)(\exp\beta(R))^{\rho^{(\gamma,\beta)}[f]}\right)\right)\right)\right)\right)$$

$$i.e.,\ \exp(\alpha(M_{g,D}^{-1}(M_{f,D}(R)))) \geq \left(\frac{\left(\overline{\sigma}_D^{(\gamma,\beta)}[f]-\varepsilon\right)(\exp\beta(R))^{\rho^{(\gamma,\beta)}[f]}}{\left(\tau_D^{(\gamma,\alpha)}[g]-\varepsilon\right)}\right)^{\frac{1}{\lambda^{(\gamma,\alpha)}[g]}}$$

$$i.e.,\ \exp(\alpha(M_{g,D}^{-1}(M_{f,D}(R)))) \geq \left[\frac{\left(\overline{\sigma}_D^{(\gamma,\beta)}[f]-\varepsilon\right)}{\left(\tau_D^{(\gamma,\alpha)}[g]+\varepsilon\right)}\right]^{\frac{1}{\lambda^{(\gamma,\alpha)}[g]}}(\exp\beta(R))^{\frac{\rho^{(\gamma,\beta)}[f]}{\lambda^{(\gamma,\alpha)}[g]}}$$

$$i.e.,\ \frac{\exp(\alpha(M_{g,D}^{-1}(M_{f,D}(R))))}{(\exp\beta(R))^{\frac{\rho^{(\gamma,\beta)}[f]}{\lambda^{(\gamma,\alpha)}[g]}}} \geq \left[\frac{\left(\overline{\sigma}_D^{(\gamma,\beta)}[f]-\varepsilon\right)}{\left(\tau_D^{(\gamma,\alpha)}[g]+\varepsilon\right)}\right]^{\frac{1}{\lambda^{(\gamma,\alpha)}[g]}}.$$

Now in view of Theorem 4.2.1, it follows that $\frac{\rho^{(\gamma,\beta)}[f]}{\lambda^{(\gamma,\alpha)}[g]} \geq \rho^{(\alpha,\beta)}[f]_g$. Since $\varepsilon\,(>0)$ is arbitrary, we get from above that

$$\limsup_{R\to\infty} \frac{\exp(\alpha(M_{g,D}^{-1}(M_{f,D}(R))))}{(\exp\beta(R))^{\rho^{(\alpha,\beta)}[f]_g}} \geq \left[\frac{\overline{\sigma}_D^{(\gamma,\beta)}[f]}{\tau_D^{(\gamma,\alpha)}[g]}\right]^{\frac{1}{\lambda^{(\gamma,\alpha)}[g]}}$$

$$i.e.,\ \sigma_D^{(\alpha,\beta)}[f]_g \geq \left[\frac{\overline{\sigma}_D^{(\gamma,\beta)}[f]}{\tau_D^{(\gamma,\alpha)}[g]}\right]^{\frac{1}{\lambda^{(\gamma,\alpha)}[g]}}. \tag{119}$$

Again in view of (115), we have from (110) for all sufficiently large values of $R$ that

$$\exp(\alpha(M_{g,D}^{-1}(M_{f,D}(R)))) \leq$$

$$\exp\left(\alpha\left(M_{g,D}^{-1}\left(\gamma^{-1}\left(\log\left(\left(\overline{\tau}_D^{(\gamma,\beta)}[f]+\varepsilon\right)(\exp\beta(R))^{\lambda^{(\gamma,\beta)}[f]}\right)\right)\right)\right)\right)$$

$$i.e., \ \exp(\alpha(M_{g,D}^{-1}(M_{f,D}(R)))) \leq \left( \frac{\left( \overline{\tau}_D^{(\gamma,\beta)}[f] + \varepsilon \right) (\exp \beta(R))^{\lambda^{(\gamma,\beta)}[f]}}{\left( \tau_D^{(\gamma,\alpha)}[g] - \varepsilon \right)} \right)^{\frac{1}{\lambda^{(\gamma,\alpha)}[g]}}$$

$$i.e., \ \exp(\alpha(M_{g,D}^{-1}(M_{f,D}(R)))) \leq \left[ \frac{\left( \overline{\tau}_D^{(\gamma,\beta)}[f] + \varepsilon \right)}{\left( \tau_D^{(\gamma,\alpha)}[g] - \varepsilon \right)} \right]^{\frac{1}{\lambda^{(\gamma,\alpha)}[g]}} (\exp \beta(R))^{\frac{\lambda^{(\gamma,\beta)}[f]}{\lambda^{(\gamma,\alpha)}[g]}}$$

$$i.e., \ \frac{\exp(\alpha(M_{g,D}^{-1}(M_{f,D}(R))))}{(\exp \beta(R))^{\frac{\lambda^{(\gamma,\beta)}[f]}{\lambda^{(\gamma,\alpha)}[g]}}} \leq \left[ \frac{\left( \overline{\tau}_D^{(\gamma,\beta)}[f] + \varepsilon \right)}{\left( \tau_D^{(\gamma,\alpha)}[g] - \varepsilon \right)} \right]^{\frac{1}{\lambda^{(\gamma,\alpha)}[g]}} .$$

Since in view of Theorem 4.2.1, we get that $\frac{\lambda^{(\gamma,\beta)}[f]}{\lambda^{(\gamma,\alpha)}[g]} \leq \rho^{(\alpha,\beta)}[f]_g$ and as $\varepsilon \ (> 0)$ is arbitrary therefore it follows from above that

$$\limsup_{R \to \infty} \frac{\exp(\alpha(M_{g,D}^{-1}(M_{f,D}(R))))}{(\exp \beta(R))^{\rho^{(\alpha,\beta)}[f]_g}} \leq \left[ \frac{\overline{\tau}_D^{(\gamma,\beta)}[f]}{\tau_D^{(\gamma,\alpha)}[g]} \right]^{\frac{1}{\lambda^{(\gamma,\alpha)}[g]}}$$

$$i.e., \ \sigma_D^{(\alpha,\beta)}[f]_g \leq \left[ \frac{\overline{\tau}_D^{(\gamma,\beta)}[f]}{\tau_D^{(\gamma,\alpha)}[g]} \right]^{\frac{1}{\lambda^{(\gamma,\alpha)}[g]}} . \tag{120}$$

Again in view of Theorem 4.2.1, it follows that $\frac{\rho^{(\gamma,\beta)}[f]}{\rho^{(\gamma,\alpha)}[g]} \leq \rho^{(\alpha,\beta)}[f]_g$. Since $\varepsilon \ (> 0)$ is arbitrary, we get from (102) that

$$\limsup_{R \to \infty} \frac{\exp(\alpha(M_{g,D}^{-1}(M_{f,D}(R))))}{(\exp \beta(R))^{\rho^{(\alpha,\beta)}[f]_g}} \leq \left[ \frac{\sigma_D^{(\gamma,\beta)}[f]}{\overline{\sigma}_D^{(\gamma,\alpha)}[g]} \right]^{\frac{1}{\rho^{(\gamma,\alpha)}[g]}}$$

$$i.e., \ \sigma_D^{(\alpha,\beta)}[f]_g \leq \left[ \frac{\sigma_D^{(\gamma,\beta)}[f]}{\overline{\sigma}_D^{(\gamma,\alpha)}[g]} \right]^{\frac{1}{\rho^{(\gamma,\alpha)}[g]}} . \tag{121}$$

Further in view of (95), we have from (110) for all sufficiently large values of $R$ that

$$\exp(\alpha(M_{g,D}^{-1}(M_{f,D}(R)))) \leq$$

$$\exp \left( \alpha \left( M_{g,D}^{-1} \left( \gamma^{-1} \left( \log \left( \left( \overline{\tau}_D^{(\gamma,\beta)}[f] + \varepsilon \right) (\exp \beta(R))^{\lambda^{(\gamma,\beta)}[f]} \right) \right) \right) \right) \right)$$

$$i.e., \ \exp(\alpha(M_{g,D}^{-1}(M_{f,D}(R)))) \leq \left( \frac{\left( \overline{\tau}_D^{(\gamma,\beta)}[f] + \varepsilon \right) (\exp \beta(R))^{\lambda^{(\gamma,\beta)}[f]}}{\left( \overline{\sigma}_D^{(\gamma,\alpha)}[g] - \varepsilon \right)} \right)^{\frac{1}{\rho^{(\gamma,\alpha)}[g]}}$$

$$i.e.,\ \exp(\alpha(M_{g,D}^{-1}(M_{f,D}(R)))) \leq \left[\frac{\left(\overline{\tau}_D^{(\gamma,\beta)}[f]+\varepsilon\right)}{\left(\overline{\sigma}_D^{(\gamma,\alpha)}[g]-\varepsilon\right)}\right]^{\frac{1}{\rho^{(\gamma,\alpha)}[g]}} (\exp\beta(R))^{\frac{\lambda^{(\gamma,\beta)}[f]}{\rho^{(\gamma,\alpha)}[g]}}$$

$$i.e.,\ \frac{\exp(\alpha(M_{g,D}^{-1}(M_{f,D}(R))))}{(\exp\beta(R))^{\frac{\lambda^{(\gamma,\beta)}[f]}{\rho^{(\gamma,\alpha)}[g]}}} \leq \left[\frac{\left(\overline{\tau}_D^{(\gamma,\beta)}[f]+\varepsilon\right)}{\left(\overline{\sigma}_D^{(\gamma,\alpha)}[g]-\varepsilon\right)}\right]^{\frac{1}{\rho^{(\gamma,\alpha)}[g]}}.$$

Since in view of Theorem 4.2.1, we get that $\frac{\lambda^{(\gamma,\beta)}[f]}{\rho^{(\gamma,\alpha)}[g]} \leq \rho^{(\alpha,\beta)}[f]_g$ and as $\varepsilon\,(>0)$ is arbitrary therefore it follows from above that

$$\limsup_{R\to\infty}\frac{\exp(\alpha(M_{g,D}^{-1}(M_{f,D}(R))))}{(\exp\beta(R))^{\rho^{(\alpha,\beta)}[f]_g}} \leq \left[\frac{\overline{\tau}_D^{(\gamma,\beta)}[f]}{\overline{\sigma}_D^{(\gamma,\alpha)}[g]}\right]^{\frac{1}{\rho^{(\gamma,\alpha)}[g]}}$$

$$i.e.,\ \sigma_D^{(\alpha,\beta)}[f]_g \leq \left[\frac{\overline{\tau}_D^{(\gamma,\beta)}[f]}{\overline{\sigma}_D^{(\gamma,\alpha)}[g]}\right]^{\frac{1}{\rho^{(\gamma,\alpha)}[g]}}. \tag{122}$$

Thus the first part of the theorem follows from $(118)$, $(119)$, $(120)$, $(121)$ and $(122)$.

Now from $(91)$ and in view of $(114)$, we get for all sufficiently large values of $R$ that

$$\exp(\alpha(M_{g,D}^{-1}(M_{f,D}(R)))) \geq$$

$$\exp\left(\alpha\left(M_{g,D}^{-1}\left(\gamma^{-1}\left(\log\left(\left(\overline{\sigma}_D^{(\gamma,\beta)}[f]-\varepsilon\right)(\exp\beta(R))^{\rho^{(\gamma,\beta)}[f]}\right)\right)\right)\right)\right)$$

$$i.e.,\ \exp(\alpha(M_{g,D}^{-1}(M_{f,D}(R)))) \geq \left(\frac{\left(\overline{\sigma}_D^{(\gamma,\beta)}[f]-\varepsilon\right)(\exp\beta(R))^{\rho^{(\gamma,\beta)}[f]}}{\left(\overline{\tau}_D^{(\gamma,\alpha)}[g]+\varepsilon\right)}\right)^{\frac{1}{\lambda^{(\gamma,\alpha)}[g]}}$$

$$i.e.,\ \exp(\alpha(M_{g,D}^{-1}(M_{f,D}(R)))) \geq \left[\frac{\left(\overline{\sigma}_D^{(\gamma,\beta)}[f]-\varepsilon\right)}{\left(\overline{\tau}_D^{(\gamma,\alpha)}[g]+\varepsilon\right)}\right]^{\frac{1}{\lambda^{(\gamma,\alpha)}[g]}} (\exp\beta(R))^{\frac{\rho^{(\gamma,\beta)}[f]}{\lambda^{(\gamma,\alpha)}[g]}}$$

$$i.e.,\ \frac{\exp(\alpha(M_{g,D}^{-1}(M_{f,D}(R))))}{(\exp\beta(R))^{\frac{\rho^{(\gamma,\beta)}[f]}{\lambda^{(\gamma,\alpha)}[g]}}} \geq \left[\frac{\left(\overline{\sigma}_D^{(\gamma,\beta)}[f]-\varepsilon\right)}{\left(\overline{\tau}_D^{(\gamma,\alpha)}[g]+\varepsilon\right)}\right]^{\frac{1}{\lambda^{(\gamma,\alpha)}[g]}}.$$

Also in view of Theorem 4.2.1, it follows that $\frac{\rho^{(\gamma,\beta)}[f]}{\lambda^{(\gamma,\alpha)}[g]} \geq \rho^{(\alpha,\beta)}[f]_g$. Since $\varepsilon\,(>0)$ is arbitrary, we get from above that

$$\liminf_{R\to\infty}\frac{\exp(\alpha(M_{g,D}^{-1}(M_{f,D}(R))))}{(\exp\beta(R))^{\rho^{(\alpha,\beta)}[f]_g}} \geq \left[\frac{\overline{\sigma}_D^{(\gamma,\beta)}[f]}{\overline{\tau}_D^{(\gamma,\alpha)}[g]}\right]^{\frac{1}{\lambda^{(\gamma,\alpha)}[g]}}$$

$$i.e.,\ \overline{\sigma}_D^{(\alpha,\beta)}[f]_g \geq \left[\frac{\overline{\sigma}_D^{(\gamma,\beta)}[f]}{\overline{\tau}_D^{(\gamma,\alpha)}[g]}\right]^{\frac{1}{\lambda^{(\gamma,\alpha)}[g]}}. \tag{123}$$

As $\frac{\rho^{(\gamma,\beta)}[f]}{\rho^{(\gamma,\alpha)}[g]} \leq \rho^{(\alpha,\beta)}[f]_g$ and $\varepsilon\,(>0)$ is arbitrary, we get from $(106)$ that

$$\liminf_{R\to\infty} \frac{\exp(\alpha(M_{g,D}^{-1}(M_{f,D}(R))))}{(\exp\beta(R))^{\rho^{(\alpha,\beta)}[f]_g}} \leq \left[\frac{\sigma_D^{(\gamma,\beta)}[f]}{\sigma_D^{(\gamma,\alpha)}[g]}\right]^{\frac{1}{\rho^{(\gamma,\alpha)}[g]}}$$

$$i.e.,\ \ \overline{\sigma}_D^{(\alpha,\beta)}[f]_g \leq \left[\frac{\sigma_D^{(\gamma,\beta)}[f]}{\sigma_D^{(\gamma,\alpha)}[g]}\right]^{\frac{1}{\rho^{(\gamma,\alpha)}[g]}}. \tag{124}$$

Analogously, we get from $(108)$ that

$$\liminf_{R\to\infty} \frac{\exp(\alpha(M_{g,D}^{-1}(M_{f,D}(R))))}{(\exp\beta(R))^{\rho^{(\alpha,\beta)}[f]_g}} \leq \left[\frac{\overline{\sigma}_D^{(\gamma,\beta)}[f]}{\overline{\sigma}_D^{(\gamma,\alpha)}[g]}\right]^{\frac{1}{\rho^{(\gamma,\alpha)}[g]}}$$

$$i.e.,\ \ \overline{\sigma}_D^{(\alpha,\beta)}[f]_g \leq \left[\frac{\overline{\sigma}_D^{(\gamma,\beta)}[f]}{\overline{\sigma}_D^{(\gamma,\alpha)}[g]}\right]^{\frac{1}{\rho^{(\gamma,\alpha)}[g]}}, \tag{125}$$

since $\frac{\rho^{(\gamma,\beta)}[f]}{\rho^{(\gamma,\alpha)}[g]} \leq \rho^{(\alpha,\beta)}[f]_g$ and $\varepsilon\,(>0)$ is arbitrary.

Also in view of $(116)$, we get from $(110)$ for a sequence of values of $R$ tending to infinity that

$$\exp(\alpha(M_{g,D}^{-1}(M_{f,D}(R)))) \leq$$
$$\exp\left(\alpha\left(M_{g,D}^{-1}\left(\gamma^{-1}\left(\log\left(\left(\overline{\tau}_D^{(\gamma,\beta)}[f]+\varepsilon\right)(\exp\beta(R))^{\lambda^{(\gamma,\beta)}[f]}\right)\right)\right)\right)\right)$$

$$i.e.,\ \ \exp(\alpha(M_{g,D}^{-1}(M_{f,D}(R)))) \leq \left(\frac{\left(\overline{\tau}_D^{(\gamma,\beta)}[f]+\varepsilon\right)(\exp\beta(R))^{\lambda^{(\gamma,\beta)}[f]}}{\left(\overline{\tau}_D^{(\gamma,\alpha)}[g]-\varepsilon\right)}\right)^{\frac{1}{\lambda^{(\gamma,\alpha)}[g]}}$$

$$i.e.,\ \ \exp(\alpha(M_{g,D}^{-1}(M_{f,D}(R)))) \leq \left[\frac{\left(\overline{\tau}_D^{(\gamma,\beta)}[f]+\varepsilon\right)}{\left(\overline{\tau}_D^{(\gamma,\alpha)}[g]-\varepsilon\right)}\right]^{\frac{1}{\lambda^{(\gamma,\alpha)}[g]}}(\exp\beta(R))^{\frac{\lambda^{(\gamma,\beta)}[f]}{\lambda^{(\gamma,\alpha)}[g]}}$$

$$i.e.,\ \ \frac{\exp(\alpha(M_{g,D}^{-1}(M_{f,D}(R))))}{(\exp\beta(R))^{\frac{\lambda^{(\gamma,\beta)}[f]}{\lambda^{(\gamma,\alpha)}[g]}}} \leq \left[\frac{\left(\overline{\tau}_D^{(\gamma,\beta)}[f]+\varepsilon\right)}{\left(\overline{\tau}_D^{(\gamma,\alpha)}[g]-\varepsilon\right)}\right]^{\frac{1}{\lambda^{(\gamma,\alpha)}[g]}}.$$

Since in view of Theorem 4.2.1, we get that $\frac{\lambda^{(\gamma,\beta)}[f]}{\lambda^{(\gamma,\alpha)}[g]} \leq \rho^{(\alpha,\beta)}[f]_g$ and as $\varepsilon\,(>0)$ is arbitrary therefore it follows from above that

$$\liminf_{R\to\infty} \frac{\exp(\alpha(M_{g,D}^{-1}(M_{f,D}(R))))}{(\exp\beta(R))^{\rho^{(\alpha,\beta)}[f]_g}} \leq \left[\frac{\overline{\tau}_D^{(\gamma,\beta)}[f]}{\overline{\tau}_D^{(\gamma,\alpha)}[g]}\right]^{\frac{1}{\lambda^{(\gamma,\alpha)}[g]}}$$

$$i.e.,\ \ \overline{\sigma}_D^{(\alpha,\beta)}[f]_g \leq \left[\frac{\overline{\tau}_D^{(\gamma,\beta)}[f]}{\overline{\tau}_D^{(\gamma,\alpha)}[g]}\right]^{\frac{1}{\lambda^{(\gamma,\alpha)}[g]}}. \tag{126}$$

Similarly from (113) and in view of (115), it follows for a sequence of values of $R$ tending to infinity that

$$\exp(\alpha(M_{g,D}^{-1}(M_{f,D}(R)))) \leq$$
$$\exp\left(\alpha\left(M_{g,D}^{-1}\left(\gamma^{-1}\left(\log\left(\left(\tau_D^{(\gamma,\beta)}[f]+\varepsilon\right)(\exp\beta(R))^{\lambda^{(\gamma,\beta)}[f]}\right)\right)\right)\right)\right)$$

$$i.e., \ \exp(\alpha(M_{g,D}^{-1}(M_{f,D}(R)))) \leq \left(\frac{\left(\tau_D^{(\gamma,\beta)}[f]+\varepsilon\right)(\exp\beta(R))^{\lambda^{(\gamma,\beta)}[f]}}{\left(\tau_D^{(\gamma,\alpha)}[g]-\varepsilon\right)}\right)^{\frac{1}{\lambda^{(\gamma,\alpha)}[g]}}$$

$$i.e., \ \exp(\alpha(M_{g,D}^{-1}(M_{f,D}(R)))) \leq \left[\frac{\left(\tau_D^{(\gamma,\beta)}[f]+\varepsilon\right)}{\left(\tau_D^{(\gamma,\alpha)}[g]-\varepsilon\right)}\right]^{\frac{1}{\lambda^{(\gamma,\alpha)}[g]}}(\exp\beta(R))^{\frac{\lambda^{(\gamma,\beta)}[f]}{\lambda^{(\gamma,\alpha)}[g]}}$$

$$i.e., \ \frac{\exp(\alpha(M_{g,D}^{-1}(M_{f,D}(R))))}{(\exp\beta(R))^{\frac{\lambda^{(\gamma,\beta)}[f]}{\lambda^{(\gamma,\alpha)}[g]}}} \leq \left[\frac{\left(\tau_D^{(\gamma,\beta)}[f]+\varepsilon\right)}{\left(\tau_D^{(\gamma,\alpha)}[g]-\varepsilon\right)}\right]^{\frac{1}{\lambda^{(\gamma,\alpha)}[g]}}.$$

As in view of Theorem 4.2.1, we get that $\frac{\lambda^{(\gamma,\beta)}[f]}{\lambda^{(\gamma,\alpha)}[g]} \leq \rho^{(\alpha,\beta)}[f]_g$ and as $\varepsilon\,(>0)$ is arbitrary, therefore it follows from above that

$$\liminf_{R\to\infty}\frac{\exp(\alpha(M_{g,D}^{-1}(M_{f,D}(R))))}{(\exp\beta(R))^{\rho^{(\alpha,\beta)}[f]_g}} \leq \left[\frac{\tau_D^{(\gamma,\beta)}[f]}{\tau_D^{(\gamma,\alpha)}[g]}\right]^{\frac{1}{\lambda^{(\gamma,\alpha)}[g]}}$$

$$i.e., \ \overline{\sigma}_D^{(\alpha,\beta)}[f]_g \leq \left[\frac{\tau_D^{(\gamma,\beta)}[f]}{\tau_D^{(\gamma,\alpha)}[g]}\right]^{\frac{1}{\lambda^{(\gamma,\alpha)}[g]}}. \tag{127}$$

Further in view of (96), we get from (110) for a sequence of values of $R$ tending to infinity that

$$\exp(\alpha(M_{g,D}^{-1}(M_{f,D}(R)))) \leq$$
$$\exp\left(\alpha\left(M_{g,D}^{-1}\left(\gamma^{-1}\left(\log\left(\left(\overline{\tau}_D^{(\gamma,\beta)}[f]+\varepsilon\right)(\exp\beta(R))^{\lambda^{(\gamma,\beta)}[f]}\right)\right)\right)\right)\right)$$

$$i.e., \ \exp(\alpha(M_{g,D}^{-1}(M_{f,D}(R)))) \leq \left(\frac{\left(\overline{\tau}_D^{(\gamma,\beta)}[f]+\varepsilon\right)(\exp\beta(R))^{\lambda^{(\gamma,\beta)}[f]}}{\left(\sigma_D^{(\gamma,\alpha)}[g]-\varepsilon\right)}\right)^{\frac{1}{\rho^{(\gamma,\alpha)}[g]}}$$

$$i.e., \ \exp(\alpha(M_{g,D}^{-1}(M_{f,D}(R)))) \leq \left[\frac{\left(\overline{\tau}_D^{(\gamma,\beta)}[f]+\varepsilon\right)}{\left(\sigma_D^{(\gamma,\alpha)}[g]-\varepsilon\right)}\right]^{\frac{1}{\rho^{(\gamma,\alpha)}[g]}}(\exp\beta(R))^{\frac{\lambda^{(\gamma,\beta)}[f]}{\rho^{(\gamma,\alpha)}[g]}}$$

$$i.e., \ \frac{\exp(\alpha(M_{g,D}^{-1}(M_{f,D}(R))))}{(\exp\beta(R))^{\frac{\lambda^{(\gamma,\beta)}[f]}{\rho^{(\gamma,\alpha)}[g]}}} \leq \left[\frac{\left(\overline{\tau}_D^{(\gamma,\beta)}[f]+\varepsilon\right)}{\left(\sigma_D^{(\gamma,\alpha)}[g]-\varepsilon\right)}\right]^{\frac{1}{\rho^{(\gamma,\alpha)}[g]}}.$$

Since in view of Theorem 4.2.1, we get that $\frac{\lambda^{(\gamma,\beta)}[f]}{\rho^{(\gamma,\alpha)}[g]} \leq \rho^{(\alpha,\beta)}[f]_g$ and as $\varepsilon\,(>0)$ is arbitrary therefore it follows from above that

$$\liminf_{R\to\infty}\frac{\exp(\alpha(M_{g,D}^{-1}(M_{f,D}(R))))}{(\exp\beta(R))^{\rho^{(\alpha,\beta)}[f]_g}} \leq \left[\frac{\overline{\tau}_D^{(\gamma,\beta)}[f]}{\sigma_D^{(\gamma,\alpha)}[g]}\right]^{\frac{1}{\rho^{(\gamma,\alpha)}[g]}}$$

$$i.e.,\ \overline{\sigma}_D^{(\alpha,\beta)}[f]_g \leq \left[\frac{\overline{\tau}_D^{(\gamma,\beta)}[f]}{\sigma_D^{(\gamma,\alpha)}[g]}\right]^{\frac{1}{\rho^{(\gamma,\alpha)}[g]}}. \tag{128}$$

Similarly from (113) and in view of (95), it follows for a sequence of values of $R$ tending to infinity that

$$\exp(\alpha(M_{g,D}^{-1}(M_{f,D}(R)))) \leq$$

$$\exp\left(\alpha\left(M_{g,D}^{-1}\left(\gamma^{-1}\left(\log\left(\left(\tau_D^{(\gamma,\beta)}[f]+\varepsilon\right)(\exp\beta(R))^{\lambda^{(\gamma,\beta)}[f]}\right)\right)\right)\right)\right)$$

$$i.e.,\ \exp(\alpha(M_{g,D}^{-1}(M_{f,D}(R)))) \leq \left(\frac{\left(\tau_D^{(\gamma,\beta)}[f]+\varepsilon\right)(\exp\beta(R))^{\lambda^{(\gamma,\beta)}[f]}}{\left(\overline{\sigma}_D^{(\gamma,\alpha)}[g]-\varepsilon\right)}\right)^{\frac{1}{\rho^{(\gamma,\alpha)}[g]}}$$

$$i.e.,\ \exp(\alpha(M_{g,D}^{-1}(M_{f,D}(R)))) \leq \left[\frac{\left(\tau_D^{(\gamma,\beta)}[f]+\varepsilon\right)}{\left(\overline{\sigma}_D^{(\gamma,\alpha)}[g]-\varepsilon\right)}\right]^{\frac{1}{\rho^{(\gamma,\alpha)}[g]}}(\exp\beta(R))^{\frac{\lambda^{(\gamma,\beta)}[f]}{\rho^{(\gamma,\alpha)}[g]}}$$

$$i.e.,\ \frac{\exp(\alpha(M_{g,D}^{-1}(M_{f,D}(R))))}{(\exp\beta(R))^{\frac{\lambda^{(\gamma,\beta)}[f]}{\rho^{(\gamma,\alpha)}[g]}}} \leq \left[\frac{\left(\tau_D^{(\gamma,\beta)}[f]+\varepsilon\right)}{\left(\overline{\sigma}_D^{(\gamma,\alpha)}[g]-\varepsilon\right)}\right]^{\frac{1}{\rho^{(\gamma,\alpha)}[g]}}.$$

As in view of Theorem 4.2.1, we get that $\frac{\lambda^{(\gamma,\beta)}[f]}{\rho^{(\gamma,\alpha)}[g]} \leq \rho^{(\alpha,\beta)}[f]_g$ and as $\varepsilon\,(>0)$ is arbitrary, therefore it follows from above that

$$\liminf_{R\to\infty}\frac{\exp(\alpha(M_{g,D}^{-1}(M_{f,D}(R))))}{(\exp\beta(R))^{\rho^{(\alpha,\beta)}[f]_g}} \leq \left[\frac{\tau_D^{(\gamma,\beta)}[f]}{\overline{\sigma}_D^{(\gamma,\alpha)}[g]}\right]^{\frac{1}{\rho^{(\gamma,\alpha)}[g]}}$$

$$i.e.,\ \overline{\sigma}_D^{(\alpha,\beta)}[f]_g \leq \left[\frac{\tau_D^{(\gamma,\beta)}[f]}{\overline{\sigma}_D^{(\gamma,\alpha)}[g]}\right]^{\frac{1}{\rho^{(\gamma,\alpha)}[g]}}. \tag{129}$$

Therefore the second part of the theorem follows from $(123)$, $(124)$, $(125)$, $(126)$, $(127)$, $(128)$ and $(129)$. ∎

**Theorem 6.3.6** *Let $f(z)$ and $g(z)$ be two entire functions of $n$ complex variables such that $0 < \rho^{(\gamma,\beta)}[f] < \infty$ and $0 < \rho^{(\gamma,\alpha)}[g] < \infty$ . Then*

$$
\max \left\{
\begin{array}{l}
\left[\dfrac{\overline{\tau}_D^{(\gamma,\beta)}[f]}{\overline{\tau}_D^{(\gamma,\alpha)}[g]}\right]^{\frac{1}{\lambda^{(\gamma,\alpha)}[g]}} , \;
\left[\dfrac{\tau_D^{(\gamma,\beta)}[f]}{\tau_D^{(\gamma,\alpha)}[g]}\right]^{\frac{1}{\lambda^{(\gamma,\alpha)}[g]}} , \;
\left[\dfrac{\overline{\sigma}_D^{(\gamma,\beta)}[f]}{\overline{\sigma}_D^{(\gamma,\alpha)}[g]}\right]^{\frac{1}{\rho^{(\gamma,\alpha)}[g]}} , \\[18pt]
\left[\dfrac{\sigma_D^{(\gamma,\beta)}[f]}{\sigma_D^{(\gamma,\alpha)}[g]}\right]^{\frac{1}{\rho^{(\gamma,\alpha)}[g]}} , \;
\left[\dfrac{\sigma_D^{(\gamma,\beta)}[f]}{\overline{\tau}_D^{(\gamma,\alpha)}[g]}\right]^{\frac{1}{\lambda^{(\gamma,\alpha)}[g]}} , \;
\left[\dfrac{\overline{\tau}_D^{(\gamma,\beta)}[f]}{\tau_D^{(\gamma,\alpha)}[g]}\right]^{\frac{1}{\lambda^{(\gamma,\alpha)}[g]}}
\end{array}
\right\}
$$

$$
\leq \overline{\tau}_D^{(\alpha,\beta)}[f]_g \leq \left[\dfrac{\overline{\tau}_D^{(\gamma,\beta)}[f]}{\overline{\sigma}_D^{(\gamma,\alpha)}[g]}\right]^{\frac{1}{\rho^{(\gamma,\alpha)}[g]}}
$$

*and*

$$
\max \left\{
\left[\dfrac{\overline{\sigma}_D^{(\gamma,\beta)}[f]}{\sigma_D^{(\gamma,\alpha)}[g]}\right]^{\frac{1}{\rho^{(\gamma,\alpha)}[g]}} , \;
\left[\dfrac{\tau_D^{(\gamma,\beta)}[f]}{\overline{\tau}_D^{(\gamma,\alpha)}[g]}\right]^{\frac{1}{\lambda^{(\gamma,\alpha)}[g]}} , \;
\left[\dfrac{\overline{\sigma}_D^{(\gamma,\beta)}[f]}{\overline{\tau}_D^{(\gamma,\alpha)}[g]}\right]^{\frac{1}{\lambda^{(\gamma,\alpha)}[g]}}
\right\} \leq \tau_D^{(\alpha,\beta)}[f]_g
$$

$$
\leq \min \left\{
\left[\dfrac{\tau_D^{(\gamma,\beta)}[f]}{\overline{\sigma}_D^{(\gamma,\alpha)}[g]}\right]^{\frac{1}{\rho^{(\gamma,\alpha)}[g]}} , \;
\left[\dfrac{\overline{\tau}_D^{(\gamma,\beta)}[f]}{\sigma_D^{(\gamma,\alpha)}[g]}\right]^{\frac{1}{\rho^{(\gamma,\alpha)}[g]}}
\right\} .
$$

**Proof.** We obtain from (112) and (114), for a sequence of values of $R$ tending to infinity that

$$
\exp(\alpha(M_{g,D}^{-1}(M_{f,D}(R)))) \geq
$$

$$
\exp\left(\alpha\left(M_{g,D}^{-1}\left(\gamma^{-1}\left(\log\left(\left(\overline{\tau}_D^{(\gamma,\beta)}[f] - \varepsilon\right)(\exp\beta(R))^{\lambda^{(\gamma,\beta)}[f]}\right)\right)\right)\right)\right)
$$

*i.e.,* $\;\; \exp(\alpha(M_{g,D}^{-1}(M_{f,D}(R)))) \geq \left(\dfrac{\left(\overline{\tau}_D^{(\gamma,\beta)}[f] - \varepsilon\right)(\exp\beta(R))^{\lambda^{(\gamma,\beta)}[f]}}{\left(\overline{\tau}_D^{(\gamma,\alpha)}[g] + \varepsilon\right)}\right)^{\frac{1}{\lambda^{(\gamma,\alpha)}[g]}}$

*i.e.,* $\;\; \exp(\alpha(M_{g,D}^{-1}(M_{f,D}(R)))) \geq \left[\dfrac{\left(\overline{\tau}_D^{(\gamma,\beta)}[f] - \varepsilon\right)}{\left(\overline{\tau}_D^{(\gamma,\alpha)}[g] + \varepsilon\right)}\right]^{\frac{1}{\lambda^{(\gamma,\alpha)}[g]}} (\exp\beta(R))^{\frac{\lambda^{(\gamma,\beta)}[f]}{\lambda^{(\gamma,\alpha)}[g]}}$

*i.e.,* $\;\; \dfrac{\exp(\alpha(M_{g,D}^{-1}(M_{f,D}(R))))}{(\exp\beta(R))^{\frac{\lambda^{(\gamma,\beta)}[f]}{\lambda^{(\gamma,\alpha)}[g]}}} \geq \left[\dfrac{\left(\overline{\tau}_D^{(\gamma,\beta)}[f] - \varepsilon\right)}{\left(\overline{\tau}_D^{(\gamma,\alpha)}[g] + \varepsilon\right)}\right]^{\frac{1}{\lambda^{(\gamma,\alpha)}[g]}} .$

Since in view of Theorem 4.2.1, we get that $\frac{\lambda^{(\gamma,\beta)}[f]}{\lambda^{(\gamma,\alpha)}[g]} \geq \lambda^{(\alpha,\beta)}[f]_g$ and as $\varepsilon \,(> 0)$ is arbitrary, therefore it follows from above that

$$\limsup_{R \to \infty} \frac{\exp(\alpha(M_{g,D}^{-1}(M_{f,D}(R))))}{(\exp \beta(R))^{\lambda^{(\alpha,\beta)}[f]_g}} \geq \left[\frac{\overline{\tau}_D^{(\gamma,\beta)}[f]}{\overline{\tau}_D^{(\gamma,\alpha)}[g]}\right]^{\frac{1}{\lambda^{(\gamma,\alpha)}[g]}}$$

$$i.e., \;\; \overline{\tau}_D^{(\alpha,\beta)}[f]_g \geq \left[\frac{\overline{\tau}_D^{(\gamma,\beta)}[f]}{\overline{\tau}_D^{(\gamma,\alpha)}[g]}\right]^{\frac{1}{\lambda^{(\gamma,\alpha)}[g]}}. \tag{130}$$

Further we obtain from (111) and (117), for a sequence of values of $R$ tending to infinity that

$$\exp(\alpha(M_{g,D}^{-1}(M_{f,D}(R)))) \geq$$
$$\exp\left(\alpha\left(M_{g,D}^{-1}\left(\gamma^{-1}\left(\log\left(\left(\tau_D^{(\gamma,\beta)}[f] - \varepsilon\right)(\exp \beta(R))^{\lambda^{(\gamma,\beta)}[f]}\right)\right)\right)\right)\right)$$

$$i.e., \;\; \exp(\alpha(M_{g,D}^{-1}(M_{f,D}(R)))) \geq \left(\frac{\left(\tau_D^{(\gamma,\beta)}[f] - \varepsilon\right)(\exp \beta(R))^{\lambda^{(\gamma,\beta)}[f]}}{\left(\tau_D^{(\gamma,\alpha)}[g] - \varepsilon\right)}\right)^{\frac{1}{\lambda^{(\gamma,\alpha)}[g]}}$$

$$i.e., \;\; \exp(\alpha(M_{g,D}^{-1}(M_{f,D}(R)))) \geq \left[\frac{\left(\tau_D^{(\gamma,\beta)}[f] - \varepsilon\right)}{\left(\tau_D^{(\gamma,\alpha)}[g] - \varepsilon\right)}\right]^{\frac{1}{\lambda^{(\gamma,\alpha)}[g]}} (\exp \beta(R))^{\frac{\lambda^{(\gamma,\beta)}[f]}{\lambda^{(\gamma,\alpha)}[g]}}$$

$$i.e., \;\; \frac{\exp(\alpha(M_{g,D}^{-1}(M_{f,D}(R))))}{(\exp \beta(R))^{\frac{\lambda^{(\gamma,\beta)}[f]}{\lambda^{(\gamma,\alpha)}[g]}}} \geq \left[\frac{\left(\tau_D^{(\gamma,\beta)}[f] - \varepsilon\right)}{\left(\tau_D^{(\gamma,\alpha)}[g] - \varepsilon\right)}\right]^{\frac{1}{\lambda^{(\gamma,\alpha)}[g]}}.$$

As in view of Theorem 4.2.1, we get that $\frac{\lambda^{(\gamma,\beta)}[f]}{\lambda^{(\gamma,\alpha)}[g]} \geq \lambda^{(\alpha,\beta)}[f]_g$ and as $\varepsilon \,(> 0)$ is arbitrary, therefore it follows from above that

$$\limsup_{R \to \infty} \frac{\exp(\alpha(M_{g,D}^{-1}(M_{f,D}(R))))}{(\exp \beta(R))^{\lambda^{(\alpha,\beta)}[f]_g}} \geq \left[\frac{\tau_D^{(\gamma,\beta)}[f]}{\tau_D^{(\gamma,\alpha)}[g]}\right]^{\frac{1}{\lambda^{(\gamma,\alpha)}[g]}}$$

$$i.e., \;\; \overline{\tau}_D^{(\alpha,\beta)}[f]_g \geq \left[\frac{\tau_D^{(\gamma,\beta)}[f]}{\tau_D^{(\gamma,\alpha)}[g]}\right]^{\frac{1}{\lambda^{(\gamma,\alpha)}[g]}}. \tag{131}$$

As $\frac{\rho^{(\gamma,\beta)}[f]}{\rho^{(\gamma,\alpha)}[g]} \geq \lambda^{(\alpha,\beta)}[f]_g$ and $\varepsilon \,(> 0)$ is arbitrary, we get from (98) that

$$\limsup_{R \to \infty} \frac{\exp(\alpha(M_{g,D}^{-1}(M_{f,D}(R))))}{(\exp \beta(R))^{\lambda^{(\alpha,\beta)}[f]_g}} \geq \left[\frac{\sigma_D^{(\gamma,\beta)}[f]}{\sigma_D^{(\gamma,\alpha)}[g]}\right]^{\frac{1}{\rho^{(\gamma,\alpha)}[g]}}$$

$$i.e., \;\; \overline{\tau}_D^{(\alpha,\beta)}[f]_g \geq \left[\frac{\sigma_D^{(\gamma,\beta)}[f]}{\sigma_D^{(\gamma,\alpha)}[g]}\right]^{\frac{1}{\rho^{(\gamma,\alpha)}[g]}}. \tag{132}$$

Analogously, we get from (100) that

$$\limsup_{R \to \infty} \frac{\exp \alpha M_{g,D}^{-1} M_{f,D}(R)}{(\exp \beta(R))^{\lambda^{(\alpha,\beta)}[f]_g}} \geq \left[ \frac{\overline{\sigma}_D^{(\gamma,\beta)}[f]}{\overline{\sigma}_D^{(\gamma,\alpha)}[g]} \right]^{\frac{1}{\rho^{(\gamma,\alpha)}[g]}}$$

$$i.e.,\ \ \overline{\tau}_D^{(\alpha,\beta)}[f]_g \geq \left[ \frac{\overline{\sigma}_D^{(\gamma,\beta)}[f]}{\overline{\sigma}_D^{(\gamma,\alpha)}[g]} \right]^{\frac{1}{\rho^{(\gamma,\alpha)}[g]}}, \tag{133}$$

since $\frac{\rho^{(\gamma,\beta)}[f]}{\rho^{(\gamma,\alpha)}[g]} \leq \lambda^{(\alpha,\beta)}[f]_g$ and $\varepsilon (> 0)$ is arbitrary.

Likewise from (92) and in view of (114), we get for a sequence of values of $R$ tending to infinity that

$$\exp(\alpha(M_{g,D}^{-1}(M_{f,D}(R)))) \geq$$

$$\exp\left(\alpha\left(M_{g,D}^{-1}\left(\gamma^{-1}\left(\log\left(\left(\sigma_D^{(\gamma,\beta)}[f] - \varepsilon\right)(\exp\beta(R))^{\rho^{(\gamma,\beta)}[f]}\right)\right)\right)\right)\right)$$

$$i.e.,\ \ \exp(\alpha(M_{g,D}^{-1}(M_{f,D}(R)))) \geq \left( \frac{\left(\sigma_D^{(\gamma,\beta)}[f] - \varepsilon\right)(\exp\beta(R))^{\rho^{(\gamma,\beta)}[f]}}{\left(\overline{\tau}_D^{(\gamma,\alpha)}[g] + \varepsilon\right)} \right)^{\frac{1}{\lambda^{(\gamma,\alpha)}[g]}}$$

$$i.e.,\ \ \exp(\alpha(M_{g,D}^{-1}(M_{f,D}(R)))) \geq \left[ \frac{\left(\sigma_D^{(\gamma,\beta)}[f] - \varepsilon\right)}{\left(\overline{\tau}_D^{(\gamma,\alpha)}[g] + \varepsilon\right)} \right]^{\frac{1}{\lambda^{(\gamma,\alpha)}[g]}} (\exp\beta(R))^{\frac{\rho^{(\gamma,\beta)}[f]}{\lambda^{(\gamma,\alpha)}[g]}}$$

$$i.e.,\ \ \frac{\exp(\alpha(M_{g,D}^{-1}(M_{f,D}(R))))}{(\exp\beta(R))^{\frac{\rho^{(\gamma,\beta)}[f]}{\lambda^{(\gamma,\alpha)}[g]}}} \geq \left[ \frac{\left(\sigma_D^{(\gamma,\beta)}[f] - \varepsilon\right)}{\left(\overline{\tau}_D^{(\gamma,\alpha)}[g] + \varepsilon\right)} \right]^{\frac{1}{\lambda^{(\gamma,\alpha)}[g]}}.$$

Since in view of Theorem 4.2.1, we get that $\frac{\rho^{(\gamma,\beta)}[f]}{\lambda^{(\gamma,\alpha)}[g]} \geq \lambda^{(\alpha,\beta)}[f]_g$ and as $\varepsilon (> 0)$ is arbitrary, therefore it follows from above that

$$\limsup_{R \to \infty} \frac{\exp(\alpha(M_{g,D}^{-1}(M_{f,D}(R))))}{(\exp\beta(R))^{\lambda^{(\alpha,\beta)}[f]_g}} \geq \left[ \frac{\sigma_D^{(\gamma,\beta)}[f]}{\overline{\tau}_D^{(\gamma,\alpha)}[g]} \right]^{\frac{1}{\lambda^{(\gamma,\alpha)}[g]}}$$

$$i.e.,\ \ \overline{\tau}_D^{(\alpha,\beta)}[f]_g \geq \left[ \frac{\sigma_D^{(\gamma,\beta)}[f]}{\overline{\tau}_D^{(\gamma,\alpha)}[g]} \right]^{\frac{1}{\lambda^{(\gamma,\alpha)}[g]}}. \tag{134}$$

Analogously from (91) and in view of (117), it follows for a sequence of values of $R$ tending to infinity that

$$\exp(\alpha(M_{g,D}^{-1}(M_{f,D}(R)))) \geq$$

$$\exp\left(\alpha\left(M_{g,D}^{-1}\left(\gamma^{-1}\left(\log\left(\left(\overline{\sigma}_D^{(\gamma,\beta)}[f]-\varepsilon\right)(\exp\beta(R))^{\rho^{(\gamma,\beta)[f]}}\right)\right)\right)\right)\right)$$

$$i.e.,\ \exp(\alpha(M_{g,D}^{-1}(M_{f,D}(R)))) \geq \left(\frac{\left(\overline{\sigma}_D^{(\gamma,\beta)}[f]-\varepsilon\right)(\exp\beta(R))^{\rho^{(\gamma,\beta)[f]}}}{\left(\tau_D^{(\gamma,\alpha)}[g]-\varepsilon\right)}\right)^{\frac{1}{\lambda^{(\gamma,\alpha)}[g]}}$$

$$i.e.,\ \exp(\alpha(M_{g,D}^{-1}(M_{f,D}(R)))) \geq \left[\frac{\left(\overline{\sigma}_D^{(\gamma,\beta)}[f]-\varepsilon\right)}{\left(\tau_D^{(\gamma,\alpha)}[g]-\varepsilon\right)}\right]^{\frac{1}{\lambda^{(\gamma,\alpha)}[g]}} (\exp\beta(R))^{\frac{\rho^{(\gamma,\beta)[f]}}{\lambda^{(\gamma,\alpha)}[g]}}$$

$$i.e.,\ \frac{\exp(\alpha(M_{g,D}^{-1}(M_{f,D}(R))))}{(\exp\beta(R))^{\frac{\rho^{(\gamma,\beta)[f]}}{\lambda^{(\gamma,\alpha)}[g]}}} \geq \left[\frac{\left(\overline{\sigma}_D^{(\gamma,\beta)}[f]-\varepsilon\right)}{\left(\tau_D^{(\gamma,\alpha)}[g]-\varepsilon\right)}\right]^{\frac{1}{\lambda^{(\gamma,\alpha)}[g]}}.$$

As in view of Theorem 4.2.1, we get that $\frac{\rho^{(\gamma,\beta)[f]}}{\lambda^{(\gamma,\alpha)}[g]} \geq \lambda^{(\alpha,\beta)}[f]_g$ and as $\varepsilon\,(>0)$ is arbitrary, therefore it follows from above that

$$\limsup_{R\to\infty}\frac{\exp(\alpha(M_{g,D}^{-1}(M_{f,D}(R))))}{(\exp\beta(R))^{\lambda^{(\alpha,\beta)}[f]_g}} \geq \left[\frac{\overline{\sigma}_D^{(\gamma,\beta)}[f]}{\tau_D^{(\gamma,\alpha)}[g]}\right]^{\frac{1}{\lambda^{(\gamma,\alpha)}[g]}}$$

$$i.e.,\ \overline{\tau}_D^{(\alpha,\beta)}[f]_g \geq \left[\frac{\overline{\sigma}_D^{(\gamma,\beta)}[f]}{\tau_D^{(\gamma,\alpha)}[g]}\right]^{\frac{1}{\lambda^{(\gamma,\alpha)}[g]}}. \tag{135}$$

Again from (95) and (110), we have for all sufficiently large values of $R$ that

$$\exp(\alpha(M_{g,D}^{-1}(M_{f,D}(R)))) \leq$$

$$\exp\left(\alpha\left(M_{g,D}^{-1}\left(\gamma^{-1}\left(\log\left(\left(\overline{\tau}_D^{(\gamma,\beta)}[f]+\varepsilon\right)(\exp\beta(R))^{\lambda^{(\gamma,\beta)[f]}}\right)\right)\right)\right)\right)$$

$$i.e.,\ \exp(\alpha(M_{g,D}^{-1}(M_{f,D}(R)))) \leq \left(\frac{\left(\overline{\tau}_D^{(\gamma,\beta)}[f]+\varepsilon\right)(\exp\beta(R))^{\lambda^{(\gamma,\beta)[f]}}}{\left(\overline{\sigma}_D^{(\gamma,\alpha)}[g]-\varepsilon\right)}\right)^{\frac{1}{\rho^{(\gamma,\alpha)}[g]}}$$

$$i.e.,\ \exp(\alpha(M_{g,D}^{-1}(M_{f,D}(R)))) \leq \left[\frac{\left(\overline{\tau}_D^{(\gamma,\beta)}[f]+\varepsilon\right)}{\left(\overline{\sigma}_D^{(\gamma,\alpha)}[g]-\varepsilon\right)}\right]^{\frac{1}{\rho^{(\gamma,\alpha)}[g]}} (\exp\beta(R))^{\frac{\lambda^{(\gamma,\beta)[f]}}{\rho^{(\gamma,\alpha)}[g]}}$$

$$i.e.,\ \frac{\exp(\alpha(M_{g,D}^{-1}(M_{f,D}(R))))}{(\exp\beta(R))^{\frac{\lambda^{(\gamma,\beta)[f]}}{\rho^{(\gamma,\alpha)}[g]}}} \leq \left[\frac{\left(\overline{\tau}_D^{(\gamma,\beta)}[f]+\varepsilon\right)}{\left(\overline{\sigma}_D^{(\gamma,\alpha)}[g]-\varepsilon\right)}\right]^{\frac{1}{\rho^{(\gamma,\alpha)}[g]}}.$$

As in view of Theorem 4.2.1, we get that $\frac{\lambda^{(\gamma,\beta)}[f]}{\rho^{(\gamma,\alpha)}[g]} \leq \lambda^{(\alpha,\beta)}[f]_g$ and as $\varepsilon\,(>0)$ is arbitrary, therefore it follows from above that

$$\limsup_{R\to\infty}\frac{\exp(\alpha(M_{g,D}^{-1}(M_{f,D}(R))))}{(\exp\beta(R))^{\lambda^{(\alpha,\beta)}[f]_g}} \leq \left[\frac{\overline{\tau}_D^{(\gamma,\beta)}[f]}{\overline{\sigma}_D^{(\gamma,\alpha)}[g]}\right]^{\frac{1}{\rho^{(\gamma,\alpha)}[g]}}$$

$$i.e.,\ \overline{\tau}_D^{(\alpha,\beta)}[f]_g \leq \left[\frac{\overline{\tau}_D^{(\gamma,\beta)}[f]}{\overline{\sigma}_D^{(\gamma,\alpha)}[g]}\right]^{\frac{1}{\rho^{(\gamma,\alpha)}[g]}}. \tag{136}$$

Thus the first part of the theorem follows from $(130)$, $(131)$, $(132)$, $(133)$, $(134)$, $(135)$ and $(136)$.

Now from $(111)$ and in view of $(114)$, we get for all sufficiently large values of $R$ that

$$\exp(\alpha(M_{g,D}^{-1}(M_{f,D}(R)))) \geq$$
$$\exp\left(\alpha\left(M_{g,D}^{-1}\left(\gamma^{-1}\left(\log\left(\left(\tau_D^{(\gamma,\beta)}[f]-\varepsilon\right)(\exp\beta(R))^{\lambda^{(\gamma,\beta)}[f]}\right)\right)\right)\right)\right)$$

$$i.e.,\ \exp(\alpha(M_{g,D}^{-1}(M_{f,D}(R)))) \geq \left(\frac{\left(\tau_D^{(\gamma,\beta)}[f]-\varepsilon\right)(\exp\beta(R))^{\lambda^{(\gamma,\beta)}[f]}}{\left(\overline{\tau}_D^{(\gamma,\alpha)}[g]+\varepsilon\right)}\right)^{\frac{1}{\lambda^{(\gamma,\alpha)}[g]}}$$

$$i.e.,\ \exp(\alpha(M_{g,D}^{-1}(M_{f,D}(R)))) \geq \left[\frac{\left(\tau_D^{(\gamma,\beta)}[f]-\varepsilon\right)}{\left(\overline{\tau}_D^{(\gamma,\alpha)}[g]+\varepsilon\right)}\right]^{\frac{1}{\lambda^{(\gamma,\alpha)}[g]}}(\exp\beta(R))^{\frac{\lambda^{(\gamma,\beta)}[f]}{\lambda^{(\gamma,\alpha)}[g]}}$$

$$i.e.,\ \frac{\exp(\alpha(M_{g,D}^{-1}(M_{f,D}(R))))}{(\exp\beta(R))^{\frac{\lambda^{(\gamma,\beta)}[f]}{\lambda^{(\gamma,\alpha)}[g]}}} \geq \left[\frac{\left(\tau_D^{(\gamma,\beta)}[f]-\varepsilon\right)}{\left(\overline{\tau}_D^{(\gamma,\alpha)}[g]+\varepsilon\right)}\right]^{\frac{1}{\lambda^{(\gamma,\alpha)}[g]}}.$$

As in view of Theorem 4.2.1, we get that $\frac{\lambda^{(\gamma,\beta)}[f]}{\lambda^{(\gamma,\alpha)}[g]} \geq \lambda^{(\alpha,\beta)}[f]_g$ and as $\varepsilon\,(>0)$ is arbitrary, therefore it follows from above that

$$\liminf_{R\to\infty}\frac{\exp(\alpha(M_{g,D}^{-1}(M_{f,D}(R))))}{(\exp\beta(R))^{\lambda^{(\alpha,\beta)}[f]_g}} \geq \left[\frac{\tau_D^{(\gamma,\beta)}[f]}{\overline{\tau}_D^{(\gamma,\alpha)}[g]}\right]^{\frac{1}{\lambda^{(\gamma,\alpha)}[g]}}$$

$$i.e.,\ \tau_D^{(\alpha,\beta)}[f]_g \geq \left[\frac{\tau_D^{(\gamma,\beta)}[f]}{\overline{\tau}_D^{(\gamma,\alpha)}[g]}\right]^{\frac{1}{\lambda^{(\gamma,\alpha)}[g]}}. \tag{137}$$

Since $\frac{\rho^{(\gamma,\beta)}[f]}{\rho^{(\gamma,\alpha)}[g]} \geq \lambda^{(\alpha,\beta)}[f]_g$ and $\varepsilon\,(>0)$ is arbitrary, we get from $(104)$ that

$$\liminf_{R\to\infty}\frac{\exp(\alpha(M_{g,D}^{-1}(M_{f,D}(R))))}{(\exp\beta(R))^{\lambda^{(\alpha,\beta)}[f]_g}} \geq \left[\frac{\overline{\sigma}_D^{(\gamma,\beta)}[f]}{\sigma_D^{(\gamma,\alpha)}[g]}\right]^{\frac{1}{\rho^{(\gamma,\alpha)}[g]}}$$

$$i.e.,\ \tau_D^{(\alpha,\beta)}[f]_g \geq \left[\frac{\overline{\sigma}_D^{(\gamma,\beta)}[f]}{\sigma_D^{(\gamma,\alpha)}[g]}\right]^{\frac{1}{\rho^{(\gamma,\alpha)}[g]}}. \tag{138}$$

Again from (91) and in view of (114), we get for all sufficiently large values of $R$ that

$$\exp(\alpha(M_{g,D}^{-1}(M_{f,D}(R)))) \geq$$

$$\exp\left(\alpha\left(M_{g,D}^{-1}\left(\gamma^{-1}\left(\log\left(\left(\overline{\sigma}_D^{(\gamma,\beta)}[f] - \varepsilon\right)(\exp\beta(R))^{\rho^{(\gamma,\beta)}[f]}\right)\right)\right)\right)\right)$$

$$i.e., \ \exp(\alpha(M_{g,D}^{-1}(M_{f,D}(R)))) \geq \left(\frac{\left(\overline{\sigma}_D^{(\gamma,\beta)}[f] - \varepsilon\right)(\exp\beta(R))^{\rho^{(\gamma,\beta)}[f]}}{\left(\overline{\tau}_D^{(\gamma,\alpha)}[g] + \varepsilon\right)}\right)^{\frac{1}{\lambda^{(\gamma,\alpha)}[g]}}$$

$$i.e., \ \exp(\alpha(M_{g,D}^{-1}(M_{f,D}(R)))) \geq \left[\frac{\left(\overline{\sigma}_D^{(\gamma,\beta)}[f] - \varepsilon\right)}{\left(\overline{\tau}_D^{(\gamma,\alpha)}[g] + \varepsilon\right)}\right]^{\frac{1}{\lambda^{(\gamma,\alpha)}[g]}} (\exp\beta(R))^{\frac{\rho^{(\gamma,\beta)}[f]}{\lambda^{(\gamma,\alpha)}[g]}}$$

$$i.e., \ \frac{\exp(\alpha(M_{g,D}^{-1}(M_{f,D}(R))))}{(\exp\beta(R))^{\frac{\rho^{(\gamma,\beta)}[f]}{\lambda^{(\gamma,\alpha)}[g]}}} \geq \left[\frac{\left(\overline{\sigma}_D^{(\gamma,\beta)}[f] - \varepsilon\right)}{\left(\overline{\tau}_D^{(\gamma,\alpha)}[g] + \varepsilon\right)}\right]^{\frac{1}{\lambda^{(\gamma,\alpha)}[g]}}.$$

Since in view of Theorem 4.2.1, we get that $\frac{\rho^{(\gamma,\beta)}[f]}{\lambda^{(\gamma,\alpha)}[g]} \geq \lambda^{(\alpha,\beta)}[f]_g$ and as $\varepsilon \ (> 0)$ is arbitrary, therefore it follows from above that

$$\liminf_{R\to\infty} \frac{\exp(\alpha(M_{g,D}^{-1}(M_{f,D}(R))))}{(\exp\beta(R))^{\lambda^{(\alpha,\beta)}[f]_g}} \geq \left[\frac{\overline{\sigma}_D^{(\gamma,\beta)}[f]}{\overline{\tau}_D^{(\gamma,\alpha)}[g]}\right]^{\frac{1}{\lambda^{(\gamma,\alpha)}[g]}}$$

$$i.e., \ \tau_D^{(\alpha,\beta)}[f]_g \geq \left[\frac{\overline{\sigma}_D^{(\gamma,\beta)}[f]}{\overline{\tau}_D^{(\gamma,\alpha)}[g]}\right]^{\frac{1}{\lambda^{(\gamma,\alpha)}[g]}}. \tag{139}$$

Further, we get from (96) and (110) for a sequence of values of $R$ tending to infinity that

$$\exp(\alpha(M_{g,D}^{-1}(M_{f,D}(R)))) \leq$$

$$\exp\left(\alpha\left(M_{g,D}^{-1}\left(\gamma^{-1}\left(\log\left(\left(\overline{\tau}_D^{(\gamma,\beta)}[f] + \varepsilon\right)(\exp\beta(R))^{\lambda^{(\gamma,\beta)}[f]}\right)\right)\right)\right)\right)$$

$$i.e., \ \exp(\alpha(M_{g,D}^{-1}(M_{f,D}(R)))) \leq \left(\frac{\left(\overline{\tau}_D^{(\gamma,\beta)}[f] + \varepsilon\right)(\exp\beta(R))^{\lambda^{(\gamma,\beta)}[f]}}{\left(\sigma_D^{(\gamma,\alpha)}[g] - \varepsilon\right)}\right)^{\frac{1}{\rho^{(\gamma,\alpha)}[g]}}$$

$$i.e., \ \exp(\alpha(M_{g,D}^{-1}(M_{f,D}(R)))) \leq \left[\frac{\left(\overline{\tau}_D^{(\gamma,\beta)}[f] + \varepsilon\right)}{\left(\sigma_D^{(\gamma,\alpha)}[g] - \varepsilon\right)}\right]^{\frac{1}{\rho^{(\gamma,\alpha)}[g]}} (\exp\beta(R))^{\frac{\lambda^{(\gamma,\beta)}[f]}{\rho^{(\gamma,\alpha)}[g]}}$$

$$i.e., \ \frac{\exp(\alpha(M_{g,D}^{-1}(M_{f,D}(R))))}{(\exp\beta(R))^{\frac{\lambda^{(\gamma,\beta)}[f]}{\rho^{(\gamma,\alpha)}[g]}}} \leq \left[\frac{\left(\overline{\tau}_D^{(\gamma,\beta)}[f] + \varepsilon\right)}{\left(\sigma_D^{(\gamma,\alpha)}[g] - \varepsilon\right)}\right]^{\frac{1}{\rho^{(\gamma,\alpha)}[g]}}.$$

As in view of Theorem 4.2.1, we get that $\frac{\lambda^{(\gamma,\beta)}[f]}{\lambda^{(\gamma,\alpha)}[g]} \leq \lambda^{(\alpha,\beta)}[f]_g$ and as $\varepsilon\,(>0)$ is arbitrary, therefore it follows from above that

$$\liminf_{R\to\infty} \frac{\exp(\alpha(M_{g,D}^{-1}(M_{f,D}(R))))}{(\exp\beta(R))^{\lambda^{(\alpha,\beta)}[f]_g}} \leq \left[\frac{\overline{\tau}_D^{(\gamma,\beta)}[f]}{\sigma_D^{(\gamma,\alpha)}[g]}\right]^{\frac{1}{\rho^{(\gamma,\alpha)}[g]}}$$

$$i.e.,\ \ \tau_D^{(\alpha,\beta)}[f]_g \leq \left[\frac{\overline{\tau}_D^{(\gamma,\beta)}[f]}{\sigma_D^{(\gamma,\alpha)}[g]}\right]^{\frac{1}{\rho^{(\gamma,\alpha)}[g]}}. \tag{140}$$

Similarly from (113) and in view of (95), it follows for a sequence of values of $R$ tending to infinity that

$$\exp(\alpha(M_{g,D}^{-1}(M_{f,D}(R)))) \leq$$

$$\exp\left(\alpha\left(M_{g,D}^{-1}\left(\gamma^{-1}\left(\log\left(\left(\tau_D^{(\gamma,\beta)}[f]+\varepsilon\right)(\exp\beta(R))^{\lambda^{(\gamma,\beta)}[f]}\right)\right)\right)\right)\right)$$

$$i.e.,\ \ \exp(\alpha(M_{g,D}^{-1}(M_{f,D}(R)))) \leq \left(\frac{\left(\tau_D^{(\gamma,\beta)}[f]+\varepsilon\right)(\exp\beta(R))^{\lambda^{(\gamma,\beta)}[f]}}{\left(\overline{\sigma}_D^{(\gamma,\alpha)}[g]-\varepsilon\right)}\right)^{\frac{1}{\rho^{(\gamma,\alpha)}[g]}}$$

$$i.e.,\ \ \exp(\alpha(M_{g,D}^{-1}(M_{f,D}(R)))) \leq \left[\frac{\left(\tau_D^{(\gamma,\beta)}[f]+\varepsilon\right)}{\left(\overline{\sigma}_D^{(\gamma,\alpha)}[g]-\varepsilon\right)}\right]^{\frac{1}{\rho^{(\gamma,\alpha)}[g]}}(\exp\beta(R))^{\frac{\lambda^{(\gamma,\beta)}[f]}{\rho^{(\gamma,\alpha)}[g]}}$$

$$i.e.,\ \ \frac{\exp(\alpha(M_{g,D}^{-1}(M_{f,D}(R))))}{(\exp\beta(R))^{\frac{\lambda^{(\gamma,\beta)}[f]}{\rho^{(\gamma,\alpha)}[g]}}} \leq \left[\frac{\left(\tau_D^{(\gamma,\beta)}[f]+\varepsilon\right)}{\left(\overline{\sigma}_D^{(\gamma,\alpha)}[g]-\varepsilon\right)}\right]^{\frac{1}{\rho^{(\gamma,\alpha)}[g]}}.$$

Since in view of Theorem 4.2.1, we get that $\frac{\lambda^{(\gamma,\beta)}[f]}{\rho^{(\gamma,\alpha)}[g]} \leq \lambda^{(\alpha,\beta)}[f]_g$ and as $\varepsilon\,(>0)$ is arbitrary, therefore it follows from above that

$$\liminf_{R\to\infty} \frac{\exp\alpha M_{g,D}^{-1}M_{f,D}(R)}{(\exp\beta(R))^{\lambda^{(\alpha,\beta)}[f]_g}} \leq \left[\frac{\tau_D^{(\gamma,\beta)}[f]}{\overline{\sigma}_D^{(\gamma,\alpha)}[g]}\right]^{\frac{1}{\rho^{(\gamma,\alpha)}[g]}}$$

$$i.e.,\ \ \tau_D^{(\alpha,\beta)}[f]_g \leq \left[\frac{\tau_D^{(\gamma,\beta)}[f]}{\overline{\sigma}_D^{(\gamma,\alpha)}[g]}\right]^{\frac{1}{\rho^{(\gamma,\alpha)}[g]}}. \tag{141}$$

Hence the second part of the theorem follows from $(137)$, $(138)$, $(139)$, $(140)$ and $(141)$.

∎

# 6.4   Conclusion.

The main aim of this chapter is to find out the limiting value of generalized relative Gol'dberg type $(\alpha, \beta)$, generalized relative Gol'dberg weak type $(\alpha, \beta)$ etc. under some different conditions. Now some more results may be thought of focusing the concepts of higher dimensions in this connection which has been briefly discussed in the next chapter.

# Chapter 7

# Generalized relative Gol'dberg order $(\alpha, \beta)$ and generalized relative Gol'dberg type $(\alpha, \beta)$ based growth measure of entire functions of several complex variables

**Abstract:** In this chapter, we intend to find out generalized relative Gol'dberg order $(\alpha, \beta)$, generalized relative Gol'dberg type $(\alpha, \beta)$ and generalized relative Gol'dberg weak type $(\alpha, \beta)$ of an entire function $f$ of several complex variables with respect to another entire function $g$ of several complex variables when generalized relative Gol'dberg order $(\gamma, \beta)$, generalized relative Gol'dberg type $(\gamma, \beta)$ and generalized relative Gol'dberg weak type $(\gamma, \beta)$ of $f$ and generalized relative Gol'dberg order $(\gamma, \alpha)$, generalized relative Gol'dberg type $(\gamma, \alpha)$ and generalized relative Gol'dberg weak type $(\gamma, \alpha)$ of $g$ with respect another entire function $h$ of several complex variables are given, where $\alpha, \beta, \gamma$ are continuous non-negative functions defined on $(-\infty, +\infty)$.

**Keywords:** Increasing function, generalized relative Gol'dberg order $(\alpha, \beta)$, generalized relative Gol'dberg type $(\alpha, \beta)$, generalized relative Gol'dberg weak type $(\alpha, \beta)$.

**Mathematics Subject Classification (2010) :** 32A15.

## 7.1   Introduction.

In continuation of the discussion of previous chapter, question may arise about the values of generalized relative Gol'dberg order $(\alpha, \beta)$, generalized relative Gol'dberg type $(\alpha, \beta)$ and generalized relative Gol'dberg weak type $(\alpha, \beta)$ of an entire function $f(z)$ of $n$ complex variables with respect to another entire function $g(z)$ of $n$ complex variables when generalized relative Gol'dberg order $(\gamma, \beta)$, generalized relative Gol'dberg type $(\gamma, \beta)$ and generalized relative Gol'dberg weak type $(\gamma, \beta)$ of $f(z)$ and generalized relative Gol'dberg order $(\gamma, \alpha)$, generalized relative Gol'dberg type $(\gamma, \alpha)$ and generalized relative Gol'dberg

**Tanmay Biswas & Chinmay Biswas**

weak type $(\gamma, \alpha)$ of $g(z)$ with respect to another entire function $h(z)$ of $n$ complex variables are given. In this chapter we intend to provide this answer. In this present chapter $\alpha, \beta$ and $\gamma$ always denote the functions belonging to $L^0$.

## 7.2   Main Results.

In this section we present the main results of the chapter.

**Theorem 7.2.1** *Let $f(z)$, $g(z)$ and $h(z)$ be three entire functions of $n$ complex variables such that $0 < \lambda^{(\gamma,\beta)}[f]_h \leq \rho^{(\gamma,\beta)}[f]_h < \infty$ and $0 < \lambda^{(\gamma,\alpha)}[g]_h \leq \rho^{(\gamma,\alpha)}[g]_h < \infty$ . Then*

$$\frac{\lambda^{(\gamma,\beta)}[f]_h}{\rho^{(\gamma,\alpha)}[g]_h} \leq \lambda^{(\alpha,\beta)}[f]_g \leq \min\left\{\frac{\lambda^{(\gamma,\beta)}[f]_h}{\lambda^{(\gamma,\alpha)}[g]_h}, \frac{\rho^{(\gamma,\beta)}[f]_h}{\rho^{(\gamma,\alpha)}[g]_h}\right\}$$

$$\leq \max\left\{\frac{\lambda^{(\gamma,\beta)}[f]_h}{\lambda^{(\gamma,\alpha)}[g]_h}, \frac{\rho^{(\gamma,\beta)}[f]_h}{\rho^{(\gamma,\alpha)}[g]_h}\right\} \leq \rho^{(\alpha,\beta)}[f]_g \leq \frac{\rho^{(\gamma,\beta)}[f]_h}{\lambda^{(\gamma,\alpha)}[g]_h}.$$

**Proof.** From the definitions of $\rho^{(\gamma,\beta)}[f]_h$ and $\lambda^{(\gamma,\beta)}[f]_h$, we have for all sufficiently large values of $R$ that

$$M_{h,D}^{-1}(M_{f,D}(R)) \leq \gamma^{-1}\left(\rho^{(\gamma,\beta)}[f]_h + \varepsilon\right)\beta(R))$$
$$i.e., \ M_{f,D}(R) \leq M_{h,D}\left(\gamma^{-1}\left(\left(\rho^{(\gamma,\beta)}[f]_h + \varepsilon\right)\beta(R)\right)\right), \tag{142}$$

$$M_{h,D}^{-1}(M_{f,D}(R)) \geq \gamma^{-1}\left(\left(\lambda^{(\gamma,\beta)}[f]_h - \varepsilon\right)\beta(R)\right)$$
$$i.e., \ M_{f,D}(R) \geq M_{h,D}\left(\gamma^{-1}\left(\left(\lambda^{(\gamma,\beta)}[f]_h - \varepsilon\right)\beta(R)\right)\right). \tag{143}$$

Also for a sequence of values of $R$ tending to infinity, we get that

$$M_{h,D}^{-1}(M_{f,D}(R)) \geq \gamma^{-1}\left(\left(\rho^{(\gamma,\beta)}[f]_h - \varepsilon\right)\beta(R)\right)$$
$$i.e., \ M_{f,D}(R) \geq M_{h,D}\left(\gamma^{-1}\left(\left(\rho^{(\gamma,\beta)}[f]_h - \varepsilon\right)\beta(R)\right)\right), \tag{144}$$
$$M_{h,D}^{-1}(M_{f,D}(R)) \leq \gamma^{-1}\left(\left(\lambda^{(\gamma,\beta)}[f]_h + \varepsilon\right)\beta(R)\right)$$
$$i.e., \ M_{f,D}(R) \leq M_{h,D}\left(\gamma^{-1}\left(\left(\lambda^{(\gamma,\beta)}[f]_h + \varepsilon\right)\beta(R)\right)\right). \tag{145}$$

Similarly from the definitions of $\rho^{(\gamma,\alpha)}[g]_h$ and $\lambda^{(\gamma,\alpha)}[g]_h$, it follows for all sufficiently large values of $R$ that

$$M_{h,D}^{-1}(M_{g,D}(R)) \leq \gamma^{-1}\left(\left(\rho^{(\gamma,\alpha)}[g]_h + \varepsilon\right)\alpha(R)\right)$$
$$i.e., \ M_{g,D}(R) \leq M_{h,D}\left(\gamma^{-1}\left(\left(\rho^{(\gamma,\alpha)}[g]_h + \varepsilon\right)\alpha(R)\right)\right)$$
$$i.e., \ M_{h,D}(R) \geq M_{g,D}\left(\alpha^{-1}\left(\frac{\gamma(R)}{(\rho^{(\gamma,\alpha)}[g]_h + \varepsilon)}\right)\right), \tag{146}$$

$$M_{h,D}^{-1}(M_{g,D}(R)) \geq \gamma^{-1}\left(\left(\lambda^{(\gamma,\alpha)}[g]_h - \varepsilon\right)\alpha(R)\right)$$
$$i.e.,\ M_{g,D}(R) \geq M_{h,D}\left(\gamma^{-1}\left(\left(\lambda^{(\gamma,\alpha)}[g]_h - \varepsilon\right)\alpha(R)\right)\right)$$
$$i.e.,\ M_{h,D}(R) \leq M_{g,D}\left(\alpha^{-1}\left(\frac{\gamma(R)}{(\lambda^{(\gamma,\alpha)}[g]_h - \varepsilon)}\right)\right) \tag{147}$$

and for a sequence of values of $R$ tending to infinity, we obtain that

$$M_{h,D}^{-1}(M_{g,D}(R)) \geq \gamma^{-1}\left(\left(\rho^{(\gamma,\alpha)}[g]_h - \varepsilon\right)\alpha(R)\right)$$
$$i.e.,\ M_{g,D}(R) \geq M_{h,D}\left(\gamma^{-1}\left(\left(\rho^{(\gamma,\alpha)}[g]_h - \varepsilon\right)\alpha(R)\right)\right)$$
$$i.e.,\ M_{h,D}(R) \leq M_{g,D}\left(\alpha^{-1}\left(\frac{\gamma(R)}{(\rho^{(\gamma,\alpha)}[g]_h - \varepsilon)}\right)\right), \tag{148}$$

$$M_{h,D}^{-1}(M_{g,D}(R)) \leq \gamma^{-1}\left(\left(\lambda^{(\gamma,\alpha)}[g]_h + \varepsilon\right)\alpha(R)\right)$$
$$i.e.,\ M_{g,D}(R) \leq M_{h,D}\left(\gamma^{-1}\left(\left(\lambda^{(\gamma,\alpha)}[g]_h + \varepsilon\right)\alpha(R)\right)\right)$$
$$i.e.,\ M_{h,D}(R) \geq M_{g,D}\left(\alpha^{-1}\left(\frac{\gamma(R)}{(\lambda^{(\gamma,\alpha)}[g]_h + \varepsilon)}\right)\right). \tag{149}$$

Now from $(144)$ and in view of $(146)$, we get for a sequence of values of $R$ tending to infinity that

$$\alpha(M_{g,D}^{-1}(M_{f,D}(R))) \geq \alpha\left(M_{g,D}^{-1}\left(M_{h,D}\left(\gamma^{-1}\left(\left(\rho^{(\gamma,\beta)}[f]_h - \varepsilon\right)\beta(R)\right)\right)\right)\right)$$

$$i.e.,\ \alpha(M_{g,D}^{-1}(M_{f,D}(R)))$$
$$\geq \alpha\left(M_{g,D}^{-1}\left(M_{g,D}\left(\alpha^{-1}\left(\frac{\gamma\left(\gamma^{-1}\left(\left(\rho^{(\gamma,\beta)}[f]_h - \varepsilon\right)\beta(R)\right)\right)}{(\rho^{(\gamma,\alpha)}[g]_h + \varepsilon)}\right)\right)\right)\right)$$

$$i.e.,\ \alpha(M_{g,D}^{-1}(M_{f,D}(R))) \geq \frac{\left(\rho^{(\gamma,\beta)}[f]_h - \varepsilon\right)}{(\rho^{(\gamma,\alpha)}[g]_h + \varepsilon)}\beta(R)$$
$$i.e.,\ \frac{\alpha(M_{g,D}^{-1}(M_{f,D}(R)))}{\beta(R)} \geq \frac{\left(\rho^{(\gamma,\beta)}[f]_h - \varepsilon\right)}{(\rho^{(\gamma,\alpha)}[g]_h + \varepsilon)}.$$

As $\varepsilon\,(>0)$ is arbitrary, it follows that

$$\limsup_{R\to\infty}\frac{\alpha(M_{g,D}^{-1}(M_{f,D}(R)))}{\beta(R)} \geq \frac{\rho^{(\gamma,\beta)}[f]_h}{\rho^{(\gamma,\alpha)}[g]_h}$$
$$i.e.,\ \rho^{(\alpha,\beta)}[f]_g \geq \frac{\rho^{(\gamma,\beta)}[f]_h}{\rho^{(\gamma,\alpha)}[g]_h}. \tag{150}$$

Analogously from $(143)$ and in view of $(149)$, it follows for a sequence of values of $R$ tending to infinity that

$$\alpha(M_{g,D}^{-1}(M_{f,D}(R))) \geq \alpha\left(M_{g,D}^{-1}\left(M_{h,D}\left(\gamma^{-1}\left(\left(\lambda^{(\gamma,\beta)}[f]_h - \varepsilon\right)\beta(R)\right)\right)\right)\right)$$

*i.e.,* $\alpha(M_{g,D}^{-1}(M_{f,D}(R)))$

$$\geq \alpha\left(M_{g,D}^{-1}\left(M_{g,D}\left(\alpha^{-1}\left(\frac{\gamma\left(\gamma^{-1}\left(\left(\lambda^{(\gamma,\beta)}[f]_h - \varepsilon\right)\beta(R)\right)\right)}{\left(\lambda^{(\gamma,\alpha)}[g]_h + \varepsilon\right)}\right)\right)\right)\right)$$

*i.e.,* $\alpha(M_{g,D}^{-1}(M_{f,D}(R))) \geq \dfrac{\left(\lambda^{(\gamma,\beta)}[f]_h - \varepsilon\right)}{\left(\lambda^{(\gamma,\alpha)}[g]_h + \varepsilon\right)}\beta(R)$

*i.e.,* $\dfrac{\alpha(M_{g,D}^{-1}(M_{f,D}(R)))}{\beta(R)} \geq \dfrac{\left(\lambda^{(\gamma,\beta)}[f]_h - \varepsilon\right)}{\left(\lambda^{(\gamma,\alpha)}[g]_h + \varepsilon\right)}.$

Since $\varepsilon\,(>0)$ is arbitrary, we get from above that

$$\limsup_{R\to\infty}\frac{\alpha(M_{g,D}^{-1}(M_{f,D}(R)))}{\beta(R)} \geq \frac{\lambda^{(\gamma,\beta)}[f]_h}{\lambda^{(\gamma,\alpha)}[g]_h}$$

$$\text{*i.e.,*} \quad \rho^{(\alpha,\beta)}[f]_g \geq \frac{\lambda^{(\gamma,\beta)}[f]_h}{\lambda^{(\gamma,\alpha)}[g]_h}. \tag{151}$$

Again in view of $(147)$, we have from $(142)$ for all sufficiently large values of $R$ that

$$\alpha(M_{g,D}^{-1}(M_{f,D}(R))) \leq \alpha\left(M_{g,D}^{-1}\left(M_{h,D}\left(\gamma^{-1}\left(\left(\rho^{(\gamma,\beta)}[f]_h + \varepsilon\right)\beta(R)\right)\right)\right)\right)$$

*i.e.,* $\alpha(M_{g,D}^{-1}(M_{f,D}(R)))$

$$\leq \alpha\left(M_{g,D}^{-1}\left(M_{g,D}\left(\alpha^{-1}\left(\frac{\gamma\left(\gamma^{-1}\left(\left(\rho^{(\gamma,\beta)}[f]_h + \varepsilon\right)\beta(R)\right)\right)}{\left(\lambda^{(\gamma,\alpha)}[g]_h - \varepsilon\right)}\right)\right)\right)\right)$$

*i.e.,* $\alpha(M_{g,D}^{-1}(M_{f,D}(R))) \leq \dfrac{\left(\rho^{(\gamma,\beta)}[f]_h + \varepsilon\right)}{\left(\lambda^{(\gamma,\alpha)}[g]_h - \varepsilon\right)}\beta(R)$

*i.e.,* $\dfrac{\alpha(M_{g,D}^{-1}(M_{f,D}(R)))}{\beta(R)} \leq \dfrac{\left(\rho^{(\gamma,\beta)}[f]_h + \varepsilon\right)}{\left(\lambda^{(\gamma,\alpha)}[g]_h - \varepsilon\right)}.$

Since $\varepsilon\,(>0)$ is arbitrary, we obtain that

$$\limsup_{R\to\infty}\frac{\alpha(M_{g,D}^{-1}(M_{f,D}(R)))}{\beta(R)} \leq \frac{\rho^{(\gamma,\beta)}[f]_h}{\lambda^{(\gamma,\alpha)}[g]_h}$$

$$\text{*i.e.,*} \quad \rho^{(\alpha,\beta)}[f]_g \leq \frac{\rho^{(\gamma,\beta)}[f]_h}{\lambda^{(\gamma,\alpha)}[g]_h}. \tag{152}$$

Again from $(143)$ and in view of $(146)$, we get for all sufficiently large values of $R$ that

$$\alpha(M_{g,D}^{-1}(M_{f,D}(R))) \geq \alpha\left(M_{g,D}^{-1}\left(M_{h,D}\left(\gamma^{-1}\left(\left(\lambda^{(\gamma,\beta)}[f]_h - \varepsilon\right)\beta(R)\right)\right)\right)\right)$$

$$i.e., \ \alpha(M_{g,D}^{-1}(M_{f,D}(R)))$$

$$\geq \alpha \left( M_{g,D}^{-1} \left( M_{g,D} \left( \alpha^{-1} \left( \frac{\gamma \left( \gamma^{-1} \left( \left( \lambda^{(\gamma,\beta)}[f]_h - \varepsilon \right) \beta(R) \right) \right)}{(\rho^{(\gamma,\alpha)}[g]_h + \varepsilon)} \right) \right) \right) \right)$$

$$i.e., \ \alpha(M_{g,D}^{-1}(M_{f,D}(R))) \geq \frac{\left( \lambda^{(\gamma,\beta)}[f]_h - \varepsilon \right)}{(\rho^{(\gamma,\alpha)}[g]_h + \varepsilon)} \beta(R)$$

$$i.e., \ \frac{\alpha(M_{g,D}^{-1}(M_{f,D}(R)))}{\beta(R)} \geq \frac{\left( \lambda^{(\gamma,\beta)}[f]_h - \varepsilon \right)}{(\rho^{(\gamma,\alpha)}[g]_h + \varepsilon)}.$$

As $\varepsilon \, (> 0)$ is arbitrary, it follows from above that

$$\liminf_{R \to \infty} \frac{\alpha(M_{g,D}^{-1}(M_{f,D}(R)))}{\beta(R)} \geq \frac{\lambda^{(\gamma,\beta)}[f]_h}{\rho^{(\gamma,\alpha)}[g]_h}$$

$$i.e., \ \lambda^{(\alpha,\beta)}[f]_g \geq \frac{\lambda^{(\gamma,\beta)}[f]_h}{\rho^{(\gamma,\alpha)}[g]_h}. \tag{153}$$

Also in view of (148), we get from (142) for a sequence of values of $R$ tending to infinity that

$$\alpha(M_{g,D}^{-1}(M_{f,D}(R))) \leq \alpha \left( M_{g,D}^{-1} \left( M_{h,D} \left( \gamma^{-1} \left( \left( \rho^{(\gamma,\beta)}[f]_h + \varepsilon \right) \beta(R) \right) \right) \right) \right)$$

$$i.e., \ \alpha(M_{g,D}^{-1}(M_{f,D}(R)))$$

$$\leq \alpha \left( M_{g,D}^{-1} \left( M_{g,D} \left( \alpha^{-1} \left( \frac{\gamma \left( \gamma^{-1} \left( \left( \rho^{(\gamma,\beta)}[f]_h + \varepsilon \right) \beta(R) \right) \right)}{(\rho^{(\gamma,\alpha)}[g]_h - \varepsilon)} \right) \right) \right) \right)$$

$$i.e., \ \alpha(M_{g,D}^{-1}(M_{f,D}(R))) \leq \frac{\left( \rho^{(\gamma,\beta)}[f]_h + \varepsilon \right)}{(\rho^{(\gamma,\alpha)}[g]_h - \varepsilon)} \beta(R)$$

$$i.e., \ \frac{\alpha(M_{g,D}^{-1}(M_{f,D}(R)))}{\beta(R)} \leq \frac{\left( \rho^{(\gamma,\beta)}[f]_h + \varepsilon \right)}{(\rho^{(\gamma,\alpha)}[g]_h - \varepsilon)}.$$

Since $\varepsilon \, (> 0)$ is arbitrary, we get from above that

$$\liminf_{R \to \infty} \frac{\alpha(M_{g,D}^{-1}(M_{f,D}(R)))}{\beta(R)} \leq \frac{\rho^{(\gamma,\beta)}[f]_h}{\rho^{(\gamma,\alpha)}[g]_h}$$

$$i.e., \ \lambda^{(\alpha,\beta)}[f]_g \leq \frac{\rho^{(\gamma,\beta)}[f]_h}{\rho^{(\gamma,\alpha)}[g]_h}. \tag{154}$$

Similarly from (145) and in view of (147), it follows for a sequence of values of $R$ tending to infinity we get that

$$\alpha(M_{g,D}^{-1}(M_{f,D}(R))) \leq \alpha \left( M_{g,D}^{-1} \left( M_{h,D} \left( \gamma^{-1} \left( \left( \lambda^{(\gamma,\beta)}[f]_h + \varepsilon \right) \beta(R) \right) \right) \right) \right)$$

i.e., $\alpha(M_{g,D}^{-1}(M_{f,D}(R)))$

$$\leq \alpha \left( M_{g,D}^{-1} \left( M_{g,D} \left( \alpha^{-1} \left( \frac{\gamma \left( \gamma^{-1} \left( \left( \lambda^{(\gamma,\beta)}[f]_h + \varepsilon \right) \beta(R) \right) \right)}{\left( \lambda^{(\gamma,\alpha)}[g]_h - \varepsilon \right)} \right) \right) \right) \right)$$

$$\text{i.e., } \alpha(M_{g,D}^{-1}(M_{f,D}(R))) \leq \frac{\left( \lambda^{(\gamma,\beta)}[f]_h + \varepsilon \right)}{\left( \lambda^{(\gamma,\alpha)}[g]_h - \varepsilon \right)} \beta(R)$$

$$\text{i.e., } \frac{\alpha(M_{g,D}^{-1}(M_{f,D}(R)))}{\beta(R)} \leq \frac{\left( \lambda^{(\gamma,\beta)}[f]_h + \varepsilon \right)}{\left( \lambda^{(\gamma,\alpha)}[g]_h - \varepsilon \right)}.$$

As $\varepsilon \, (> 0)$ is arbitrary, we obtain from above that

$$\liminf_{R \to \infty} \frac{\alpha(M_{g,D}^{-1}(M_{f,D}(R)))}{\beta(R)} \leq \frac{\lambda^{(\gamma,\beta)}[f]_h}{\lambda^{(\gamma,\alpha)}[g]_h}$$

$$\text{i.e., } \lambda^{(\alpha,\beta)}[f]_g \leq \frac{\lambda^{(\gamma,\beta)}[f]_h}{\lambda^{(\gamma,\alpha)}[g]_h}. \tag{155}$$

Thus the theorem follows from $(150)$, $(151)$, $(152)$, $(153)$, $(154)$ and $(155)$. ∎

In view of Theorem 7.2.1, one can easily verify the following corollaries:

**Corollary 7.2.1** *Let $f(z)$, $g(z)$ and $h(z)$ be three entire functions of $n$ complex variables such that $0 < \lambda^{(\gamma,\beta)}[f]_h = \rho^{(\gamma,\beta)}[f]_h < \infty$ and $0 < \lambda^{(\gamma,\alpha)}[g]_h \leq \rho^{(\gamma,\alpha)}[g]_h < \infty$. Then*

$$\lambda^{(\alpha,\beta)}[f]_g = \frac{\rho^{(\gamma,\beta)}[f]_h}{\rho^{(\gamma,\alpha)}[g]_h} \quad \text{and} \quad \rho^{(\alpha,\beta)}[f]_g = \frac{\rho^{(\gamma,\beta)}[f]_h}{\lambda^{(\gamma,\alpha)}[g]_h}.$$

*In addition, if $\rho^{(\gamma,\beta)}[f]_h = \rho^{(\gamma,\alpha)}[g]_h$, then*

$$\lambda^{(\alpha,\beta)}[f]_g = \rho^{(\beta,\alpha)}[g]_f = 1.$$

**Corollary 7.2.2** *Let $f(z)$, $g(z)$ and $h(z)$ be three entire functions of $n$ complex variables such that $0 < \lambda^{(\gamma,\beta)}[f]_h \leq \rho^{(\gamma,\beta)}[f]_h < \infty$ and $0 < \lambda^{(\gamma,\alpha)}[g]_h = \rho^{(\gamma,\alpha)}[g]_h < \infty$. Then*

$$\lambda^{(\alpha,\beta)}[f]_g = \frac{\lambda^{(\gamma,\beta)}[f]_h}{\rho^{(\gamma,\alpha)}[g]_h} \quad \text{and} \quad \rho^{(\alpha,\beta)}[f]_g = \frac{\rho^{(\gamma,\beta)}[f]_h}{\rho^{(\gamma,\alpha)}[g]_h}.$$

*In addition, if $\rho^{(\gamma,\beta)}[f]_h = \rho^{(\gamma,\alpha)}[g]_h$, then*

$$\rho^{(\alpha,\beta)}[f]_g = \lambda^{(\beta,\alpha)}[g]_f = 1.$$

**Corollary 7.2.3** *Let $f(z)$, $g(z)$ and $h(z)$ be three entire functions of $n$ complex variables such that $0 < \lambda^{(\gamma,\beta)}[f]_h = \rho^{(\gamma,\beta)}[f]_h < \infty$ and $0 < \lambda^{(\gamma,\alpha)}[g]_h = \rho^{(\gamma,\alpha)}[g]_h < \infty$. Then*

$$\lambda^{(\alpha,\beta)}[f]_g = \rho^{(\alpha,\beta)}[f]_g = \frac{\rho^{(\gamma,\beta)}[f]_h}{\rho^{(\gamma,\alpha)}[g]_h}.$$

**Corollary 7.2.4** *Let $f(z)$, $g(z)$ and $h(z)$ be three entire functions of $n$ complex variables such that $0 < \lambda^{(\gamma,\beta)}[f]_h = \rho^{(\gamma,\beta)}[f]_h < \infty$ and $0 < \lambda^{(\gamma,\alpha)}[g]_h = \rho^{(\gamma,\alpha)}[g]_h < \infty$. Also suppose that $\rho^{(\gamma,\beta)}[f]_h = \rho^{(\gamma,\alpha)}[g]_h$. Then*

$$\lambda^{(\alpha,\beta)}[f]_g = \rho^{(\alpha,\beta)}[f]_g = \lambda^{(\beta,\alpha)}[g]_f = \rho^{(\beta,\alpha)}[g]_f = 1.$$

**Corollary 7.2.5** *Let $f(z)$, $g(z)$ and $h(z)$ be three entire functions of $n$ complex variables and either $0 < \lambda^{(\gamma,\beta)}[f]_h < \rho^{(\gamma,\beta)}[f]_h < \infty$ or $0 < \lambda^{(\gamma,\alpha)}[g]_h < \rho^{(\gamma,\alpha)}[g]_h < \infty$. Then*

$$\rho^{(\alpha,\beta)}[f]_g \cdot \rho^{(\beta,\alpha)}[g]_f \geq 1.$$

*If $0 < \lambda^{(\gamma,\beta)}[f]_h = \rho^{(\gamma,\beta)}[f]_h < \infty$ and $0 < \lambda^{(\gamma,\alpha)}[g]_h = \rho^{(\gamma,\alpha)}[g]_h < \infty$, then*

$$\rho^{(\alpha,\beta)}[f]_g \cdot \rho^{(\beta,\alpha)}[g]_f = 1.$$

**Corollary 7.2.6** *Let $f(z)$, $g(z)$ and $h(z)$ be three entire functions of $n$ complex variables and either $0 < \lambda^{(\gamma,\beta)}[f]_h < \rho^{(\gamma,\beta)}[f]_h < \infty$ or $0 < \lambda^{(\gamma,\alpha)}[g]_h < \rho^{(\gamma,\alpha)}[g]_h < \infty$. Then*

$$\lambda^{(\alpha,\beta)}[f]_g \cdot \lambda^{(\beta,\alpha)}[g]_f \leq 1.$$

*If $0 < \lambda^{(\gamma,\beta)}[f]_h = \rho^{(\gamma,\beta)}[f]_h < \infty$ and $0 < \lambda^{(\gamma,\alpha)}[g]_h = \rho^{(\gamma,\alpha)}[g]_h < \infty$, then*

$$\lambda^{(\alpha,\beta)}[f]_g \cdot \lambda^{(\beta,\alpha)}[g]_f = 1.$$

**Corollary 7.2.7** *Let $f(z)$, $g(z)$ and $h(z)$ be three entire functions of $n$ complex variables such that $0 < \lambda^{(\gamma,\beta)}[f]_h \leq \rho^{(\gamma,\beta)}[f]_h < \infty$ and $\lambda^{(\gamma,\alpha)}[g]_h \leq \rho^{(\gamma,\alpha)}[g]_h$. Then*

$$(i) \quad \lambda^{(\alpha,\beta)}[f]_g = \infty \text{ when } \rho^{(\gamma,\alpha)}[g]_h = 0,$$
$$(ii) \quad \rho^{(\alpha,\beta)}[f]_g = \infty \text{ when } \lambda^{(\gamma,\alpha)}[g]_h = 0,$$
$$(iii) \quad \lambda^{(\alpha,\beta)}[f]_g = 0 \text{ when } \rho^{(\gamma,\alpha)}[g]_h = \infty$$

*and*

$$(iv) \quad \rho^{(\alpha,\beta)}[f]_g = 0 \text{ when } \lambda^{(\gamma,\alpha)}[g]_h = \infty.$$

**Corollary 7.2.8** *Let $f(z)$, $g(z)$ and $h(z)$ be three entire functions of $n$ complex variables such that $\lambda^{(\gamma,\beta)}[f]_h \leq \rho^{(\gamma,\beta)}[f]_h$ and $0 < \lambda^{(\gamma,\alpha)}[g]_h \leq \rho^{(\gamma,\alpha)}[g]_h < \infty$. Then*

$$(i) \quad \rho^{(\alpha,\beta)}[f]_g = 0 \text{ when } \rho^{(\gamma,\beta)}[f]_h = 0,$$
$$(ii) \quad \lambda^{(\alpha,\beta)}[f]_g = 0 \text{ when } \lambda^{(\gamma,\beta)}[f]_h = 0,$$
$$(iii) \quad \rho^{(\alpha,\beta)}[f]_g = \infty \text{ when } \rho^{(\gamma,\beta)}[f]_h = \infty$$

*and*

$$(iv) \quad \lambda^{(\alpha,\beta)}[f]_g = \infty \text{ when } \lambda^{(\gamma,\beta)}[f]_h = \infty.$$

**Corollary 7.2.9** *Let $f(z)$, $g(z)$ and $h(z)$ be three entire functions of $n$ complex variables such that $0 < \lambda^{(\gamma,\beta)}[f]_h \leq \rho^{(\gamma,\beta)}[f]_h < \infty$ and $0 < \lambda^{(\gamma,\alpha)}[g]_h \leq \rho^{(\gamma,\alpha)}[g]_h < \infty$. Then*

$$\rho^{(\alpha,\beta)}[f]_g = \frac{\rho^{(\gamma,\beta)}[f]_h}{\rho^{(\gamma,\alpha)}[g]_h} \quad and \quad \lambda^{(\alpha,\beta)}[f]_g = \frac{\lambda^{(\gamma,\beta)}[f]_h}{\lambda^{(\gamma,\alpha)}[g]_h}$$

*when $\lambda^{(\gamma,\alpha)}[g]_h = \rho^{(\gamma,\alpha)}[g]_h$.*

**Corollary 7.2.10** *Let $f(z)$, $g(z)$ and $h(z)$ be three entire functions of $n$ complex variables such that $0 < \lambda^{(\gamma,\beta)}[f]_h \leq \rho^{(\gamma,\beta)}[f]_h < \infty$ and $0 < \lambda^{(\gamma,\alpha)}[g]_h \leq \rho^{(\gamma,\alpha)}[g]_h < \infty$. Then*

$$\rho^{(\alpha,\beta)}[f]_g = \frac{\lambda^{(\gamma,\beta)}[f]_h}{\lambda^{(\gamma,\alpha)}[g]_h} \quad and \quad \lambda^{(\alpha,\beta)}[f]_g = \frac{\rho^{(\gamma,\beta)}[f]_h}{\rho^{(\gamma,\alpha)}[g]_h}$$

*when $\lambda^{(\gamma,\beta)}[f]_h = \rho^{(\gamma,\beta)}[f]_h$.*

Next we prove our results for generalized relative Gol'dberg type $(\alpha,\beta)$, generalized relative Gol'dberg weak type $(\alpha,\beta)$ etc.. We include the proof of the first main Theorem 7.2.2 for the sake of completeness. The others are basically omitted since they are easily proven with the same techniques or with some easy reasoning.

**Theorem 7.2.2** *Let $f(z)$, $g(z)$ and $h(z)$ be three entire functions of $n$ complex variables such that $0 < \lambda^{(\gamma,\beta)}[f]_h \leq \rho^{(\gamma,\beta)}[f]_h < \infty$ and $0 < \lambda^{(\gamma,\alpha)}[g]_h \leq \rho^{(\gamma,\alpha)}[g]_h < \infty$. Also let $\lambda^{(\gamma,\alpha)}[g]_h = \rho^{(\gamma,\alpha)}[g]_h$. Then*

$$\left[ \frac{\overline{\sigma}_D^{(\gamma,\beta)}[f]_h}{\sigma_D^{(\gamma,\alpha)}[g]_h} \right]^{\frac{1}{\rho^{(\gamma,\alpha)}[g]_h}} \leq \overline{\sigma}_D^{(\alpha,\beta)}[f]_g \leq \min \left\{ \left[ \frac{\overline{\sigma}_D^{(\gamma,\beta)}[f]_h}{\overline{\sigma}_D^{(\gamma,\alpha)}[g]_h} \right]^{\frac{1}{\rho^{(\gamma,\alpha)}[g]_h}}, \left[ \frac{\sigma_D^{(\gamma,\beta)}[f]_h}{\sigma_D^{(\gamma,\alpha)}[g]_h} \right]^{\frac{1}{\rho^{(\gamma,\alpha)}[g]_h}} \right\}$$

$$\leq \max \left\{ \left[ \frac{\overline{\sigma}_D^{(\gamma,\beta)}[f]_h}{\overline{\sigma}_D^{(\gamma,\alpha)}[g]_h} \right]^{\frac{1}{\rho^{(\gamma,\alpha)}[g]_h}}, \left[ \frac{\sigma_D^{(\gamma,\beta)}[f]_h}{\sigma_D^{(\gamma,\alpha)}[g]_h} \right]^{\frac{1}{\rho^{(\gamma,\alpha)}[g]_h}} \right\} \leq \sigma_D^{(\alpha,\beta)}[f]_g \leq \left[ \frac{\sigma_D^{(\gamma,\beta)}[f]_h}{\overline{\sigma}_D^{(\gamma,\alpha)}[g]_h} \right]^{\frac{1}{\rho^{(\gamma,\alpha)}[g]_h}}.$$

**Proof.** From the definitions of $\sigma_D^{(\gamma,\beta)}[f]_h$ and $\overline{\sigma}_D^{(\gamma,\beta)}[f]_h$, we have for all sufficiently large values of $R$ that

$$M_{h,D}^{-1}(M_{f,D}(R)) \leq \gamma^{-1}\left( \log\left( \left( \sigma_D^{(\gamma,\beta)}[f]_h + \varepsilon \right) (\exp(\beta(R)))^{\rho^{(\gamma,\beta)}[f]_h} \right) \right),$$

$$i.e., \ M_{f,D}(R) \leq M_{h,D}\left( \gamma^{-1}\left( \log\left( \left( \sigma_D^{(\gamma,\beta)}[f]_h + \varepsilon \right) (\exp(\beta(R)))^{\rho^{(\gamma,\beta)}[f]_h} \right) \right) \right) \quad (156)$$

and

$$M_{h,D}^{-1}(M_{f,D}(R)) \geq \gamma^{-1}\left( \log\left( \left( \overline{\sigma}_D^{(\gamma,\beta)}[f]_h - \varepsilon \right) (\exp(\beta(R)))^{\rho^{(\gamma,\beta)}[f]_h} \right) \right)$$

$$i.e., \ M_{f,D}(R) \geq M_{h,D}\left( \gamma^{-1}\left( \log\left( \left( \overline{\sigma}_D^{(\gamma,\beta)}[f]_h - \varepsilon \right) (\exp(\beta(R)))^{\rho^{(\gamma,\beta)}[f]_h} \right) \right) \right). \quad (157)$$

Also for a sequence of values of $R$ tending to infinity, we get that

$$M_{h,D}^{-1}(M_{f,D}(R)) \geq \gamma^{-1}\left(\log\left(\left(\sigma_D^{(\gamma,\beta)}[f]_h - \varepsilon\right)(\exp(\beta(R)))^{\rho^{(\gamma,\beta)}[f]_h}\right)\right)$$

$$i.e., \ M_{f,D}(R) \geq M_{h,D}\left(\gamma^{-1}\left(\log\left(\left(\sigma_D^{(\gamma,\beta)}[f]_h - \varepsilon\right)(\exp(\beta(R)))^{\rho^{(\gamma,\beta)}[f]_h}\right)\right)\right) \quad (158)$$

and

$$M_{h,D}^{-1}(M_{f,D}(R)) \leq \gamma^{-1}\left(\log\left(\left(\overline{\sigma}_D^{(\gamma,\beta)}[f]_h + \varepsilon\right)(\exp(\beta(R)))^{\rho^{(\gamma,\beta)}[f]_h}\right)\right)$$

$$i.e., \ M_{f,D}(R) \leq M_{h,D}\left(\gamma^{-1}\left(\log\left(\left(\overline{\sigma}_D^{(\gamma,\beta)}[f]_h + \varepsilon\right)(\exp(\beta(R)))^{\rho^{(\gamma,\beta)}[f]_h}\right)\right)\right). \quad (159)$$

Similarly from the definitions of $\sigma_D^{(\gamma,\alpha)}[g]_h$ and $\overline{\sigma}_D^{(\gamma,\alpha)}[g]_h$, it follows for all sufficiently large values of $R$ that

$$M_{h,D}^{-1}(M_{g,D}(R)) \leq \gamma^{-1}\left(\log\left(\left(\sigma_D^{(\gamma,\alpha)}[g]_h + \varepsilon\right)(\exp\alpha(R))^{\rho^{(\gamma,\alpha)}[g]_h}\right)\right)$$

$$i.e., \ M_{g,D}(R) \leq M_{h,D}\left(\gamma^{-1}\left(\log\left(\left(\sigma_D^{(\gamma,\alpha)}[g]_h + \varepsilon\right)(\exp\alpha(R))^{\rho^{(\gamma,\alpha)}[g]_h}\right)\right)\right)$$

$$i.e., \ M_{h,D}(R) \geq M_{g,D}\left(\alpha^{-1}\left(\log\left(\left(\frac{\exp(\gamma(R))}{\left(\sigma_D^{(\gamma,\alpha)}[g]_h + \varepsilon\right)}\right)^{\frac{1}{\rho^{(\gamma,\alpha)}[g]_h}}\right)\right)\right), \quad (160)$$

$$M_{h,D}^{-1}(M_{g,D}(R)) \geq \gamma^{-1}\left(\log\left(\left(\overline{\sigma}_D^{(\gamma,\alpha)}[g]_h - \varepsilon\right)(\exp\alpha(R))^{\rho^{(\gamma,\alpha)}[g]_h}\right)\right)$$

$$i.e., \ M_{g,D}(R) \geq M_{h,D}\left(\gamma^{-1}\left(\log\left(\left(\overline{\sigma}_D^{(\gamma,\alpha)}[g]_h - \varepsilon\right)(\exp\alpha(R))^{\rho^{(\gamma,\alpha)}[g]_h}\right)\right)\right)$$

$$i.e., \ M_{h,D}(R) \leq M_{g,D}\left(\alpha^{-1}\left(\log\left(\left(\frac{\exp(\gamma(R))}{\left(\overline{\sigma}_D^{(\gamma,\alpha)}[g]_h - \varepsilon\right)}\right)^{\frac{1}{\rho^{(\gamma,\alpha)}[g]_h}}\right)\right)\right) \quad (161)$$

and for a sequence of values of $R$ tending to infinity we obtain that

$$M_{h,D}^{-1}(M_{g,D}(R)) \geq \gamma^{-1}\left(\log\left(\left(\sigma_D^{(\gamma,\alpha)}[g]_h - \varepsilon\right)(\exp\alpha(R))^{\rho^{(\gamma,\alpha)}[g]_h}\right)\right)$$

$$i.e., \ M_{g,D}(R) \geq M_{h,D}\left(\gamma^{-1}\left(\log\left(\left(\sigma_D^{(\gamma,\alpha)}[g]_h - \varepsilon\right)(\exp\alpha(R))^{\rho^{(\gamma,\alpha)}[g]_h}\right)\right)\right)$$

$$i.e., \ M_{h,D}(R) \leq M_{g,D}\left(\alpha^{-1}\left(\log\left(\left(\frac{\exp(\gamma(R))}{\left(\sigma_D^{(\gamma,\alpha)}[g]_h - \varepsilon\right)}\right)^{\frac{1}{\rho^{(\gamma,\alpha)}[g]_h}}\right)\right)\right), \quad (162)$$

$$M_{h,D}^{-1}(M_{g,D}(R)) \leq \gamma^{-1}\left(\log\left(\left(\overline{\sigma}_D^{(\gamma,\alpha)}[g]_h + \varepsilon\right)(\exp\alpha(R))^{\rho^{(\gamma,\alpha)}[g]_h}\right)\right)$$

$$i.e., \ M_{g,D}(R) \leq M_{h,D}\left(\gamma^{-1}\left(\log\left(\left(\overline{\sigma}_D^{(\gamma,\alpha)}[g]_h + \varepsilon\right)(\exp\alpha(R))^{\rho^{(\gamma,\alpha)}[g]_h}\right)\right)\right)$$

$$i.e., \ M_{h,D}(R) \geq M_{g,D}\left(\alpha^{-1}\left(\log\left(\left(\frac{\exp(\gamma(R))}{\left(\overline{\sigma}_D^{(\gamma,\alpha)}[g]_h + \varepsilon\right)}\right)^{\frac{1}{\rho^{(\gamma,\alpha)}[g]_h}}\right)\right)\right). \quad (163)$$

Now from (158) and in view of (160), we get for a sequence of values of $R$ tending to infinity we get that

$$\exp(\alpha(M_{g,D}^{-1}(M_{f,D}(R)))) \geq$$

$$\exp\left(\alpha\left(M_{g,D}^{-1}\left(M_{h,D}\left(\gamma^{-1}\left(\log\left(\left(\sigma_D^{(\gamma,\beta)}[f]_h - \varepsilon\right)(\exp(\beta(R)))^{\rho^{(\gamma,\beta)}[f]_h}\right)\right)\right)\right)\right)$$

$$i.e., \ \exp(\alpha(M_{g,D}^{-1}(M_{f,D}(R)))) \geq \left(\frac{\left(\left(\sigma_D^{(\gamma,\beta)}[f]_h - \varepsilon\right)(\exp(\beta(R)))^{\rho^{(\gamma,\beta)}[f]_h}\right)}{\left(\sigma_D^{(\gamma,\alpha)}[g]_h + \varepsilon\right)}\right)^{\frac{1}{\rho^{(\gamma,\alpha)}[g]_h}}$$

$$i.e., \ \exp(\alpha(M_{g,D}^{-1}(M_{f,D}(R)))) \geq \left[\frac{\left(\sigma_D^{(\gamma,\beta)}[f]_h - \varepsilon\right)}{\left(\sigma_D^{(\gamma,\alpha)}[g]_h + \varepsilon\right)}\right]^{\frac{1}{\rho^{(\gamma,\alpha)}[g]_h}} (\exp(\beta(R)))^{\frac{\rho^{(\gamma,\beta)}[f]_h}{\rho^{(\gamma,\alpha)}[g]_h}}$$

$$i.e., \ \frac{\exp(\alpha(M_{g,D}^{-1}(M_{f,D}(R))))}{(\exp(\beta(R)))^{\frac{\rho^{(\gamma,\beta)}[f]_h}{\rho^{(\gamma,\alpha)}[g]_h}}} \geq \left[\frac{\left(\sigma_D^{(\gamma,\beta)}[f]_h - \varepsilon\right)}{\left(\sigma_D^{(\gamma,\alpha)}[g]_h + \varepsilon\right)}\right]^{\frac{1}{\rho^{(\gamma,\alpha)}[g]_h}}.$$

As $\varepsilon \, (> 0)$ is arbitrary, in view of Corollary 7.2.9 it follows that

$$\limsup_{R\to\infty} \frac{\exp(\alpha(M_{g,D}^{-1}(M_{f,D}(R))))}{(\exp(\beta(R)))^{\rho^{(\alpha,\beta)}[f]_g}} \geq \left[\frac{\sigma_D^{(\gamma,\beta)}[f]_h}{\sigma_D^{(\gamma,\alpha)}[g]_h}\right]^{\frac{1}{\rho^{(\gamma,\alpha)}[g]_h}}$$

$$i.e., \ \sigma_D^{(\alpha,\beta)}[f]_g \geq \left[\frac{\sigma_D^{(\gamma,\beta)}[f]_h}{\sigma_D^{(\gamma,\alpha)}[g]_h}\right]^{\frac{1}{\rho^{(\gamma,\alpha)}[g]_h}}. \tag{164}$$

Analogously from (157) and in view of (163), it follows for a sequence of values of $R$ tending to infinity that

$$\exp(\alpha(M_{g,D}^{-1}(M_{f,D}(R)))) \geq$$

$$\exp\left(\alpha\left(M_{g,D}^{-1}\left(M_{h,D}\left(\gamma^{-1}\left(\log\left(\left(\overline{\sigma}_D^{(\gamma,\beta)}[f]_h - \varepsilon\right)(\exp(\beta(R)))^{\rho^{(\gamma,\beta)}[f]_h}\right)\right)\right)\right)\right)$$

$$i.e., \ \exp(\alpha(M_{g,D}^{-1}(M_{f,D}(R)))) \geq \left(\frac{\left(\left(\overline{\sigma}_D^{(\gamma,\beta)}[f]_h - \varepsilon\right)(\exp(\beta(R)))^{\rho^{(\gamma,\beta)}[f]_h}\right)}{\left(\overline{\sigma}_D^{(\gamma,\alpha)}[g]_h + \varepsilon\right)}\right)^{\frac{1}{\rho^{(\gamma,\alpha)}[g]_h}}$$

$$i.e., \ \exp(\alpha(M_{g,D}^{-1}(M_{f,D}(R)))) \geq \left[\frac{\left(\overline{\sigma}_D^{(\gamma,\beta)}[f]_h - \varepsilon\right)}{\left(\overline{\sigma}_D^{(\gamma,\alpha)}[g]_h + \varepsilon\right)}\right]^{\frac{1}{\rho^{(\gamma,\alpha)}[g]_h}} (\exp(\beta(R)))^{\frac{\rho^{(\gamma,\beta)}[f]_h}{\rho^{(\gamma,\alpha)}[g]_h}}$$

$$i.e., \ \frac{\exp(\alpha(M_{g,D}^{-1}(M_{f,D}(R))))}{(\exp(\beta(R)))^{\frac{\rho^{(\gamma,\beta)}[f]_h}{\rho^{(\gamma,\alpha)}[g]_h}}} \geq \left[\frac{\left(\overline{\sigma}_D^{(\gamma,\beta)}[f]_h - \varepsilon\right)}{\left(\overline{\sigma}_D^{(\gamma,\alpha)}[g]_h + \varepsilon\right)}\right]^{\frac{1}{\rho^{(\gamma,\alpha)}[g]_h}}.$$

Since $\varepsilon\,(>0)$ is arbitrary, we get from above and Corollary 7.2.9 that

$$\limsup_{R\to\infty}\frac{\exp(\alpha(M_{g,D}^{-1}(M_{f,D}(R))))}{(\exp(\beta(R)))^{\rho^{(\alpha,\beta)}[f]_g}} \geq \left[\frac{\overline{\sigma}_D^{(\gamma,\beta)}[f]_h}{\overline{\sigma}_D^{(\gamma,\alpha)}[g]_h}\right]^{\frac{1}{\rho^{(\gamma,\alpha)}[g]_h}}$$

$$i.e.,\ \sigma_D^{(\alpha,\beta)}[f]_g \geq \left[\frac{\overline{\sigma}_D^{(\gamma,\beta)}[f]_h}{\overline{\sigma}_D^{(\gamma,\alpha)}[g]_h}\right]^{\frac{1}{\rho^{(\gamma,\alpha)}[g]_h}}. \tag{165}$$

Again in view of (161), we have from (156) for all sufficiently large values of $R$ that

$$\exp(\alpha(M_{g,D}^{-1}(M_{f,D}(R)))) \leq$$

$$\exp\left(\alpha\left(M_{g,D}^{-1}\left(M_{h,D}\left(\gamma^{-1}\left(\log\left(\left(\sigma_D^{(\gamma,\beta)}[f]_h + \varepsilon\right)(\exp(\beta(R)))^{\rho^{(\gamma,\beta)}[f]_h}\right)\right)\right)\right)\right)$$

$$i.e.,\ \exp(\alpha(M_{g,D}^{-1}(M_{f,D}(R)))) \leq \left(\frac{\left(\left(\sigma_D^{(\gamma,\beta)}[f]_h + \varepsilon\right)(\exp(\beta(R)))^{\rho^{(\gamma,\beta)}[f]_h}\right)}{\left(\overline{\sigma}_D^{(\gamma,\alpha)}[g]_h - \varepsilon\right)}\right)^{\frac{1}{\rho^{(\gamma,\alpha)}[g]_h}}$$

$$i.e.,\ \exp(\alpha(M_{g,D}^{-1}(M_{f,D}(R)))) \leq \left[\frac{\left(\sigma_D^{(\gamma,\beta)}[f]_h + \varepsilon\right)}{\left(\overline{\sigma}_D^{(\gamma,\alpha)}[g]_h - \varepsilon\right)}\right]^{\frac{1}{\rho^{(\gamma,\alpha)}[g]_h}}(\exp(\beta(R)))^{\frac{\rho^{(\gamma,\beta)}[f]_h}{\rho^{(\gamma,\alpha)}[g]_h}}$$

$$i.e.,\ \frac{\exp(\alpha(M_{g,D}^{-1}(M_{f,D}(R))))}{(\exp(\beta(R)))^{\frac{\rho^{(\gamma,\beta)}[f]_h}{\rho^{(\gamma,\alpha)}[g]_h}}} \leq \left[\frac{\left(\sigma_D^{(\gamma,\beta)}[f]_h + \varepsilon\right)}{\left(\overline{\sigma}_D^{(\gamma,\alpha)}[g]_h - \varepsilon\right)}\right]^{\frac{1}{\rho^{(\gamma,\alpha)}[g]_h}}.$$

Since $\varepsilon\,(>0)$ is arbitrary, we obtain in view of Corollary 7.2.9 that

$$\limsup_{R\to\infty}\frac{\exp(\alpha(M_{g,D}^{-1}(M_{f,D}(R))))}{(\exp(\beta(R)))^{\rho^{(\alpha,\beta)}[f]_g}} \leq \left[\frac{\sigma_D^{(\gamma,\beta)}[f]_h}{\overline{\sigma}_D^{(\gamma,\alpha)}[g]_h}\right]^{\frac{1}{\rho^{(\gamma,\alpha)}[g]_h}}$$

$$i.e.,\ \sigma_D^{(\alpha,\beta)}[f]_g \leq \left[\frac{\sigma_D^{(\gamma,\beta)}[f]_h}{\overline{\sigma}_D^{(\gamma,\alpha)}[g]_h}\right]^{\frac{1}{\rho^{(\gamma,\alpha)}[g]_h}}. \tag{166}$$

Again from (157) and in view of (160), we get for all sufficiently large values of $R$ that

$$\exp(\alpha(M_{g,D}^{-1}(M_{f,D}(R)))) \geq$$

$$\exp\left(\alpha\left(M_{g,D}^{-1}\left(M_{h,D}\left(\gamma^{-1}\left(\log\left(\left(\overline{\sigma}_D^{(\gamma,\beta)}[f]_h - \varepsilon\right)(\exp(\beta(R)))^{\rho^{(\gamma,\beta)}[f]_h}\right)\right)\right)\right)\right)$$

$$i.e.,\ \exp(\alpha(M_{g,D}^{-1}(M_{f,D}(R)))) \geq \left(\frac{\left(\left(\overline{\sigma}_D^{(\gamma,\beta)}[f]_h - \varepsilon\right)(\exp(\beta(R)))^{\rho^{(\gamma,\beta)}[f]_h}\right)}{\left(\sigma_D^{(\gamma,\alpha)}[g]_h + \varepsilon\right)}\right)^{\frac{1}{\rho^{(\gamma,\alpha)}[g]_h}}$$

$$i.e., \quad \exp(\alpha(M_{g,D}^{-1}(M_{f,D}(R)))) \geq \left[\frac{\left(\overline{\sigma}_D^{(\gamma,\beta)}[f]_h - \varepsilon\right)}{\left(\sigma_D^{(\gamma,\alpha)}[g]_h + \varepsilon\right)}\right]^{\frac{1}{\rho^{(\gamma,\alpha)}[g]_h}} (\exp(\beta(R)))^{\frac{\rho^{(\gamma,\beta)}[f]_h}{\rho^{(\gamma,\alpha)}[g]_h}}$$

$$i.e., \quad \frac{\exp(\alpha(M_{g,D}^{-1}(M_{f,D}(R))))}{(\exp(\beta(R)))^{\frac{\rho^{(\gamma,\beta)}[f]_h}{\rho^{(\gamma,\alpha)}[g]_h}}} \geq \left[\frac{\left(\overline{\sigma}_D^{(\gamma,\beta)}[f]_h - \varepsilon\right)}{\left(\sigma_D^{(\gamma,\alpha)}[g]_h + \varepsilon\right)}\right]^{\frac{1}{\rho^{(\gamma,\alpha)}[g]_h}}.$$

As $\varepsilon \, (> 0)$ is arbitrary, it follows from above and Corollary 7.2.9 that

$$\liminf_{R\to\infty} \frac{\exp(\alpha(M_{g,D}^{-1}(M_{f,D}(R))))}{(\exp(\beta(R)))^{\rho^{(\alpha,\beta)}[f]_g}} \geq \left[\frac{\overline{\sigma}_D^{(\gamma,\beta)}[f]_h}{\sigma_D^{(\gamma,\alpha)}[g]_h}\right]^{\frac{1}{\rho^{(\gamma,\alpha)}[g]_h}}$$

$$i.e., \quad \overline{\sigma}_D^{(\alpha,\beta)}[f]_g \geq \left[\frac{\overline{\sigma}_D^{(\gamma,\beta)}[f]_h}{\sigma_D^{(\gamma,\alpha)}[g]_h}\right]^{\frac{1}{\rho^{(\gamma,\alpha)}[g]_h}}. \tag{167}$$

Also in view of $(162)$, we get from $(156)$ for a sequence of values of $R$ tending to infinity that

$$\exp(\alpha(M_{g,D}^{-1}(M_{f,D}(R)))) \leq$$

$$\exp\left(\alpha\left(M_{g,D}^{-1}\left(M_{h,D}\left(\gamma^{-1}\left(\log\left(\left(\sigma_D^{(\gamma,\beta)}[f]_h + \varepsilon\right)(\exp(\beta(R)))^{\rho^{(\gamma,\beta)}[f]_h}\right)\right)\right)\right)\right)\right)$$

$$i.e., \quad \exp(\alpha(M_{g,D}^{-1}(M_{f,D}(R)))) \leq \left(\frac{\left(\left(\sigma_D^{(\gamma,\beta)}[f]_h + \varepsilon\right)(\exp(\beta(R)))^{\rho^{(\gamma,\beta)}[f]_h}\right)}{\left(\sigma_D^{(\gamma,\alpha)}[g]_h - \varepsilon\right)}\right)^{\frac{1}{\rho^{(\gamma,\alpha)}[g]_h}}$$

$$i.e., \quad \exp(\alpha(M_{g,D}^{-1}(M_{f,D}(R)))) \leq \left[\frac{\left(\sigma_D^{(\gamma,\beta)}[f]_h + \varepsilon\right)}{\left(\sigma_D^{(\gamma,\alpha)}[g]_h - \varepsilon\right)}\right]^{\frac{1}{\rho^{(\gamma,\alpha)}[g]_h}} (\exp(\beta(R)))^{\frac{\rho^{(\gamma,\beta)}[f]_h}{\rho^{(\gamma,\alpha)}[g]_h}}$$

$$i.e., \quad \frac{\exp(\alpha(M_{g,D}^{-1}(M_{f,D}(R))))}{(\exp(\beta(R)))^{\frac{\rho^{(\gamma,\beta)}[f]_h}{\rho^{(\gamma,\alpha)}[g]_h}}} \leq \left[\frac{\left(\sigma_D^{(\gamma,\beta)}[f]_h + \varepsilon\right)}{\left(\sigma_D^{(\gamma,\alpha)}[g]_h - \varepsilon\right)}\right]^{\frac{1}{\rho^{(\gamma,\alpha)}[g]_h}}.$$

Since $\varepsilon \, (> 0)$ is arbitrary, we get from Corollary 7.2.9 and above that

$$\liminf_{R\to\infty} \frac{\exp(\alpha(M_{g,D}^{-1}(M_{f,D}(R))))}{(\exp(\beta(R)))^{\rho^{(\alpha,\beta)}[f]_g}} \leq \left[\frac{\sigma_D^{(\gamma,\beta)}[f]_h}{\sigma_D^{(\gamma,\alpha)}[g]_h}\right]^{\frac{1}{\rho^{(\gamma,\alpha)}[g]_h}}$$

$$i.e., \quad \overline{\sigma}_D^{(\alpha,\beta)}[f]_g \leq \left[\frac{\sigma_D^{(\gamma,\beta)}[f]_h}{\sigma_D^{(\gamma,\alpha)}[g]_h}\right]^{\frac{1}{\rho^{(\gamma,\alpha)}[g]_h}}. \tag{168}$$

Similarly from (159) and in view of (161), it follows for a sequence of values of $R$ tending to infinity that

$$\exp(\alpha(M_{g,D}^{-1}(M_{f,D}(R)))) \leq$$

$$\exp\left(\alpha\left(M_{g,D}^{-1}\left(M_{h,D}\left(\gamma^{-1}\left(\log\left(\left(\overline{\sigma}_D^{(\gamma,\beta)}[f]_h + \varepsilon\right)(\exp(\beta(R)))^{\rho^{(\gamma,\beta)}[f]_h}\right)\right)\right)\right)\right)\right)$$

$$i.e.,\ \exp(\alpha(M_{g,D}^{-1}(M_{f,D}(R)))) \leq \left(\frac{\left(\left(\overline{\sigma}_D^{(\gamma,\beta)}[f]_h + \varepsilon\right)(\exp(\beta(R)))^{\rho^{(\gamma,\beta)}[f]_h}\right)}{\left(\overline{\sigma}_D^{(\gamma,\alpha)}[g]_h - \varepsilon\right)}\right)^{\frac{1}{\rho^{(\gamma,\alpha)}[g]_h}}$$

$$i.e.,\ \exp(\alpha(M_{g,D}^{-1}(M_{f,D}(R)))) \leq \left[\frac{\left(\overline{\sigma}_D^{(\gamma,\beta)}[f]_h + \varepsilon\right)}{\left(\overline{\sigma}_D^{(\gamma,\alpha)}[g]_h - \varepsilon\right)}\right]^{\frac{1}{\rho^{(\gamma,\alpha)}[g]_h}}(\exp(\beta(R)))^{\frac{\rho^{(\gamma,\beta)}[f]_h}{\rho^{(\gamma,\alpha)}[g]_h}}$$

$$i.e.,\ \frac{\exp(\alpha(M_{g,D}^{-1}(M_{f,D}(R))))}{(\exp(\beta(R)))^{\frac{\rho^{(\gamma,\beta)}[f]_h}{\rho^{(\gamma,\alpha)}[g]_h}}} \leq \left[\frac{\left(\overline{\sigma}_D^{(\gamma,\beta)}[f]_h + \varepsilon\right)}{\left(\overline{\sigma}_D^{(\gamma,\alpha)}[g]_h - \varepsilon\right)}\right]^{\frac{1}{\rho^{(\gamma,\alpha)}[g]_h}}.$$

As $\varepsilon\ (> 0)$ is arbitrary, we obtain from Corollary 7.2.9 and above that

$$\liminf_{R\to\infty}\frac{\exp(\alpha(M_{g,D}^{-1}(M_{f,D}(R))))}{(\exp(\beta(R)))^{\rho^{(\alpha,\beta)}[f]_g}} \leq \left[\frac{\overline{\sigma}_D^{(\gamma,\beta)}[f]_h}{\overline{\sigma}_D^{(\gamma,\alpha)}[g]_h}\right]^{\frac{1}{\rho^{(\gamma,\alpha)}[g]_h}}$$

$$i.e.,\ \overline{\sigma}_D^{(\alpha,\beta)}[f]_g \leq \left[\frac{\overline{\sigma}_D^{(\gamma,\beta)}[f]_h}{\overline{\sigma}_D^{(\gamma,\alpha)}[g]_h}\right]^{\frac{1}{\rho^{(\gamma,\alpha)}[g]_h}}. \tag{169}$$

Thus the theorem follows from $(164)$, $(165)$, $(166)$, $(167)$, $(168)$ and $(169)$. ∎

In view of Theorem 7.2.2, one can easily verify the following corollaries :

**Corollary 7.2.11** *Let $f(z)$, $g(z)$ and $h(z)$ be three entire functions of $n$ complex variables such that $0 < \lambda^{(\gamma,\beta)}[f]_h \leq \rho^{(\gamma,\beta)}[f]_h < \infty$ and $0 < \lambda^{(\gamma,\alpha)}[g]_h \leq \rho^{(\gamma,\alpha)}[g]_h < \infty$ . Also let $\lambda^{(\gamma,\alpha)}[g]_h = \rho^{(\gamma,\alpha)}[g]_h$. Then*

$$\sigma_D^{(\alpha,\beta)}[f]_g = \left[\frac{\sigma_D^{(\gamma,\beta)}[f]_h}{\sigma_D^{(\gamma,\alpha)}[g]_h}\right]^{\frac{1}{\rho^{(\gamma,\alpha)}[g]_h}} \quad and \quad \overline{\sigma}_D^{(\alpha,\beta)}[f]_g = \left[\frac{\overline{\sigma}_D^{(\gamma,\beta)}[f]_h}{\sigma_D^{(\gamma,\alpha)}[g]_h}\right]^{\frac{1}{\rho^{(\gamma,\alpha)}[g]_h}}.$$

*In addition, if $\sigma_D^{(\gamma,\beta)}[f]_h = \sigma_D^{(\gamma,\alpha)}[g]_h$ and $0 < \lambda^{(\gamma,\beta)}[f]_h = \rho^{(\gamma,\beta)}[f]_h < \infty$, then*

$$\sigma_D^{(\alpha,\beta)}[f]_g = \overline{\sigma}_D^{(\beta,\alpha)}[g]_f = 1.$$

**Corollary 7.2.12** *Let $f(z)$, $g(z)$ and $h(z)$ be three entire functions of $n$ complex variables such that $\sigma_D^{(\gamma,\beta)}[f]_h = \overline{\sigma}_D^{(\gamma,\beta)}[f]_h$ and $0 < \lambda^{(\gamma,\alpha)}[g]_h = \rho^{(\gamma,\alpha)}[g]_h < \infty$. Then*

$$\sigma_D^{(\alpha,\beta)}[f]_g = \overline{\sigma}_D^{(\alpha,\beta)}[f]_g = \left[\frac{\sigma_D^{(\gamma,\beta)}[f]_h}{\sigma_D^{(\gamma,\alpha)}[g]_h}\right]^{\frac{1}{\rho^{(\gamma,\alpha)}[g]_h}}.$$

*In addition, if $\sigma_D^{(\gamma,\beta)}[f]_h = \sigma_D^{(\gamma,\alpha)}[g]_h$ and $0 < \lambda^{(\gamma,\beta)}[f]_h = \rho^{(\gamma,\beta)}[f]_h < \infty$, then*

$$\sigma_D^{(\alpha,\beta)}[f]_g = \overline{\sigma}_D^{(\alpha,\beta)}[f]_g = \sigma_D^{(\beta,\alpha)}[g]_f = \overline{\sigma}_D^{(\beta,\alpha)}[g]_f = 1.$$

**Corollary 7.2.13** *Let $f(z)$, $g(z)$ and $h(z)$ be three entire functions of $n$ complex variables such that $0 < \lambda^{(\gamma,\beta)}[f]_h \leq \rho^{(\gamma,\beta)}[f]_h < \infty$ and $0 < \lambda^{(\gamma,\alpha)}[g]_h = \rho^{(\gamma,\alpha)}[g]_h < \infty$. Then*

$$(i)\quad \sigma_D^{(\alpha,\beta)}[f]_g = \overline{\sigma}_D^{(\alpha,\beta)}[f]_g = \infty \text{ when } \sigma_D^{(\gamma,\alpha)}[g]_h = 0$$

*and*

$$(ii)\quad \sigma_D^{(\alpha,\beta)}[f]_g = \overline{\sigma}_D^{(\alpha,\beta)}[f]_g = 0 \text{ when } \sigma_D^{(\gamma,\alpha)}[g]_h = \infty.$$

**Corollary 7.2.14** *Let $f(z)$, $g(z)$ and $h(z)$ be three entire functions of $n$ complex variables such that $\lambda^{(\gamma,\beta)}[f]_h \leq \rho^{(\gamma,\beta)}[f]_h$ and $0 < \lambda^{(\gamma,\alpha)}[g]_h = \rho^{(\gamma,\alpha)}[g]_h < \infty$. Then*

$$(i)\quad \sigma_D^{(\alpha,\beta)}[f]_g = 0 \text{ when } \sigma_D^{(\gamma,\beta)}[f]_h = 0,$$

$$(ii)\quad \overline{\sigma}_D^{(\alpha,\beta)}[f]_g = 0 \text{ when } \overline{\sigma}_D^{(\alpha,\beta)}[f]_g = 0,$$

$$(iii)\quad \sigma_D^{(\alpha,\beta)}[f]_g = \infty \text{ when } \sigma_D^{(\gamma,\beta)}[f]_h = \infty$$

*and*

$$(iv)\quad \overline{\sigma}_D^{(\alpha,\beta)}[f]_g = \infty \text{ when } \overline{\sigma}_D^{(\alpha,\beta)}[f]_g = \infty.$$

In the line of Theorem 7.2.2 and with the help of Corollary 7.2.10, one can easily establish the following theorem and therefore its proof is omitted:

**Theorem 7.2.3** *Let $f(z)$, $g(z)$ and $h(z)$ be three entire functions of $n$ complex variables such that $0 < \lambda^{(\gamma,\beta)}[f]_h = \rho^{(\gamma,\beta)}[f]_h < \infty$ and $0 < \lambda^{(\gamma,\alpha)}[g]_h \leq \rho^{(\gamma,\alpha)}[g]_h < \infty$. Then*

$$\left[\frac{\overline{\sigma}_D^{(\gamma,\beta)}[f]_h}{\sigma_D^{(\gamma,\alpha)}[g]_h}\right]^{\frac{1}{\rho^{(\gamma,\alpha)}[g]_h}} \leq \tau_D^{(\alpha,\beta)}[f]_g \leq \min\left\{\left[\frac{\overline{\sigma}_D^{(\gamma,\beta)}[f]_h}{\overline{\sigma}_D^{(\gamma,\alpha)}[g]_h}\right]^{\frac{1}{\rho^{(\gamma,\alpha)}[g]_h}}, \left[\frac{\sigma_D^{(\gamma,\beta)}[f]_h}{\sigma_D^{(\gamma,\alpha)}[g]_h}\right]^{\frac{1}{\rho^{(\gamma,\alpha)}[g]_h}}\right\}$$

$$\leq \max\left\{\left[\frac{\overline{\sigma}_D^{(\gamma,\beta)}[f]_h}{\overline{\sigma}_D^{(\gamma,\alpha)}[g]_h}\right]^{\frac{1}{\rho^{(\gamma,\alpha)}[g]_h}}, \left[\frac{\sigma_D^{(\gamma,\beta)}[f]_h}{\sigma_D^{(\gamma,\alpha)}[g]_h}\right]^{\frac{1}{\rho^{(\gamma,\alpha)}[g]_h}}\right\} \leq \overline{\tau}_D^{(\alpha,\beta)}[f]_g \leq \left[\frac{\sigma_D^{(\gamma,\beta)}[f]_h}{\overline{\sigma}_D^{(\gamma,\alpha)}[g]_h}\right]^{\frac{1}{\rho^{(\gamma,\alpha)}[g]_h}}.$$

In view of Theorem 7.2.3, one can easily derive the following corollaries:

**Corollary 7.2.15** *Let* $f(z)$, $g(z)$ *and* $h(z)$ *be three entire functions of* $n$ *complex variables such that* $0 < \lambda^{(\gamma,\beta)}[f]_h = \rho^{(\gamma,\beta)}[f]_h < \infty$ *and* $0 < \lambda^{(\gamma,\alpha)}[g]_h \leq \rho^{(\gamma,\alpha)}[g]_h < \infty$. *Then*

$$\overline{\tau}_D^{(\alpha,\beta)}[f]_g = \left[\frac{\sigma_D^{(\gamma,\beta)}[f]_h}{\overline{\sigma}_D^{(\gamma,\alpha)}[g]_h}\right]^{\frac{1}{\rho^{(\gamma,\alpha)}[g]_h}} \quad and \quad \tau_D^{(\alpha,\beta)}[f]_g = \left[\frac{\sigma_D^{(\gamma,\beta)}[f]_h}{\sigma_D^{(\gamma,\alpha)}[g]_h}\right]^{\frac{1}{\rho^{(\gamma,\alpha)}[g]_h}}.$$

*In addition, if* $\sigma_D^{(\gamma,\beta)}[f]_h = \overline{\sigma}_D^{(\gamma,\alpha)}[g]_h$ *and* $< \lambda^{(\gamma,\alpha)}[g]_h = \rho^{(\gamma,\alpha)}[g]_h < \infty$, *then*

$$\overline{\tau}_D^{(\alpha,\beta)}[f]_g = \tau_D^{(\beta,\alpha)}[g]_f = 1.$$

**Corollary 7.2.16** *Let* $f(z)$, $g(z)$ *and* $h(z)$ *be three entire functions of* $n$ *complex variables such that* $0 < \lambda^{(\gamma,\beta)}[f]_h = \rho^{(\gamma,\beta)}[f]_h < \infty$ *and* $\sigma_D^{(\gamma,\alpha)}[g]_h = \overline{\sigma}_D^{(\gamma,\alpha)}[g]_h$. *Then*

$$\overline{\tau}_D^{(\alpha,\beta)}[f]_g = \tau_D^{(\alpha,\beta)}[f]_g = \left[\frac{\sigma_D^{(\gamma,\beta)}[f]_h}{\sigma_D^{(\gamma,\alpha)}[g]_h}\right]^{\frac{1}{\rho^{(\gamma,\alpha)}[g]_h}}.$$

*In addition, if* $\sigma_D^{(\gamma,\beta)}[f]_h = \overline{\sigma}_D^{(\gamma,\alpha)}[g]_h$ *and* $0 < \lambda^{(\gamma,\alpha)}[g]_h = \rho^{(\gamma,\alpha)}[g]_h < \infty$, *then*

$$\overline{\tau}_D^{(\alpha,\beta)}[f]_g = \tau_D^{(\alpha,\beta)}[f]_g = \overline{\tau}_D^{(\beta,\alpha)}[g]_f = \tau_D^{(\beta,\alpha)}[g]_f = 1.$$

**Corollary 7.2.17** *Let* $f(z)$, $g(z)$ *and* $h(z)$ *be three entire functions of* $n$ *complex variables such that* $0 < \lambda^{(\gamma,\beta)}[f]_h = \rho^{(\gamma,\beta)}[f]_h < \infty$ *and* $0 < \lambda^{(\gamma,\alpha)}[g]_h \leq \rho^{(\gamma,\alpha)}[g]_h < \infty$. *Then*

$$(i) \quad \overline{\tau}_D^{(\alpha,\beta)}[f]_g = \tau_D^{(\alpha,\beta)}[f]_g = \infty \ when \ \sigma_D^{(\gamma,\alpha)}[g]_h = 0$$

*and*

$$(ii) \quad \overline{\tau}_D^{(\alpha,\beta)}[f]_g = \tau_D^{(\alpha,\beta)}[f]_g = 0 \ when \ \sigma_D^{(\gamma,\alpha)}[g]_h = \infty.$$

**Corollary 7.2.18** *Let* $f(z)$, $g(z)$ *and* $h(z)$ *be three entire functions of* $n$ *complex variables such that* $0 < \lambda^{(\gamma,\beta)}[f]_h = \rho^{(\gamma,\beta)}[f]_h < \infty$ *and* $\lambda^{(\gamma,\alpha)}[g]_h \leq \rho^{(\gamma,\alpha)}[g]_h$. *Then*

$$(i) \quad \overline{\tau}_D^{(\alpha,\beta)}[f]_g = 0 \ when \ \sigma_D^{(\gamma,\beta)}[f]_h = 0,$$
$$(ii) \quad \tau_D^{(\alpha,\beta)}[f]_g = 0 \ when \ \overline{\sigma}_D^{(\gamma,\beta)}[f]_h = 0,$$
$$(iii) \quad \overline{\tau}_D^{(\alpha,\beta)}[f]_g = \infty \ when \ \sigma_D^{(\gamma,\beta)}[f]_h = \infty$$

*and*

$$(iv) \quad \tau_D^{(\alpha,\beta)}[f]_g = \infty \ when \ \overline{\sigma}_D^{(\gamma,\beta)}[f]_h = \infty.$$

Similarly in the line of Theorem 7.2.2 and Theorem 7.2.3 and with the help of Corollary 7.2.9 and Corollary 7.2.10, one may easily prove the following two theorems and therefore their proofs are omitted:

**Theorem 7.2.4** *Let $f(z)$, $g(z)$ and $h(z)$ be three entire functions of $n$ complex variables such that $0 < \lambda^{(\gamma,\beta)}[f]_h \le \rho^{(\gamma,\beta)}[f]_h < \infty$ and $0 < \lambda^{(\gamma,\alpha)}[g]_h = \rho^{(\gamma,\alpha)}[g]_h < \infty$ . Then*

$$\left[\frac{\tau_D^{(\gamma,\beta)}[f]_h}{\overline{\tau}_D^{(\gamma,\alpha)}[g]_h}\right]^{\frac{1}{\lambda^{(\gamma,\alpha)}[g]_h}} \le \tau_D^{(\alpha,\beta)}[f]_g \le \min\left\{\left[\frac{\tau_D^{(\gamma,\beta)}[f]_h}{\tau_D^{(\gamma,\alpha)}[g]_h}\right]^{\frac{1}{\lambda^{(\gamma,\alpha)}[g]_h}}, \left[\frac{\overline{\tau}_D^{(\gamma,\beta)}[f]_h}{\overline{\tau}_D^{(\gamma,\alpha)}[g]_h}\right]^{\frac{1}{\lambda^{(\gamma,\alpha)}[g]_h}}\right\}$$

$$\le \max\left\{\left[\frac{\tau_D^{(\gamma,\beta)}[f]_h}{\tau_D^{(\gamma,\alpha)}[g]_h}\right]^{\frac{1}{\lambda^{(\gamma,\alpha)}[g]_h}}, \left[\frac{\overline{\tau}_D^{(\gamma,\beta)}[f]_h}{\overline{\tau}_D^{(\gamma,\alpha)}[g]_h}\right]^{\frac{1}{\lambda^{(\gamma,\alpha)}[g]_h}}\right\} \le \overline{\tau}_D^{(\alpha,\beta)}[f]_g \le \left[\frac{\overline{\tau}_D^{(\gamma,\beta)}[f]_h}{\tau_D^{(\gamma,\alpha)}[g]_h}\right]^{\frac{1}{\lambda^{(\gamma,\alpha)}[g]_h}} .$$

In view of Theorem 7.2.4, the following corollaries may also be obtained:

**Corollary 7.2.19** *Let $f(z)$, $g(z)$ and $h(z)$ be three entire functions of $n$ complex variables such that $0 < \lambda^{(\gamma,\beta)}[f]_h \le \rho^{(\gamma,\beta)}[f]_h < \infty$ and $0 < \lambda^{(\gamma,\alpha)}[g]_h = \rho^{(\gamma,\alpha)}[g]_h < \infty$ . Then*

$$\overline{\tau}_D^{(\alpha,\beta)}[f]_g = \left[\frac{\overline{\tau}_D^{(\gamma,\beta)}[f]_h}{\overline{\tau}_D^{(\gamma,\alpha)}[g]_h}\right]^{\frac{1}{\lambda^{(\gamma,\alpha)}[g]_h}} \quad and \quad \tau_D^{(\alpha,\beta)}[f]_g = \left[\frac{\tau_D^{(\gamma,\beta)}[f]_h}{\tau_D^{(\gamma,\alpha)}[g]_h}\right]^{\frac{1}{\lambda^{(\gamma,\alpha)}[g]_h}} .$$

*In addition, if $\overline{\tau}_D^{(\gamma,\beta)}[f]_h = \overline{\tau}_D^{(\gamma,\alpha)}[g]_h$ and $0 < \lambda^{(\gamma,\beta)}[f]_h = \rho^{(\gamma,\beta)}[f]_h < \infty$, then*

$$\overline{\tau}_D^{(\alpha,\beta)}[f]_g = \tau_D^{(\beta,\alpha)}[g]_f = 1.$$

**Corollary 7.2.20** *Let $f(z)$, $g(z)$ and $h(z)$ be three entire functions of $n$ complex variables such that $\overline{\tau}_D^{(\gamma,\beta)}[f]_h = \tau_D^{(\gamma,\beta)}[f]_h$ and $0 < \lambda^{(\gamma,\alpha)}[g]_h = \rho^{(\gamma,\alpha)}[g]_h < \infty$ . Then*

$$\overline{\tau}_D^{(\alpha,\beta)}[f]_g = \tau_D^{(\alpha,\beta)}[f]_g = \left[\frac{\overline{\tau}_D^{(\gamma,\beta)}[f]_h}{\overline{\tau}_D^{(\gamma,\alpha)}[g]_h}\right]^{\frac{1}{\lambda^{(\gamma,\alpha)}[g]_h}} .$$

*In addition, if $\overline{\tau}_D^{(\gamma,\beta)}[f]_h = \overline{\tau}_D^{(\gamma,\alpha)}[g]_h$ and $0 < \lambda^{(\gamma,\beta)}[f]_h = \rho^{(\gamma,\beta)}[f]_h < \infty$, then*

$$\overline{\tau}_D^{(\alpha,\beta)}[f]_g = \tau_D^{(\alpha,\beta)}[f]_g = \overline{\tau}_D^{(\beta,\alpha)}[g]_f = \tau_D^{(\beta,\alpha)}[g]_f = 1.$$

**Corollary 7.2.21** *Let $f(z)$, $g(z)$ and $h(z)$ be three entire functions of $n$ complex variables such that $0 < \lambda^{(\gamma,\beta)}[f]_h \le \rho^{(\gamma,\beta)}[f]_h < \infty$ and $0 < \lambda^{(\gamma,\alpha)}[g]_h = \rho^{(\gamma,\alpha)}[g]_h < \infty$ . Then*

$$(i) \quad \overline{\tau}_D^{(\alpha,\beta)}[f]_g = \tau_D^{(\alpha,\beta)}[f]_g = \infty \ \ when \ \overline{\tau}_D^{(\gamma,\alpha)}[g]_h = 0$$

*and*

$$(ii) \quad \overline{\tau}_D^{(\alpha,\beta)}[f]_g = \tau_D^{(\alpha,\beta)}[f]_g = 0 \ \ when \ \overline{\tau}_D^{(\gamma,\alpha)}[g]_h = \infty.$$

**Corollary 7.2.22** *Let $f(z)$, $g(z)$ and $h(z)$ be three entire functions of $n$ complex variables such that $\lambda^{(\gamma,\beta)}[f]_h \le \rho^{(\gamma,\beta)}[f]_h$ and $0 < \lambda^{(\gamma,\alpha)}[g]_h = \rho^{(\gamma,\alpha)}[g]_h < \infty$ . Then*

$$(i) \quad \overline{\tau}_D^{(\alpha,\beta)}[f]_g = 0 \ \ when \ \overline{\tau}_D^{(\gamma,\beta)}[f]_h = 0,$$

$$(ii) \quad \tau_D^{(\alpha,\beta)}[f]_g = 0 \ \ when \ \tau_D^{(\gamma,\beta)}[f]_h = 0,$$

$$(iii) \quad \overline{\tau}_D^{(\alpha,\beta)}[f]_g = \infty \ \ when \ \overline{\tau}_D^{(\gamma,\beta)}[f]_h = \infty$$

*and*

$$(iv) \quad \tau_D^{(\alpha,\beta)}[f]_g = \infty \quad when \quad \tau_D^{(\gamma,\beta)}[f]_h = \infty.$$

**Theorem 7.2.5** *Let* $f(z)$, $g(z)$ *and* $h(z)$ *be three entire functions of* $n$ *complex variables such that* $0 < \lambda^{(\gamma,\beta)}[f]_h = \rho^{(\gamma,\beta)}[f]_h < \infty$ *and* $0 < \lambda^{(\gamma,\alpha)}[g]_h \le \rho^{(\gamma,\alpha)}[g]_h < \infty$ . *Then*

$$\left[\frac{\tau_D^{(\gamma,\beta)}[f]_h}{\overline{\tau}_D^{(\gamma,\alpha)}[g]_h}\right]^{\frac{1}{\lambda^{(\gamma,\alpha)}[g]_h}} \le \overline{\sigma}_D^{(\alpha,\beta)}[f]_g \le \min\left\{\left[\frac{\tau_D^{(\gamma,\beta)}[f]_h}{\tau_D^{(\gamma,\alpha)}[g]_h}\right]^{\frac{1}{\lambda^{(\gamma,\alpha)}[g]_h}}, \left[\frac{\overline{\tau}_D^{(\gamma,\beta)}[f]_h}{\overline{\tau}_D^{(\gamma,\alpha)}[g]_h}\right]^{\frac{1}{\lambda^{(\gamma,\alpha)}[g]_h}}\right\}$$

$$\le \max\left\{\left[\frac{\tau_D^{(\gamma,\beta)}[f]_h}{\tau_D^{(\gamma,\alpha)}[g]_h}\right]^{\frac{1}{\lambda^{(\gamma,\alpha)}[g]_h}}, \left[\frac{\overline{\tau}_D^{(\gamma,\beta)}[f]_h}{\overline{\tau}_D^{(\gamma,\alpha)}[g]_h}\right]^{\frac{1}{\lambda^{(\gamma,\alpha)}[g]_h}}\right\} \le \sigma_D^{(\alpha,\beta)}[f]_g \le \left[\frac{\tau_D^{(\gamma,\beta)}[f]_h}{\tau_D^{(\gamma,\alpha)}[g]_h}\right]^{\frac{1}{\lambda^{(\gamma,\alpha)}[g]_h}}.$$

From Theorem 7.2.5, the following corollaries are immediate:

**Corollary 7.2.23** *Let* $f(z)$, $g(z)$ *and* $h(z)$ *be three entire functions of* $n$ *complex variables such that* $0 < \lambda^{(\gamma,\beta)}[f]_h = \rho^{(\gamma,\beta)}[f]_h < \infty$ *and* $0 < \lambda^{(\gamma,\alpha)}[g]_h \le \rho^{(\gamma,\alpha)}[g]_h < \infty$. *Then*

$$\sigma_D^{(\alpha,\beta)}[f]_g = \left[\frac{\overline{\tau}_D^{(\gamma,\beta)}[f]_h}{\tau_D^{(\gamma,\alpha)}[g]_h}\right]^{\frac{1}{\lambda^{(\gamma,\alpha)}[g]_h}} \quad and \quad \overline{\sigma}_D^{(\alpha,\beta)}[f]_g = \left[\frac{\overline{\tau}_D^{(\gamma,\beta)}[f]_h}{\overline{\tau}_D^{(\gamma,\alpha)}[g]_h}\right]^{\frac{1}{\lambda^{(\gamma,\alpha)}[g]_h}}.$$

*In addition, if* $\overline{\tau}_D^{(\gamma,\beta)}[f]_h = \tau_D^{(\gamma,\alpha)}[g]_h$ *and* $0 < \lambda^{(\gamma,\alpha)}[g]_h = \rho^{(\gamma,\alpha)}[g]_h < \infty$, *then*

$$\sigma_D^{(\alpha,\beta)}[f]_g = \overline{\sigma}_D^{(\beta,\alpha)}[g]_f = 1.$$

**Corollary 7.2.24** *Let* $f(z)$, $g(z)$ *and* $h(z)$ *be three entire functions of* $n$ *complex variables such that* $0 < \lambda^{(\gamma,\beta)}[f]_h = \rho^{(\gamma,\beta)}[f]_h < \infty$ *and* $\overline{\tau}_D^{(\gamma,\alpha)}[g]_h = \tau_D^{(\gamma,\alpha)}[g]_h$. *Then*

$$\sigma_D^{(\alpha,\beta)}[f]_g = \overline{\sigma}_D^{(\alpha,\beta)}[f]_g = \left[\frac{\overline{\tau}_D^{(\gamma,\beta)}[f]_h}{\overline{\tau}_D^{(\gamma,\alpha)}[g]_h}\right]^{\frac{1}{\lambda^{(\gamma,\alpha)}[g]_h}}.$$

*In addition, if* $\overline{\tau}_D^{(\gamma,\beta)}[f]_h = \tau_D^{(\gamma,\alpha)}[g]_h$ *and* $0 < \lambda^{(\gamma,\alpha)}[g]_h = \rho^{(\gamma,\alpha)}[g]_h < \infty$, *then*

$$\sigma_D^{(\alpha,\beta)}[f]_g = \overline{\sigma}_D^{(\alpha,\beta)}[f]_g = \sigma_D^{(\beta,\alpha)}[g]_f = \overline{\sigma}_D^{(\beta,\alpha)}[g]_f = 1.$$

**Corollary 7.2.25** *Let* $f(z)$, $g(z)$ *and* $h(z)$ *be three entire functions of* $n$ *complex variables such that* $0 < \lambda^{(\gamma,\beta)}[f]_h = \rho^{(\gamma,\beta)}[f]_h < \infty$ *and* $0 < \lambda^{(\gamma,\alpha)}[g]_h \le \rho^{(\gamma,\alpha)}[g]_h < \infty$. *Then*

$$(i) \quad \sigma_D^{(\alpha,\beta)}[f]_g = \overline{\sigma}_D^{(\alpha,\beta)}[f]_g = \infty \quad when \quad \overline{\tau}_D^{(\gamma,\alpha)}[g]_h = 0$$

*and*

$$(ii) \quad \sigma_D^{(\alpha,\beta)}[f]_g = \overline{\sigma}_D^{(\alpha,\beta)}[f]_g = 0 \quad when \quad \overline{\tau}_D^{(\gamma,\alpha)}[g]_h = \infty.$$

**Corollary 7.2.26** *Let $f(z)$, $g(z)$ and $h(z)$ be three entire functions of $n$ complex variables such that $0 < \lambda^{(\gamma,\beta)}[f]_h = \rho^{(\gamma,\beta)}[f]_h < \infty$ and $\lambda^{(\gamma,\alpha)}[g]_h \leq \rho^{(\gamma,\alpha)}[g]_h$. Then*

$$(i) \quad \sigma_D^{(\alpha,\beta)}[f]_g = 0 \text{ when } \overline{\tau}_D^{(\gamma,\beta)}[f]_h = 0,$$

$$(ii) \quad \overline{\sigma}_D^{(\alpha,\beta)}[f]_g = 0 \text{ when } \tau_D^{(\gamma,\beta)}[f]_h = 0,$$

$$(iii) \quad \sigma_D^{(\alpha,\beta)}[f]_g = \infty \text{ when } \overline{\tau}_D^{(\gamma,\beta)}[f]_h = \infty$$

*and*

$$(iv) \quad \overline{\sigma}_D^{(\alpha,\beta)}[f]_g = \infty \text{ when } \tau_D^{(\gamma,\beta)}[f]_h = \infty.$$

Now we state the following two theorems without their proofs as those can easily be carried out in the line of Theorem 7.2.2 and with the help of Theorem 7.2.1:

**Theorem 7.2.6** *Let $f(z)$, $g(z)$ and $h(z)$ be three entire functions of $n$ complex variables such that $0 < \lambda^{(\gamma,\beta)}[f]_h \leq \rho^{(\gamma,\beta)}[f]_h < \infty$ and $0 < \lambda^{(\gamma,\alpha)}[g]_h \leq \rho^{(\gamma,\alpha)}[g]_h < \infty$. Then*

$$\max\left\{ \left[\frac{\overline{\sigma}_D^{(\gamma,\beta)}[f]_h}{\tau_D^{(\gamma,\alpha)}[g]_h}\right]^{\frac{1}{\lambda^{(\gamma,\alpha)}[g]_h}}, \left[\frac{\sigma_D^{(\gamma,\beta)}[f]_h}{\overline{\tau}_D^{(\gamma,\alpha)}[g]_h}\right]^{\frac{1}{\lambda^{(\gamma,\alpha)}[g]_h}} \right\} \leq \sigma_D^{(\alpha,\beta)}[f]_g$$

$$\leq \min\left\{ \left[\frac{\overline{\tau}_D^{(\gamma,\beta)}[f]_h}{\tau_D^{(\gamma,\alpha)}[g]_h}\right]^{\frac{1}{\lambda^{(\gamma,\alpha)}[g]_h}}, \left[\frac{\sigma_D^{(\gamma,\beta)}[f]_h}{\overline{\sigma}_D^{(\gamma,\alpha)}[g]_h}\right]^{\frac{1}{\rho^{(\gamma,\alpha)}[g]_h}}, \left[\frac{\overline{\tau}_D^{(\gamma,\beta)}[f]_h}{\overline{\sigma}_D^{(\gamma,\alpha)}[g]_h}\right]^{\frac{1}{\rho^{(\gamma,\alpha)}[g]_h}} \right\}$$

*and*

$$\left[\frac{\overline{\sigma}_D^{(\gamma,\beta)}[f]_h}{\overline{\tau}_D^{(\gamma,\alpha)}[g]_h}\right]^{\frac{1}{\lambda^{(\gamma,\alpha)}[g]_h}} \leq \overline{\sigma}_D^{(\alpha,\beta)}[f]_g$$

$$\leq \min\left\{ \begin{array}{cccc} \left[\dfrac{\overline{\sigma}_D^{(\gamma,\beta)}[f]_h}{\overline{\sigma}_D^{(\gamma,\alpha)}[g]_h}\right]^{\frac{1}{\rho^{(\gamma,\alpha)}[g]_h}}, & \left[\dfrac{\sigma_D^{(\gamma,\beta)}[f]_h}{\sigma_D^{(\gamma,\alpha)}[g]_h}\right]^{\frac{1}{\rho^{(\gamma,\alpha)}[g]_h}}, & \left[\dfrac{\tau_D^{(\gamma,\beta)}[f]_h}{\tau_D^{(\gamma,\alpha)}[g]_h}\right]^{\frac{1}{\lambda^{(\gamma,\alpha)}[g]_h}}, \\[3ex] \left[\dfrac{\overline{\tau}_D^{(\gamma,\beta)}[f]_h}{\overline{\tau}_D^{(\gamma,\alpha)}[g]_h}\right]^{\frac{1}{\lambda^{(\gamma,\alpha)}[g]_h}}, & \left[\dfrac{\overline{\tau}_D^{(\gamma,\beta)}[f]_h}{\sigma_D^{(\gamma,\alpha)}[g]_h}\right]^{\frac{1}{\rho^{(\gamma,\alpha)}[g]_h}}, & \left[\dfrac{\tau_D^{(\gamma,\beta)}[f]_h}{\overline{\sigma}_D^{(\gamma,\alpha)}[g]_h}\right]^{\frac{1}{\rho^{(\gamma,\alpha)}[g]_h}} \end{array} \right\}.$$

**Theorem 7.2.7** *Let $f(z)$, $g(z)$ and $h(z)$ be three entire functions of $n$ complex variables such that $0 < \lambda^{(\gamma,\beta)}[f]_h \leq \rho^{(\gamma,\beta)}[f]_h < \infty$ and $0 < \lambda^{(\gamma,\alpha)}[g]_h \leq \rho^{(\gamma,\alpha)}[g]_h < \infty$. Then*

$$\max\left\{ \begin{array}{cccc} \left[\dfrac{\overline{\tau}_D^{(\gamma,\beta)}[f]_h}{\overline{\tau}_D^{(\gamma,\alpha)}[g]_h}\right]^{\frac{1}{\lambda^{(\gamma,\alpha)}[g]_h}}, & \left[\dfrac{\tau_D^{(\gamma,\beta)}[f]_h}{\tau_D^{(\gamma,\alpha)}[g]_h}\right]^{\frac{1}{\lambda^{(\gamma,\alpha)}[g]_h}}, & \left[\dfrac{\overline{\sigma}_D^{(\gamma,\beta)}[f]_h}{\overline{\sigma}_D^{(\gamma,\alpha)}[g]_h}\right]^{\frac{1}{\rho^{(\gamma,\alpha)}[g]_h}}, \\[3ex] \left[\dfrac{\sigma_D^{(\gamma,\beta)}[f]_h}{\sigma_D^{(\gamma,\alpha)}[g]_h}\right]^{\frac{1}{\rho^{(\gamma,\alpha)}[g]_h}}, & \left[\dfrac{\sigma_D^{(\gamma,\beta)}[f]_h}{\overline{\tau}_D^{(\gamma,\alpha)}[g]_h}\right]^{\frac{1}{\lambda^{(\gamma,\alpha)}[g]_h}}, & \left[\dfrac{\overline{\sigma}_D^{(\gamma,\beta)}[f]_h}{\tau_D^{(\gamma,\alpha)}[g]_h}\right]^{\frac{1}{\lambda^{(\gamma,\alpha)}[g]_h}} \end{array} \right\}$$

$$\leq \overline{\tau}_D^{(\alpha,\beta)}[f]_g \leq \left[\frac{\tau_D^{(\gamma,\beta)}[f]_h}{\overline{\sigma}_D^{(\gamma,\alpha)}[g]_h}\right]^{\frac{1}{\rho^{(\gamma,\alpha)}[g]_h}}$$

*and*

$$\max\left\{\left[\frac{\overline{\sigma}_D^{(\gamma,\beta)}[f]_h}{\sigma_D^{(\gamma,\alpha)}[g]_h}\right]^{\frac{1}{\rho^{(\gamma,\alpha)}[g]_h}}, \left[\frac{\tau_D^{(\gamma,\beta)}[f]_h}{\overline{\tau}_D^{(\gamma,\alpha)}[g]_h}\right]^{\frac{1}{\lambda^{(\gamma,\alpha)}[g]_h}}, \left[\frac{\overline{\sigma}_D^{(\gamma,\beta)}[f]_h}{\overline{\tau}_D^{(\gamma,\alpha)}[g]_h}\right]^{\frac{1}{\lambda^{(\gamma,\alpha)}[g]_h}}\right\} \le \tau_D^{(\alpha,\beta)}[f]_g$$

$$\le \min\left\{\left[\frac{\tau_D^{(\gamma,\beta)}[f]_h}{\overline{\sigma}_D^{(\gamma,\alpha)}[g]_h}\right]^{\frac{1}{\rho^{(\gamma,\alpha)}[g]_h}}, \left[\frac{\overline{\tau}_D^{(\gamma,\beta)}[f]_h}{\sigma_D^{(\gamma,\alpha)}[g]_h}\right]^{\frac{1}{\rho^{(\gamma,\alpha)}[g]_h}}\right\}.$$

# 7.3   Conclusion.

The ultimate focus of this chapter is to find out the limiting value of generalized relative Gol'dberg order $(\alpha, \beta)$, generalized relative Gol'dberg type $(\alpha, \beta)$ and generalized relative Gol'dberg weak type $(\alpha, \beta)$ under some different conditions. But some basic results involving the same need to be investigated in this area of research which have been reflected in the theories as proved in the next chapter.

# Chapter 8

# Sum and product theorems depending on the generalized relative Gol'dberg order $(\alpha, \beta)$ and generalized relative Gol'dberg type $(\alpha, \beta)$

**Abstract:** In this chapter, we proved some results about sum and product theorems depending on the generalized relative Gol'dberg order $(\alpha, \beta)$ ,generalized relative Gol'dberg lower order $(\alpha, \beta)$, generalized relative Gol'dberg type $(\alpha, \beta)$ and generalized relative Gol'dberg weak type $(\alpha, \beta)$ of entire function of $n$ complex variables with respect to another entire function of $n$ complex variables, where $\alpha, \beta$ are continuous non-negative functions defined on $(-\infty, +\infty)$.
**Keywords:** Generalized relative Gol'dberg order $(\alpha, \beta)$ ,generalized relative Gol'dberg lower order $(\alpha, \beta)$, generalized relative Gol'dberg type $(\alpha, \beta)$, generalized relative Gol'dberg weak type $(\alpha, \beta)$ , increasing function, Property (G), Property (X).
**Mathematics Subject Classification (2010) :** 32A15.

## 8.1   Introduction.

First of all, we just recall the following well known inequalities for all sufficiently large $R$ relating to any two entire functions $f_1(z)$ and $f_2(z)$ of $n$ complex variables:

$$M_{f_1 \pm f_2, D}(R) \leq M_{f_1, D}(R) + M_{f_2, D}(R), \tag{170}$$

$$M_{f_1 \pm f_2, D}(R) \geq M_{f_1, D}(R) - M_{f_2, D}(R) \tag{171}$$

and

$$M_{f_1 \cdot f_2, D}(R) \leq M_{f_1, D}(R) \cdot M_{f_2, D}(R) . \tag{172}$$

**Tanmay Biswas & Chinmay Biswas**

Now let $L$ be a class of continuous non-negative on $(-\infty, +\infty)$ function $\alpha$ such that $\alpha(x) = \alpha(x_0) \geq 0$ for $x \leq x_0$ with $\alpha(x) \uparrow +\infty$ as $x \to +\infty$. For any $\alpha \in L$, we say that $\alpha \in L^0$, if $\alpha(cx) = (1 + o(1))\alpha(x)$ as $x_0 \leq x \to +\infty$ for each $c \in (0, +\infty)$. Clearly, $L^0 \subset L$.

Further detailed investigations on the properties of $(p,q)$-$\varphi$ relative Gol'dberg order and the $(p,q)$-$\varphi$ relative Gol'dberg lower order have been made in [1]. In this connection we just state the following theorems which are introduced by Datta et al. [1] .

**Theorem 8.1.1** *Let us consider $f_1(z)$, $f_2(z)$ and $g_1(z)$ are any three entire functions of $n$ complex variables. Also let at least $f_1(z)$ or $f_2(z)$ is of regular $(p,q)$-$\varphi$ relative Gol'dberg growth with respect to $g_1(z)$. Then*

$$\lambda_{g_1}^{(p,q)}(f_1 \pm f_2, \varphi) \leq \max\{\lambda_{g_1}^{(p,q)}(f_1, \varphi), \lambda_{g_1}^{(p,q)}(f_2, \varphi)\}.$$

*The equality holds when any one of $\lambda_{g_1}^{(p,q)}(f_i, \varphi) > \lambda_{g_1}^{(p,q)}(f_j, \varphi)$ hold with at least $f_j(z)$ is of regular $(p,q)$-$\varphi$ relative Gol'dberg growth with respect to $g_1(z)$ where $i, j = 1, 2$ and $i \neq j$.*

**Theorem 8.1.2** *Let us consider $f_1(z)$, $f_2(z)$ and $g_1(z)$ are any three entire functions of $n$ complex variables such that $\rho_{g_1}^{(p,q)}(f_1, \varphi)$ and $\rho_{g_1}^{(p,q)}(f_2, \varphi)$ exists. Then*

$$\rho_{g_1}^{(p,q)}(f_1 \pm f_2, \varphi) \leq \max\{\rho_{g_1}^{(p,q)}(f_1, \varphi), \rho_{g_1}^{(p,q)}(f_2, \varphi)\}.$$

*The equality holds when $\rho_{g_1}^{(p,q)}(f_1, \varphi) \neq \rho_{g_1}^{(p,q)}(f_2, \varphi)$.*

**Theorem 8.1.3** *Let $f_1(z)$, $g_1(z)$ and $g_2(z)$ be any three entire functions of $n$ complex variables such that $\lambda_{g_1}^{(p,q)}(f_1, \varphi)$ and $\lambda_{g_2}^{(p,q)}(f_1, \varphi)$ exists. Then*

$$\lambda_{g_1 \pm g_2}^{(p,q)}(f_1, \varphi) \geq \min\{\lambda_{g_1}^{(p,q)}(f_1, \varphi), \lambda_{g_2}^{(p,q)}(f_1, \varphi)\}.$$

*The equality holds when $\lambda_{g_1}^{(p,q)}(f_1, \varphi) \neq \lambda_{g_2}^{(p,q)}(f_1, \varphi)$.*

**Theorem 8.1.4** *Let $f_1(z)$, $g_1(z)$ and $g_2(z)$ be any three entire functions of $n$ complex variables. Also let $f_1(z)$ is of regular $(p,q)$-$\varphi$ relative Gol'dberg growth with respect to at least any one of $g_1(z)$ or $g_2(z)$. Then*

$$\rho_{g_1 \pm g_2}^{(p,q)}(f_1, \varphi) \geq \min\{\rho_{g_1}^{(p,q)}(f_1, \varphi), \rho_{g_2}^{(p,q)}(f_1, \varphi)\}.$$

*The equality holds when any one of $\rho_{g_i}^{(p,q)}(f_1, \varphi) < \rho_{g_j}^{(p,q)}(f_1, \varphi)$ hold with at least $f_1(z)$ is of regular $(p,q)$-$\varphi$ relative Gol'dberg growth with respect to $g_j(z)$ where $i, j = 1, 2$ and $i \neq j$.*

**Theorem 8.1.5** *Let $f_1(z)$, $g_1(z)$ and $g_2(z)$ be any three entire functions of $n$ complex variables. Then*

$$\rho_{g_1 \pm g_2}^{(p,q)}(f_1 \pm f_2, \varphi) \leq \max[\min\{\rho_{g_1}^{(p,q)}(f_1, \varphi), \rho_{g_2}^{(p,q)}(f_1, \varphi)\}, \min\{\rho_{g_1}^{(p,q)}(f_2, \varphi), \rho_{g_2}^{(p,q)}(f_2, \varphi)\}]$$

*when the following two conditions holds:*

(i) $\rho_{g_i}^{(p,q)}(f_1, \varphi) < \rho_{g_j}^{(p,q)}(f_1, \varphi)$ *with at least* $f_1(z)$ *is of regular* $(p,q)$-$\varphi$ *relative Gol'dberg growth with respect to* $g_j(z)$ *for* $i = 1, 2, j = 1, 2$ *and* $i \neq j$; *and*

(ii) $\rho_{g_i}^{(p,q)}(f_2, \varphi) < \rho_{g_j}^{(p,q)}(f_2, \varphi)$ *with at least* $f_2(z)$ *is of regular* $(p,q)$-$\varphi$ *relative Gol'dberg growth with respect to* $g_j(z)$ *for* $i = 1, 2, j = 1, 2$ *and* $i \neq j$.

*The equality holds when any one of* $\rho_{g_1}^{(p,q)}(f_i, \varphi) < \rho_{g_1}^{(p,q)}(f_j, \varphi)$ *and any one of* $\rho_{g_2}^{(p,q)}(f_i, \varphi) < \rho_{g_2}^{(p,q)}(f_j, \varphi)$ *hold simultaneously for* $i = 1, 2; j = 1, 2$ *and* $i \neq j$.

**Theorem 8.1.6** *Let* $f_1(z)$, $g_1(z)$ *and* $g_2(z)$ *be any three entire functions of* $n$ *complex variables. Then*

$$\lambda_{g_1 \pm g_2}^{(p,q)}(f_1 \pm f_2, \varphi) \geq \min[\max\{\lambda_{g_1}^{(p,q)}(f_1, \varphi), \lambda_{g_1}^{(p,q)}(f_2, \varphi)\}, \max\{\lambda_{g_2}^{(p,q)}(f_1, \varphi), \lambda_{g_2}^{(p,q)}(f_2, \varphi)\}]$$

*when the following two conditions holds:*

(i) $\lambda_{g_1}^{(p,q)}(f_i, \varphi) > \lambda_{g_1}^{(p,q)}(f_j, \varphi)$ *with at least* $f_j(z)$ *is of regular* $(p,q)$-$\varphi$ *relative Gol'dberg growth with respect to* $g_1(z)$ *for* $i = 1, 2, j = 1, 2$ *and* $i \neq j$; *and*

(ii) $\lambda_{g_2}^{(p,q)}(f_i, \varphi) > \lambda_{g_2}^{(p,q)}(f_j, \varphi)$ *with at least* $f_j(z)$ *is of regular* $(p,q)$-$\varphi$ *relative Gol'dberg growth with respect to* $g_2(z)$ *for* $i = 1, 2, j = 1, 2$ *and* $i \neq j$.

*The equality holds when any one of* $\lambda_{g_i}^{(p,q)}(f_1, \varphi) < \lambda_{g_j}^{(p,q)}(f_1, \varphi)$ *and any one of* $\lambda_{g_i}^{(p,q)}(f_2, \varphi) < \lambda_{g_j}^{(p,q)}(f_2, \varphi)$ *hold simultaneously for* $i = 1, 2; j = 1, 2$ *and* $i \neq j$.

**Theorem 8.1.7** *Let us consider* $f_1(z)$, $f_2(z)$ *and* $g_1(z)$ *are any three entire functions of* $n$ *complex variables. Also let at least* $f_1(z)$ *or* $f_2(z)$ *is of regular* $(p,q)$-$\varphi$ *relative Gol'dberg growth with respect to* $g_1(z)$ *and* $g_1(z)$ *satisfy the Property (G). Then*

$$\lambda_{g_1}^{(p,q)}(f_1 \cdot f_2, \varphi) \leq \max\left\{\lambda_{g_1}^{(p,q)}(f_1, \varphi), \lambda_{g_1}^{(p,q)}(f_2, \varphi)\right\}.$$

*The equality holds when* $f_1(z)$ *and* $f_2(z)$ *satisfy Property (X).*

**Theorem 8.1.8** *Let us consider* $f_1(z)$, $f_2(z)$ *and* $g_1(z)$ *are any three entire functions of* $n$ *complex variables such that* $\rho_{g_1}^{(p,q)}(f_1, \varphi)$ *and* $\rho_{g_1}^{(p,q)}(f_2, \varphi)$ *exists and* $g_1(z)$ *satisfy the Property (G). Then*

$$\rho_{g_1}^{(p,q)}(f_1 \cdot f_2, \varphi) \leq \max\{\rho_{g_1}^{(p,q)}(f_1, \varphi), \rho_{g_1}^{(p,q)}(f_2, \varphi)\}.$$

*The equality holds when* $f_1$ *and* $f_2$ *satisfy Property (X).*

**Theorem 8.1.9** *Let* $f_1(z)$, $g_1(z)$ *and* $g_2(z)$ *be any three entire functions of* $n$ *complex variables such that* $\lambda_{g_1}^{(p,q)}(f_1, \varphi)$ *and* $\lambda_{g_2}^{(p,q)}(f_1, \varphi)$ *exists and* $g_1 \cdot g_2(z)$ *satisfy the Property (G). Then*

$$\lambda_{g_1 \cdot g_2}^{(p,q)}(f_1, \varphi) \geq \min\{\lambda_{g_1}^{(p,q)}(f_1, \varphi), \lambda_{g_2}^{(p,q)}(f_1, \varphi)\}.$$

*The equality holds when* $g_1(z)$ *and* $g_2(z)$ *satisfy Property (X).*

**Theorem 8.1.10** *Let $f_1(z)$, $g_1(z)$ and $g_2(z)$ be any three entire functions of $n$ complex variables. Also let $f_1(z)$ is of regular $(p,q)$-$\varphi$ relative Gol'dberg growth with respect to at least any one of $g_1(z)$ or $g_2(z)$ and $g_1 \cdot g_2(z)$ satisfy the Property (G). Then*

$$\rho_{g_1 \cdot g_2}^{(p,q)}(f_1, \varphi) \geq \min\left\{\rho_{g_1}^{(p,q)}(f_1, \varphi), \rho_{g_2}^{(p,q)}(f_1, \varphi)\right\}.$$

*The equality holds when $g_1(z)$ and $g_2(z)$ satisfy Property (X).*

**Theorem 8.1.11** *Let $f_1(z)$, $f_2(z)$, $g_1(z)$ and $g_2(z)$ be any four entire functions of $n$ complex variables. Also let $g_1 \cdot g_2(z)$ be satisfy the Property (G). Then*

$$\rho_{g_1 \cdot g_2}^{(p,q)}(f_1 \cdot f_2, \varphi)$$
$$= \max[\min\{\rho_{g_1}^{(p,q)}(f_1, \varphi), \rho_{g_1}^{(p,q)}(f_2, \varphi)\}, \min\{\rho_{g_2}^{(p,q)}(f_1, \varphi), \rho_{g_2}^{(p,q)}(f_2, \varphi)\}]$$

*when the following four conditions holds:*
*(i) $f_1(z)$ is of regular $(p,q)$-$\varphi$ relative Gol'dberg growth with respect to at least any one of $g_1(z)$ or $g_2(z)$;*
*(ii) $f_2(z)$ is of regular $(p,q)$-$\varphi$ relative Gol'dberg growth with respect to at least any one of $g_1(z)$ or $g_2(z)$;*
*(iii) $f_1(z)$ and $f_2(z)$ satisfy Property (X); and*
*(iv) $g_1(z)$ and $g_2(z)$ satisfy Property (X).*

**Theorem 8.1.12** *Let $f_1(z)$, $f_2(z)$, $g_1(z)$ and $g_2(z)$ be any four entire functions of $n$ complex variables.. Also let $g_1 \cdot g_2(z)$, $g_1(z)$ and $g_2(z)$ satisfy the Property (G). Then*

$$\lambda_{g_1 \cdot g_2}^{(p,q)}(f_1 \cdot f_2, \varphi)$$
$$= \min[\max\{\lambda_{g_1}^{(p,q)}(f_1, \varphi), \lambda_{g_1}^{(p,q)}(f_2, \varphi)\}, \max\{\lambda_{g_2}^{(p,q)}(f_1, \varphi), \lambda_{g_2}^{(p,q)}(f_2, \varphi)\}]$$

*when the following four conditions holds:*
*(i) At least $f_1(z)$ or $f_2(z)$ is of regular $(p,q)$-$\varphi$ relative Gol'dberg growth with respect to $g_1(z)$;*
*(ii) At least $f_1(z)$ or $f_2(z)$ is of regular $(p,q)$-$\varphi$ relative Gol'dberg growth with respect to $g_2(z)$;*
*(iii) $f_1(z)$ and $f_2(z)$ satisfy Property (X); and*
*(iv) $g_1(z)$ and $g_2(z)$ satisfy Property (X).*

In the cases of generalized relative Gol'dberg order $(\alpha, \beta)$ and generalized relative Gol'dberg type $(\alpha, \beta)$, it therefore seems natural to make parallel investigation of their basic properties.

Now extending the notion of addition and multiplication theorems, in this chapter we wish to extend the above results in the light of generalized relative Gol'dberg order $(\alpha, \beta)$ and generalized relative Gol'dberg type $(\alpha, \beta)$ where $\alpha$ and $\beta$ always denote the functions belonging to $L^0$.

In this connection, we finally remind the following definitions from which are needed in the sequel.

**Definition 8.1.1** *(cf. [2]) A non-constant entire function $f(z)$ of $n$ complex variables is said to have Property (G), if for any $\delta > 1$,*

$$(M_{f,D}(R))^2 \leq M_{f,D}(R^\delta).$$

**Definition 8.1.2** *A pair of entire functions $f(z)$ and $g(z)$ of $n$ complex variables are mutually said to have Property (X) if for all sufficiently large values of $R$, both*

$$M_{f\cdot g,D}(R) > M_{f,D}(R)$$

*and*

$$M_{f\cdot g,D}(R) > M_{g,D}(R)$$

*hold simultaneously.*

## 8.2   Main Results.

In this section we present the main results of this chapter.

**Theorem 8.2.1** *Let us consider $f_1(z)$, $f_2(z)$ and $g_1(z)$ are any three entire functions of $n$ complex variables. Also let at least $f_1(z)$ or $f_2(z)$ is of regular generalized relative Gol'dberg $(\alpha, \beta)$ growth with respect to $g_1(z)$. Then*

$$\lambda^{(\alpha,\beta)}[f_1 \pm f_2]_{g_1} \leq \max\left\{\lambda^{(\alpha,\beta)}[f_1]_{g_1}, \lambda^{(\alpha,\beta)}[f_2]_{g_1}\right\}.$$

*The equality holds when $\lambda^{(\alpha,\beta)}[f_i]_{g_1} > \lambda^{(\alpha,\beta)}[f_j]_{g_1}$ with at least $f_j(z)$ is of regular generalized relative Gol'dberg $(\alpha, \beta)$ growth with respect to $g_1(z)$ where $i, j = 1, 2$ and $i \neq j$.*

**Proof.** If $\lambda^{(\alpha,\beta)}[f_1 \pm f_2]_{g_1} = 0$ then the result is obvious. So we suppose that $\lambda^{(\alpha,\beta)}[f_1 \pm f_2]_{g_1} > 0$. We can clearly assume that $\lambda^{(\alpha,\beta)}[f_k]_{g_1}$ is finite for $k = 1, 2$. Also let

$$\max\{\lambda^{(\alpha,\beta)}[f_1]_{g_1}, \lambda^{(\alpha,\beta)}[f_2]_{g_1}\} = \Psi$$

and $f_2(z)$ is of regular relative generalized Gol'dberg growth $(\alpha, \beta)$ with respect to $g_1(z)$. Now for any arbitrary $\varepsilon > 0$ from the definition of $\lambda^{(\alpha,\beta)}[f_1]_{g_1}$, we obtain for a sequence values of $R$ tending to infinity that

$$M_{f_1,D}(R) \leq M_{g_1,D}\left(\alpha^{-1}\left(\log\left((\exp(\beta(R)))^{\left(\lambda^{(\alpha,\beta)}[f_1]_{g_1}+\varepsilon\right)}\right)\right)\right)$$

$$i.e., \ M_{f_1,D}(R) \leq M_{g_1,D}\left(\alpha^{-1}\left(\log\left((\exp(\beta(R)))^{(\Psi+\varepsilon)}\right)\right)\right). \tag{173}$$

Also for any arbitrary $\varepsilon > 0$ from the definition of $\rho^{(\alpha,\beta)}[f_2]_{g_1}\left(= \lambda^{(\alpha,\beta)}[f_2]_{g_1}\right)$, we obtain for all sufficiently large values of $R$ that

$$M_{f_2,D}(R) \leq M_{g_1,D}\left(\alpha^{-1}\left(\log\left((\exp(\beta(R)))^{\left(\lambda^{(\alpha,\beta)}[f_2]_{g_1}+\varepsilon\right)}\right)\right)\right) \tag{174}$$

$$i.e., \ M_{f_2,D}(R) \leq M_{g_1,D}\left(\alpha^{-1}\left(\log\left((\exp(\beta(R)))^{(\Psi+\varepsilon)}\right)\right)\right). \tag{175}$$

Therefore we obtain from (173) and (175) for a sequence values of $R$ tending to infinity that

$$M_{f_1\pm f_2,D}(R) < 2M_{g_1,D}\left(\alpha^{-1}\left(\log\left((\exp(\beta(R)))^{(\Psi+\varepsilon)}\right)\right)\right)$$

$$i.e., \ M_{f_1\pm f_2,D}(R) < M_{g_1,D}\left(\alpha^{-1}\left(\log\left((\exp(\beta(R)))^{(\Psi+2\varepsilon)}\right)\right)\right).$$

As $\varepsilon > 0$ is arbitrary, it follows from above that

$$i.e., \ \frac{\alpha\left(M_{g_1,D}^{-1}\left(M_{f_1\pm f_2,D}(R)\right)\right)}{\beta(R)} < (\Psi+2\varepsilon) \ .$$

Since $\varepsilon > 0$ is arbitrary, we get from above that

$$\lambda^{(\alpha,\beta)}[f_1 \pm f_2]_{g_1} \leq \Psi = \max\left\{\lambda^{(\alpha,\beta)}[f_1]_{g_1}, \lambda^{(\alpha,\beta)}[f_2]_{g_1}\right\}.$$

Similarly, if we consider that $f_1(z)$ is of regular generalized relative Gol'dberg $(\alpha, \beta)$ growth with respect to $g_1(z)$ or both $f_1(z)$ and $f_2(z)$ are of regular generalized relative Gol'dberg $(\alpha, \beta)$ growth with respect to $g_1(z)$, then one can easily verify that

$$\lambda^{(\alpha,\beta)}[f_1 \pm f_2]_{g_1} \leq \Psi = \max\left\{\lambda^{(\alpha,\beta)}[f_1]_{g_1}, \lambda^{(\alpha,\beta)}[f_2]_{g_1}\right\}. \tag{176}$$

Further without loss of any generality, let $\lambda^{(\alpha,\beta)}[f_1]_{g_1} < \lambda^{(\alpha,\beta)}[f_2]_{g_1}$ and $f(z) = f_1(z) \pm f_2(z)$.

So in view of (176) we get that $\lambda^{(\alpha,\beta)}[f]_{g_1} \leq \lambda^{(\alpha,\beta)}[f_2]_{g_1}$.

As, $f_2(z) = \pm(f(z) - f_1(z))$ and in this case we obtain that $\lambda^{(\alpha,\beta)}[f_2]_{g_1} \leq \max\left\{\lambda^{(\alpha,\beta)}[f]_{g_1}, \lambda^{(\alpha,\beta)}[f_1]_{g_1}\right\}$.

As we assume that $\lambda^{(\alpha,\beta)}[f_1]_{g_1} < \lambda^{(\alpha,\beta)}[f_2]_{g_1}$, therefore we have $\lambda^{(\alpha,\beta)}[f_2]_{g_1} \leq \lambda^{(\alpha,\beta)}[f]_{g_1}$ and hence $\lambda^{(\alpha,\beta)}[f]_{g_1} = \lambda^{(\alpha,\beta)}[f_2]_{g_1} = \max\left\{\lambda^{(\alpha,\beta)}[f_1]_{g_1}, \lambda^{(\alpha,\beta)}[f_2]_{g_1}\right\}$.

Therefore, $\lambda^{(\alpha,\beta)}[f_1 \pm f_2]_{g_1} = \lambda^{(\alpha,\beta)}[f_i]_{g_1} \mid i = 1,2$ provided $\lambda^{(\alpha,\beta)}[f_1]_{g_1} \neq \lambda^{(\alpha,\beta)}[f_2]_{g_1}$. Thus the theorem is established. ■

Now we state the following theorem without its proof as it can easily be carried out in the line of Theorem 8.2.1.

**Theorem 8.2.2** *Let us consider $f_1(z)$, $f_2(z)$ and $g_1(z)$ are any three entire functions of $n$ complex variables. Also let $\rho^{(\alpha,\beta)}[f_1]_{g_1}$ and $\rho^{(\alpha,\beta)}[f_2]_{g_1}$ exists, then*

$$\rho^{(\alpha,\beta)}[f_1 \pm f_2]_{g_1} \leq \max\left\{\rho^{(\alpha,\beta)}[f_1]_{g_1}, \rho^{(\alpha,\beta)}[f_2]_{g_1}\right\}.$$

*The equality holds when $\rho^{(\alpha,\beta)}[f_1]_{g_1} \neq \rho^{(\alpha,\beta)}[f_2]_{g_1}$.*

**Theorem 8.2.3** *Let $f_1(z)$, $g_1(z)$ and $g_2(z)$ be any three entire functions of $n$ complex variables. Also let $\lambda^{(\alpha,\beta)}[f_1]_{g_1}$ and $\lambda^{(\alpha,\beta)}[f_1]_{g_2}$ exists, then*

$$\lambda^{(\alpha,\beta)}[f_1]_{g_1 \pm g_2} \geq \min\left\{\lambda^{(\alpha,\beta)}[f_1]_{g_1}, \lambda^{(\alpha,\beta)}[f_1]_{g_2}\right\}.$$

*The sign of equality holds when $\lambda^{(\alpha,\beta)}[f_1]_{g_1} \neq \lambda^{(\alpha,\beta)}[f_1]_{g_2}$.*

**Proof.** If $\lambda^{(\alpha,\beta)}[f_1]_{g_1 \pm g_2} = \infty$, then the result is obvious. So let $\lambda^{(\alpha,\beta)}[f_1]_{g_1 \pm g_2} < \infty$. We can clearly assume that $\lambda^{(\alpha,\beta)}[f_1]_{g_k}$ is finite for $k = 1, 2$. Further let

$$\Psi = \min\left\{\lambda^{(\alpha,\beta)}[f_1]_{g_1}, \lambda^{(\alpha,\beta)}[f_1]_{g_2}\right\}.$$

Now for any arbitrary $\varepsilon > 0$ from the definition of $\lambda^{(\alpha,\beta)}[f_1]_{g_k}$ where $k = 1, 2$, we have for all sufficiently large values of $R$ that

$$M_{g_k,D}\left(\alpha^{-1}\left(\log\left((\exp(\beta(R)))^{\left(\lambda^{(\alpha,\beta)}[f_1]_{g_k} - \varepsilon\right)}\right)\right)\right) \leq M_{f_1,D}(R)$$

$$i.e, \ M_{g_k,D}\left(\alpha^{-1}\left(\log\left((\exp(\beta(R)))^{(\Psi - \varepsilon)}\right)\right)\right) \leq M_{f_1,D}(R) \tag{177}$$

Now we obtain from above for all sufficiently large values of $R$ that

$$M_{g_1 \pm g_2,D}\left(\alpha^{-1}\left(\log\left((\exp(\beta(R)))^{(\Psi - \varepsilon)}\right)\right)\right) < M_{g_1,D}\left(\alpha^{-1}\left(\log\left((\exp(\beta(R)))^{(\Psi - \varepsilon)}\right)\right)\right)$$

$$+ M_{g_2,D}\left(\alpha^{-1}\left(\log\left((\exp(\beta(R)))^{(\Psi - \varepsilon)}\right)\right)\right)$$

$$i.e., \ M_{g_1 \pm g_2,D}\left(\alpha^{-1}\left(\log\left((\exp(\beta(R)))^{(\Psi - \varepsilon)}\right)\right)\right) < 2M_{f_1,D}(R)$$

$$i.e., \ \frac{1}{2}M_{g_1 \pm g_2,D}\left(\alpha^{-1}\left(\log\left((\exp(\beta(R)))^{(\Psi - \varepsilon)}\right)\right)\right) < M_{f_1,D}(R)$$

$$i.e., \ M_{g_1 \pm g_2,D}\left(\alpha^{-1}\left(\log\left((\exp(\beta(R)))^{(\Psi - 2\varepsilon)}\right)\right)\right) < M_{f_1,D}(R)$$

$$i.e., \ \frac{\alpha\left(M_{g_1 \pm g_2,D}^{-1}(M_{f_1,D}(R))\right)}{\beta(R)} > \Psi - 2\varepsilon.$$

Since $\varepsilon > 0$ is arbitrary, we get from above that

$$\lambda^{(\alpha,\beta)}[f_1]_{g_1 \pm g_2} \geq \Psi = \min\left\{\lambda^{(\alpha,\beta)}[f_1]_{g_1}, \lambda^{(\alpha,\beta)}[f_1]_{g_2}\right\}. \tag{178}$$

Now without loss of any generality, we may consider that $\lambda^{(\alpha,\beta)}[f_1]_{g_1} < \lambda^{(\alpha,\beta)}[f_1]_{g_2}$ and $g = g_1 \pm g_2$. Then in view of (178) we get that $\lambda^{(\alpha,\beta)}[f_1]_g \geq \lambda^{(\alpha,\beta)}[f_1]_{g_1}$.

Further, $g_1 = (g \pm g_2)$ and in this case we obtain that

$$\lambda^{(\alpha,\beta)}[f_1]_{g_1} \geq \min\left\{\lambda^{(\alpha,\beta)}[f_1]_g, \lambda^{(\alpha,\beta)}[f_1]_{g_2}\right\}.$$

As we assume that $\lambda^{(\alpha,\beta)}[f_1]_{g_1} < \lambda^{(\alpha,\beta)}[f_1]_{g_2}$, therefore we have $\lambda^{(\alpha,\beta)}[f_1]_{g_1} \geq \lambda^{(\alpha,\beta)}[f_1]_g$ and hence $\lambda^{(\alpha,\beta)}[f_1]_g = \lambda^{(\alpha,\beta)}[f_1]_{g_1} = \min\{\lambda^{(\alpha,\beta)}[f_1]_{g_1}, \lambda^{(\alpha,\beta)}[f_1]_{g_2}\}$.

Hence the theorem follows. ∎

**Theorem 8.2.4** *Let $f_1(z)$, $g_1(z)$ and $g_2(z)$ be any three entire functions of $n$ complex variables. Also let $f_1(z)$ is of regular generalized relative Gol'dberg $(\alpha, \beta)$ growth with respect to at least any one of $g_1(z)$ or $g_2(z)$. Then*

$$\rho^{(\alpha,\beta)}[f_1]_{g_1 \pm g_2} \geq \min\left\{\rho^{(\alpha,\beta)}[f_1]_{g_1}, \rho^{(\alpha,\beta)}[f_1]_{g_2}\right\}.$$

*The sign of equality holds when $\rho^{(\alpha,\beta)}[f_1]_{g_i} < \rho^{(\alpha,\beta)}[f_1]_{g_j}$ with at least $f_1(z)$ is of regular generalized relative Gol'dberg $(\alpha, \beta)$ growth with respect to $g_j(z)$ where $i, j = 1, 2$ and $i \neq j$.*

We omit the proof of Theorem 8.2.4 as it can easily be carried out in the line of Theorem 8.2.3.

**Theorem 8.2.5** *Let $f_1(z)$, $g_1(z)$ and $g_2(z)$ be any three entire functions of $n$ complex variables. Then*

$$\rho^{(\alpha,\beta)}[f_1 \pm f_2]_{g_1 \pm g_2}$$
$$\leq \max[\min\{\rho^{(\alpha,\beta)}[f_1]_{g_1}, \rho^{(\alpha,\beta)}[f_1]_{g_2}\}, \min\{\rho^{(\alpha,\beta)}[f_2]_{g_1}, \rho^{(\alpha,\beta)}[f_2]_{g_2}\}]$$

*when the following two conditions holds:*
*(i) $\rho^{(\alpha,\beta)}[f_1]_{g_i} < \rho^{(\alpha,\beta)}[f_1]_{g_j}$ with at least $f_1(z)$ is of regular generalized relative Gol'dberg $(\alpha, \beta)$ growth with respect to $g_j(z)$ for $i = 1, 2$, $j = 1, 2$ and $i \neq j$; and*
*(ii) $\rho^{(\alpha,\beta)}[f_2]_{g_i} < \rho^{(\alpha,\beta)}[f_2]_{g_j}$ with at least $f_2(z)$ is of regular generalized relative Gol'dberg $(\alpha, \beta)$ growth with respect to $g_j(z)$ for $i = 1, 2$, $j = 1, 2$ and $i \neq j$.*
*The sign of equality holds when $\rho^{(\alpha,\beta)}[f_i]_{g_1} < \rho^{(\alpha,\beta)}[f_j]_{g_1}$ and $\rho^{(\alpha,\beta)}[f_i]_{g_2} < \rho^{(\alpha,\beta)}[f_j]_{g_2}$ holds simultaneously for $i = 1, 2$; $j = 1, 2$ and $i \neq j$.*

**Proof.** In view of Theorem 8.2.2 and Theorem 8.2.4 we get that

$$\max[\min\{\rho^{(\alpha,\beta)}[f_1]_{g_1}, \rho^{(\alpha,\beta)}[f_1]_{g_2}\}, \min\{\rho^{(\alpha,\beta)}[f_2]_{g_1}, \rho^{(\alpha,\beta)}[f_2]_{g_2}\}]$$
$$= \max[\rho^{(\alpha,\beta)}[f_1]_{g_1 \pm g_2}, \rho^{(\alpha,\beta)}[f_2]_{g_1 \pm g_2}]$$
$$\geq \rho^{(\alpha,\beta)}[f_1 \pm f_2]_{g_1 \pm g_2}. \tag{179}$$

Since $\rho^{(\alpha,\beta)}[f_i]_{g_1} < \rho^{(\alpha,\beta)}[f_j]_{g_1}$ and $\rho^{(\alpha,\beta)}[f_i]_{g_2} < \rho^{(\alpha,\beta)}[f_j]_{g_2}$ hold simultaneously for $i = 1, 2$; $j = 1, 2$ and $i \neq j$, we obtain that

$$\text{either } \min\{\rho^{(\alpha,\beta)}[f_1]_{g_1}, \rho^{(\alpha,\beta)}[f_1]_{g_2}\} > \min\{\rho^{(\alpha,\beta)}[f_2]_{g_1}, \rho^{(\alpha,\beta)}[f_2]_{g_2}\} \text{ or}$$

$$\min\{\rho^{(\alpha,\beta)}[f_2]_{g_1}, \rho^{(\alpha,\beta)}[f_2]_{g_2}\} > \min\{\rho^{(\alpha,\beta)}[f_1]_{g_1}, \rho^{(\alpha,\beta)}[f_1]_{g_2}\} \text{ holds.}$$

Therefore in view of the conditions $(i)$ and $(ii)$ of the theorem, it follows from above that

$$\text{either } \rho^{(\alpha,\beta)}[f_1]_{g_1 \pm g_2} > \rho^{(\alpha,\beta)}[f_2]_{g_1 \pm g_2} \text{ or}$$
$$\rho^{(\alpha,\beta)}[f_2]_{g_1 \pm g_2} > \rho^{(\alpha,\beta)}[f_1]_{g_1 \pm g_2},$$

which is the condition for holding equality in (179).

Hence the theorem follows. ∎

**Theorem 8.2.6** *Let $f_1(z)$, $g_1(z)$ and $g_2(z)$ be any three entire functions of $n$ complex variables. Then*

$$\lambda^{(\alpha,\beta)}\left[f_1 \pm f_2\right]_{g_1 \pm g_2}$$
$$\geq \min[\max\{\lambda^{(\alpha,\beta)}\left[f_1\right]_{g_1}, \lambda^{(\alpha,\beta)}\left[f_2\right]_{g_1}\}, \max\{\lambda^{(\alpha,\beta)}\left[f_1\right]_{g_2}, \lambda^{(\alpha,\beta)}\left[f_2\right]_{g_2}\}]$$

*when the following two conditions holds:*
*(i) $\lambda^{(\alpha,\beta)}\left[f_i\right]_{g_1} > \lambda^{(\alpha,\beta)}\left[f_j\right]_{g_1}$ with at least $f_j(z)$ is of regular generalized relative Gol'dberg $(\alpha, \beta)$ growth with respect to $g_1(z)$ for $i = 1, 2$, $j = 1, 2$ and $i \neq j$; and*
*(ii) $\lambda^{(\alpha,\beta)}\left[f_i\right]_{g_2} > \lambda^{(\alpha,\beta)}\left[f_j\right]_{g_2}$ with at least $f_j(z)$ is of regular generalized relative Gol'dberg $(\alpha, \beta)$ growth with respect to $g_2(z)$ for $i = 1, 2$, $j = 1, 2$ and $i \neq j$.*
*The sign of equality holds when $\lambda^{(\alpha,\beta)}\left[f_1\right]_{g_i} < \lambda^{(\alpha,\beta)}\left[f_1\right]_{g_j}$ and $\lambda^{(\alpha,\beta)}\left[f_2\right]_{g_i} < \lambda^{(\alpha,\beta)}\left[f_2\right]_{g_j}$ hold simultaneously for $i = 1, 2$; $j = 1, 2$ and $i \neq j$.*

**Proof.** In view of Theorem 8.2.1 and Theorem 8.2.3, we obtain that

$$\min[\max\{\lambda^{(\alpha,\beta)}\left[f_1\right]_{g_1}, \lambda^{(\alpha,\beta)}\left[f_2\right]_{g_1}\}, \max\{\lambda^{(\alpha,\beta)}\left[f_1\right]_{g_2}, \lambda^{(\alpha,\beta)}\left[f_2\right]_{g_2}\}]$$
$$= \min[\lambda^{(\alpha,\beta)}\left[f_1 \pm f_2\right]_{g_1}, \lambda^{(\alpha,\beta)}\left[f_1 \pm f_2\right]_{g_2}]$$
$$\geq \lambda^{(\alpha,\beta)}\left[f_1 \pm f_2\right]_{g_1 \pm g_2}. \tag{180}$$

Since $\lambda^{(\alpha,\beta)}\left[f_1\right]_{g_i} < \lambda^{(\alpha,\beta)}\left[f_1\right]_{g_j}$ and $\lambda^{(\alpha,\beta)}\left[f_2\right]_{g_i} < \lambda^{(\alpha,\beta)}\left[f_2\right]_{g_j}$ hold simultaneously for $i = 1, 2$; $j = 1, 2$ and $i \neq j$, we get that

$$\text{either } \max\{\lambda^{(\alpha,\beta)}\left[f_1\right]_{g_1}, \lambda^{(\alpha,\beta)}\left[f_2\right]_{g_1}\} < \max\{\lambda^{(\alpha,\beta)}\left[f_1\right]_{g_2}, \lambda^{(\alpha,\beta)}\left[f_2\right]_{g_2}\} \text{ or}$$

$$\max\{\lambda^{(\alpha,\beta)}\left[f_1\right]_{g_2}, \lambda^{(\alpha,\beta)}\left[f_2\right]_{g_2}\} < \max\{\lambda^{(\alpha,\beta)}\left[f_1\right]_{g_1}, \lambda^{(\alpha,\beta)}\left[f_2\right]_{g_1}\} \text{ holds.}$$

Since condition $(i)$ and $(ii)$ of the theorem hold, it follows from above that

$$\text{either } \lambda^{(\alpha,\beta)}\left[f_1 \pm f_2\right]_{g_1} < \lambda^{(\alpha,\beta)}\left[f_1 \pm f_2\right]_{g_2} \text{ or}$$
$$\lambda^{(\alpha,\beta)}\left[f_1 \pm f_2\right]_{g_2} < \lambda^{(\alpha,\beta)}\left[f_1 \pm f_2\right]_{g_1},$$

which is the condition for holding equality in (180).
    Hence the theorem follows. ∎

**Theorem 8.2.7** *Let us consider $f_1(z)$, $f_2(z)$ and $g_1(z)$ are any three entire functions of $n$ complex variables. Also let at least $f_1(z)$ or $f_2(z)$ is of regular generalized relative Gol'dberg $(\alpha, \beta)$ growth with respect to $g_1(z)$ and $g_1(z)$ satisfy the Property (G). Then*

$$\lambda^{(\alpha,\beta)}\left[f_1 \cdot f_2\right]_{g_1} \leq \max\left\{\lambda^{(\alpha,\beta)}\left[f_1\right]_{g_1}, \lambda^{(\alpha,\beta)}\left[f_2\right]_{g_1}\right\}.$$

*The equality holds when $f_1(z)$ and $f_2(z)$ satisfy Property (X).*

**Proof.** Let $\lambda^{(\alpha,\beta)}\left[f_1 \cdot f_2\right]_{g_1} > 0$. Otherwise if $\lambda^{(\alpha,\beta)}\left[f_1 \cdot f_2\right]_{g_1} = 0$ then the result is obvious. Also let $f_2(z)$ is of regular generalized relative Gol'dberg $(\alpha,\beta)$ growth with respect to $g_1(z)$ and $\max\left\{\lambda^{(\alpha,\beta)}\left[f_1\right]_{g_1}, \lambda^{(\alpha,\beta)}\left[f_2\right]_{g_1}\right\} = \Psi$. We can clearly assume that $\lambda^{(\alpha,\beta)}\left[f_k\right]_{g_1}$ is finite for $k = 1, 2$. Therefore we obtain from (173) and (175) for a sequence values of $R$ tending to infinity that

$$M_{f_1 \cdot f_2, D}(R) < \left[M_{g_1, D}\left(\alpha^{-1}\log\left(\left(\exp\left(\beta\left(R\right)\right)\right)^{(\Psi+\varepsilon)}\right)\right)\right]^2.$$

Now in view of Definition 8.1.1, we obtain from above for a sequence values of $R$ tending to infinity that

$$M_{f_1 \cdot f_2, D}(R) < M_{g_1, D}\left(\alpha^{-1}\log\left(\left(\exp\left(\beta\left(R\right)\right)\right)^{\delta(\Psi+\varepsilon)}\right)\right),$$

since $g_1(z)$ has the Property (G) and $\delta > 1$. As $\varepsilon > 0$ is arbitrary, we obtain from above by letting $\delta \to 1^+$ that

$$\lambda^{(\alpha,\beta)}\left[f_1 \cdot f_2\right]_{g_1} \leq \Psi = \max\left\{\lambda^{(\alpha,\beta)}\left[f_1\right]_{g_1}, \lambda^{(\alpha,\beta)}\left[f_2\right]_{g_1}\right\}.$$

Similarly if we consider that $f_1(z)$ is of regular generalized relative Gol'dberg $(\alpha,\beta)$ growth with respect to $g_1(z)$ or both $f_1(z)$ and $f_2(z)$ are of regular generalized relative Gol'dberg $(\alpha,\beta)$ growth with respect to $g_1(z)$, then one can also easily get the above conclusion.

Now let $f_1(z)$ and $f_2(z)$ satisfy Property (X), then of course we have $M_{f_1 \cdot f_2, D}(R) > M_{f_1, D}(R)$ and $M_{f_1 \cdot f_2, D}(R) > M_{f_2, D}(R)$ for all sufficiently large values of $R$. Therefore for all sufficiently large values of $R$ we get that

$$\frac{\alpha\left(M_{g_1, D}^{-1}\left(M_{f_1 \cdot f_2, D}(R)\right)\right)}{\beta(R)} > \frac{\alpha\left(M_{g_1, D}^{-1}\left(M_{f_1, D}(R)\right)\right)}{\beta(R)}.$$

So $\lambda^{(\alpha,\beta)}\left[f_1 \cdot f_2\right]_{g_1} \geq \lambda^{(\alpha,\beta)}\left[f_1\right]_{g_1}$ and similarly, $\lambda^{(\alpha,\beta)}\left[f_1 \cdot f_2\right]_{g_1} \geq \lambda^{(\alpha,\beta)}\left[f_2\right]_{g_1}$. Hence the theorem follows. ∎

Now we state the following theorem without its proof as it can easily be carried out in the line of Theorem 8.2.7.

**Theorem 8.2.8** *Let us consider $f_1(z)$, $f_2(z)$ and $g_1(z)$ are any three entire functions of $n$ complex variables such that $\rho^{(\alpha,\beta)}\left[f_1\right]_{g_1}$ and $\rho^{(\alpha,\beta)}\left[f_2\right]_{g_1}$ exists and $g_1(z)$ satisfy the Property (G). Then*

$$\rho^{(\alpha,\beta)}\left[f_1 \cdot f_2\right]_{g_1} \leq \max\left\{\rho^{(\alpha,\beta)}\left[f_1\right]_{g_1}, \rho^{(\alpha,\beta)}\left[f_2\right]_{g_1}\right\}.$$

*The equality holds when $f_1(z)$ and $f_2(z)$ satisfy Property (X).*

**Theorem 8.2.9** *Let $f_1(z)$, $g_1(z)$ and $g_2(z)$ be any three entire functions of $n$ complex variables such that $\lambda^{(\alpha,\beta)}\left[f_1\right]_{g_1}$ and $\lambda^{(\alpha,\beta)}\left[f_1\right]_{g_2}$ exists and $g_1 \cdot g_2(z)$ satisfy the Property (G). Then*

$$\lambda^{(\alpha,\beta)}\left[f_1\right]_{g_1 \cdot g_2} \geq \min\left\{\lambda^{(\alpha,\beta)}\left[f_1\right]_{g_1}, \lambda^{(\alpha,\beta)}\left[f_1\right]_{g_2}\right\}.$$

*The equality holds when $g_1(z)$ and $g_2(z)$ satisfy Property (X).*

**Proof.** Let $\lambda^{(\alpha,\beta)} [f_1]_{g_1 \cdot g_2} < \infty$. Otherwise if $\lambda^{(\alpha,\beta)} [f_1]_{g_1 \cdot g_2} = \infty$ then the result is obvious. Also let $\min \left\{ \lambda^{(\alpha,\beta)} [f_1]_{g_1}, \lambda^{(\alpha,\beta)} [f_1]_{g_2} \right\} = \Psi$ .We can clearly assume that $\lambda^{(\alpha,\beta)} [f_1]_{g_k}$ is finite for $k = 1, 2$. Now for any arbitrary $\varepsilon > 0$, with $\varepsilon < \Psi$, we obtain from (177) for all sufficiently large values of $R$ that

$$M_{g_1 \cdot g_2, D} \left( \alpha^{-1} \left( \log \left( (\exp (\beta (R)))^{(\Psi - \varepsilon)} \right) \right) \right) < [M_{f_1, D} (R)]^2$$

$$i.e., \left[ M_{g_1 \cdot g_2, D} \left( \alpha^{-1} \left( \log \left( (\exp (\beta (R)))^{(\Psi - \varepsilon)} \right) \right) \right) \right]^{\frac{1}{2}} < M_{f_1, D} (R).$$

Now in view of Definition 8.1.1, we obtain from above for all sufficiently large values of $R$ that

$$M_{g_1 \cdot g_2, D} \left( \alpha^{-1} \left( \log \left( (\exp (\beta (R)))^{\frac{(\Psi - \varepsilon)}{\delta}} \right) \right) \right) < M_{f_1, D} (R)$$

since $g_1 \cdot g_2 (z)$ has the Property (G) and $\delta > 1$. As $\varepsilon > 0$ is arbitrary, we obtain from above by letting $\delta \to 1^+$ that

$$\lambda^{(\alpha,\beta)} [f_1]_{g_1 \cdot g_2} \geq \Psi = \min \left\{ \lambda^{(\alpha,\beta)} [f_1]_{g_1}, \lambda^{(\alpha,\beta)} [f_1]_{g_2} \right\}.$$

Now let $g_1 (z)$ and $g_2 (z)$ satisfy Property (X), then of course we have $M_{g_1 \cdot g_2, D} (R) > M_{g_1, D} (R)$ and $M_{g_1 \cdot g_2, D} (R) > M_{g_2, D} (R)$ for all sufficiently large values of $R$. Therefore for all sufficiently large values of $R$, we obtain that $M_{g_1 \cdot g_2, D}^{-1} (R) \leq M_{g_1, D}^{-1} (R)$ and $M_{g_1 \cdot g_2, D}^{-1} (R) \leq M_{g_2, D}^{-1} (R)$. Hence it follows that for all sufficiently large values of $R$

$$\frac{\alpha \left( M_{g_1 \cdot g_2, D}^{-1} (M_{f_1, D} (R)) \right)}{\beta (R)} < \frac{\alpha \left( M_{g_1, D}^{-1} (M_{f_1, D} (R)) \right)}{\beta (R)}.$$

So $\lambda^{(\alpha,\beta)} [f_1]_{g_1 \cdot g_2} \leq \lambda^{(\alpha,\beta)} [f_1]_{g_1}$ and similarly, $\lambda^{(\alpha,\beta)} [f_1]_{g_1 \cdot g_2} \leq \lambda^{(\alpha,\beta)} [f_1]_{g_2}$. Thus the theorem follows. ∎

**Theorem 8.2.10** *Let $f_1 (z)$, $g_1 (z)$ and $g_2 (z)$ be any three entire functions of $n$ complex variables. Also let $f_1 (z)$ is of regular generalized relative Gol'dberg $(\alpha, \beta)$ growth with respect to at least any one of $g_1 (z)$ or $g_2 (z)$ and $g_1 \cdot g_2 (z)$ satisfy the Property (G). Then*

$$\rho^{(\alpha,\beta)} [f_1]_{g_1 \cdot g_2} \geq \min \left\{ \rho^{(\alpha,\beta)} [f_1]_{g_1}, \rho^{(\alpha,\beta)} [f_1]_{g_2} \right\}.$$

*The equality holds when $g_1 (z)$ and $g_2 (z)$ satisfy Property (X).*

Now we state the following two theorems without their proofs as those can easily be carried out with the help of Theorem 8.2.7, Theorem 8.2.8, Theorem 8.2.9 and Theorem 8.2.10 and in the line of Theorem 8.2.5 and Theorem 8.2.6 respectively.

**Theorem 8.2.11** *Let* $f_1(z)$, $f_2(z)$, $g_1(z)$ *and* $g_2(z)$ *be any four entire functions of* $n$ *complex variables. Also let* $g_1 \cdot g_2(z)$ *be satisfy the Property (G). Then*

$$\rho^{(\alpha,\beta)}[f_1 \cdot f_2]_{g_1 \cdot g_2}$$
$$= \max[\min\{\rho^{(\alpha,\beta)}[f_1]_{g_1}, \rho^{(\alpha,\beta)}[f_2]_{g_1}\}, \min\{\rho^{(\alpha,\beta)}[f_1]_{g_2}, \rho^{(\alpha,\beta)}[f_2]_{g_2}\}]$$

*when the following four conditions hold:*
*(i)* $f_1(z)$ *is of regular generalized relative Gol'dberg* $(\alpha,\beta)$ *growth with respect to at least any one of* $g_1(z)$ *or* $g_2(z)$;
*(ii)* $f_2(z)$ *is of regular generalized relative Gol'dberg* $(\alpha,\beta)$ *growth with respect to at least any one of* $g_1(z)$ *or* $g_2(z)$;
*(iii)* $f_1(z)$ *and* $f_2(z)$ *satisfy Property (X); and*
*(iv)* $g_1(z)$ *and* $g_2(z)$ *satisfy Property (X).*

**Theorem 8.2.12** *Let* $f_1(z)$, $f_2(z)$, $g_1(z)$ *and* $g_2(z)$ *be any four entire functions of* $n$ *complex variables. Also let* $g_1 \cdot g_2(z)$, $g_1(z)$ *and* $g_2(z)$ *satisfy the Property (G). Then*

$$\lambda^{(\alpha,\beta)}[f_1 \cdot f_2]_{g_1 \cdot g_2}$$
$$= \min[\max\{\lambda^{(\alpha,\beta)}[f_1]_{g_1}, \lambda^{(\alpha,\beta)}[f_2]_{g_1}\}, \max\{\lambda^{(\alpha,\beta)}[f_1]_{g_2}, \lambda^{(\alpha,\beta)}[f_2]_{g_2}\}]$$

*when the following four conditions hold:*
*(i) At least* $f_1(z)$ *or* $f_2(z)$ *is of regular generalized relative Gol'dberg* $(\alpha,\beta)$ *growth with respect to* $g_1(z)$;
*(ii) At least* $f_1(z)$ *or* $f_2(z)$ *is of regular generalized relative Gol'dberg* $(\alpha,\beta)$ *growth with respect to* $g_2(z)$;
*(iii)* $f_1(z)$ *and* $f_2(z)$ *satisfy Property (X); and*
*(iv)* $g_1(z)$ *and* $g_2(z)$ *satisfy Property (X).*

Next we intend to find out some theorems of relating to generalized relative Gol'dberg type $(\alpha,\beta)$ and generalized relative Gol'dberg weak type $(\alpha,\beta)$ of entire functions of of $n$ complex variables with respect to another one taking into consideration of the above theorems.

**Theorem 8.2.13** *Let* $f_1(z)$, $f_2(z)$, $g_1(z)$ *and* $g_2(z)$ *be any four entire functions of* $n$ *complex variables. Also let* $\rho^{(\alpha,\beta)}[f_1]_{g_1}$, $\rho^{(\alpha,\beta)}[f_2]_{g_1}$, $\rho^{(\alpha,\beta)}[f_1]_{g_2}$ *and* $\rho^{(\alpha,\beta)}[f_2]_{g_2}$ *are all non-zero and finite.*
*(A) If* $\rho^{(\alpha,\beta)}[f_i]_{g_1} > \rho^{(\alpha,\beta)}[f_j]_{g_1}$ *for* $i = j = 1, 2$ *and* $i \neq j$, *then*

$$\sigma_D^{(\alpha,\beta)}[f_1 \pm f_2]_{g_1} = \sigma_D^{(\alpha,\beta)}[f_i]_{g_1} \quad and \quad \overline{\sigma}_D^{(\alpha,\beta)}[f_1 \pm f_2]_{g_1} = \overline{\sigma}_D^{(\alpha,\beta)}[f_i]_{g_1}.$$

*(B) If* $\rho^{(\alpha,\beta)}[f_1]_{g_i} < \rho^{(\alpha,\beta)}[f_1]_{g_j}$ *with at least* $f_1(z)$ *is of regular generalized relative Gol'dberg* $(\alpha,\beta)$ *growth with respect to* $g_j(z)$ *for* $i = j = 1, 2$ *and* $i \neq j$, *then*

$$\sigma_D^{(\alpha,\beta)}[f_1]_{g_1 \pm g_2} = \sigma_D^{(\alpha,\beta)}[f_1]_{g_i} \quad and \quad \overline{\sigma}_D^{(\alpha,\beta)}[f_1]_{g_1 \pm g_2} = \overline{\sigma}_D^{(\alpha,\beta)}[f_1]_{g_i}.$$

*(C)Assume the functions* $f_1(z)$, $f_2(z)$, $g_1(z)$ *and* $g_2(z)$ *satisfy the following conditions:*
*(i)* $\rho^{(\alpha,\beta)}[f_1]_{g_i} < \rho^{(\alpha,\beta)}[f_1]_{g_j}$ *with at least* $f_1(z)$ *is of regular generalized relative Gol'dberg* $(\alpha,\beta)$ *growth with respect to* $g_j(z)$ *for* $i = 1, 2$, $j = 1, 2$ *and* $i \neq j$;
*(ii)* $\rho^{(\alpha,\beta)}[f_2]_{g_i} < \rho^{(\alpha,\beta)}[f_2]_{g_j}$ *with at least* $f_2(z)$ *is of regular generalized relative Gol'dberg* $(\alpha,\beta)$ *growth with respect to* $g_j(z)$ *for* $i = 1, 2$, $j = 1, 2$ *and* $i \neq j$;
*(iii)* $\rho^{(\alpha,\beta)}[f_1]_{g_i} > \rho^{(\alpha,\beta)}[f_1]_{g_j}$ *and* $\rho^{(\alpha,\beta)}[f_2]_{g_i} > \rho^{(\alpha,\beta)}[f_2]_{g_j}$ *holds simultaneously for* $i = 1, 2$; $j = 1, 2$ *and* $i \neq j$;
*(iv)* $\rho^{(\alpha,\beta)}[f_l]_{g_m} = \max[\min\{\rho^{(\alpha,\beta)}[f_1]_{g_1}, \rho^{(\alpha,\beta)}[f_1]_{g_2}\}, \min\{\rho^{(\alpha,\beta)}[f_2]_{g_1}, \rho^{(\alpha,\beta)}[f_2]_{g_2}\}] \mid l = m = 1, 2$;
*then we have*

$$\sigma_D^{(\alpha,\beta)}[f_1 \pm f_2]_{g_1 \pm g_2} = \sigma_D^{(\alpha,\beta)}[f_l]_{g_m} \quad and \quad \overline{\sigma}_D^{(\alpha,\beta)}[f_1 \pm f_2]_{g_1 \pm g_2} = \overline{\sigma}_D^{(\alpha,\beta)}[f_l]_{g_m}.$$

**Proof.** We obtain for all sufficiently large values of $R$ that

$$M_{f_k,D}(R) \leq M_{g_l,D}\left(\alpha^{-1}(\log\{(\sigma_D^{(\alpha,\beta)}[f_k]_{g_l} + \varepsilon)(\exp(\beta(R)))^{\rho^{(\alpha,\beta)}[f_k]_{g_l}}\})\right), \tag{181}$$

$$M_{f_k,D}(R) \geq M_{g_l,D}\left(\alpha^{-1}(\log\{(\overline{\sigma}_D^{(\alpha,\beta)}[f_k]_{g_l} - \varepsilon)(\exp(\beta(R)))^{\rho^{(\alpha,\beta)}[f_k]_{g_l}}\})\right) \tag{182}$$

and for a sequence of values of $R$ tending to infinity, we obtain that

$$M_{f_k,D}(R) \geq M_{g_l,D}\left(\alpha^{-1}(\log\{(\sigma_D^{(\alpha,\beta)}[f_k]_{g_l} - \varepsilon)(\exp(\beta(R)))^{\rho^{(\alpha,\beta)}[f_k]_{g_l}}\})\right) \tag{183}$$

and

$$M_{f_k,D}(R) \leq M_{g_l,D}\left(\alpha^{-1}(\log\{(\overline{\sigma}_D^{(\alpha,\beta)}[f_k]_{g_l} + \varepsilon)(\exp(\beta(R)))^{\rho^{(\alpha,\beta)}[f_k]_{g_l}}\})\right), \quad / \tag{184}$$

where $\varepsilon > 0$ is any arbitrary positive number $k = 1, 2$ and $l = 1, 2$.

**Case I.** Suppose that $\rho^{(\alpha,\beta)}[f_1]_{g_1} > \rho^{(\alpha,\beta)}[f_2]_{g_1}$ hold. Then for arbitrary $\varepsilon (> 0)$ and for all sufficiently large values of $R$, we get in view of (181) that

$$M_{f_1 \pm f_2,D}(R) \leq M_{f_1,D}(R) + M_{f_2,D}(R)$$

$$\leq M_{g_1,D}(\alpha^{-1}(\log\{(\sigma_D^{(\alpha,\beta)}[f_1]_{g_1} + \varepsilon)(\exp(\beta(R)))^{\rho^{(\alpha,\beta)}[f_1]_{g_1}}\}))(1 + A) \tag{185}$$

where $A = \dfrac{M_{g_1,D}(\alpha^{-1}(\log\{(\sigma_D^{(\alpha,\beta)}[f_2]_{g_1} + \varepsilon)(\exp(\beta(R)))^{\rho^{(\alpha,\beta)}[f_2]_{g_1}}\}))}{M_{g_1,D}(\alpha^{-1}(\log\{(\sigma_D^{(\alpha,\beta)}[f_1]_{g_1} + \varepsilon)(\exp(\beta(R)))^{\rho^{(\alpha,\beta)}[f_1]_{g_1}}\}))}$ and in view of $\rho^{(\alpha,\beta)}[f_1]_{g_1} > \rho^{(\alpha,\beta)}[f_2]_{g_1}$, and for all sufficiently large values of $R$, we can make the term $A$ sufficiently small, i.e. $A < \varepsilon_1$. Hence for any $\delta = 1 + \varepsilon_1$, it follows from (185) for all sufficiently large values of $R$ that

$$M_{f_1 \pm f_2,D}(R)$$
$$\leq M_{g_1,D}(\alpha^{-1}(\log\{(\sigma_D^{(\alpha,\beta)}[f_1]_{g_1} + \varepsilon)(\exp(\beta(R)))^{\rho^{(\alpha,\beta)}[f_1]_{g_1}}\}))(1 + \varepsilon_1)$$

$i.e.,\ M_{f_1 \pm f_2, D}(R)$

$$\leq M_{g_1, D}(\alpha^{-1}(\log\{(\sigma_D^{(\alpha,\beta)}[f_1]_{g_1} + \varepsilon)(\exp(\beta(R)))^{\rho^{(\alpha,\beta)}[f_1]_{g_1}}\})) \cdot \delta.$$

Since in view of Theorem 8.2.2, $\rho^{(\alpha,\beta)}[f_1]_{g_1} \leq \max\{\rho^{(\alpha,\beta)}[f_1]_{g_1}, \rho^{(\alpha,\beta)}[f_2]_{g_1} = \rho^{(\alpha,\beta)}[f_1]_{g_1}$, letting $\delta \to 1+$, and for all sufficiently large values of $R$, we get

$$\limsup_{R \to +\infty} \frac{\exp\left(\alpha\left(M_{g_1, D}^{-1}(M_{f_1 \pm f_2, D}(R))\right)\right)}{(\exp(\beta(R)))^{\rho^{(\alpha,\beta)}[f_1 \pm f_2]_{g_1}}} \leq \sigma_D^{(\alpha,\beta)}[f_1]_{g_1}$$

$$i.e., \sigma_D^{(\alpha,\beta)}[f_1 \pm f_2]_{g_1} \leq \sigma_D^{(\alpha,\beta)}[f_1]_{g_1}. \tag{186}$$

Next we take $f(z) = f_1(z) \pm f_2(z)$. Since $\rho^{(\alpha,\beta)}[f_1]_{g_1} > \rho^{(\alpha,\beta)}[f_2]_{g_1}$ hold, then $\sigma_D^{(\alpha,\beta)}[f]_{g_1} = \sigma_D^{(\alpha,\beta)}[f_1 \pm f_2]_{g_1} \leq \sigma_D^{(\alpha,\beta)}[f_1]_{g_1}$. Further, let $f_1(z) = (f(z) \pm f_2(z))$. Now, in view of Theorem 8.2.2 and $\rho^{(\alpha,\beta)}[f_1]_{g_1} > \rho^{(\alpha,\beta)}[f_2]_{g_1}$, we obtain that $\rho^{(\alpha,\beta)}[f]_{g_1} > \rho^{(\alpha,\beta)}[f_2]_{g_1}$ holds. Hence in view of (186) $\sigma_D^{(\alpha,\beta)}[f_1]_{g_1} \leq \sigma_D^{(\alpha,\beta)}[f]_{g_1} = \sigma_D^{(\alpha,\beta)}[f_1 \pm f_2]_{g_1}$. Therefore $\sigma_D^{(\alpha,\beta)}[f]_{g_1} = \sigma_D^{(\alpha,\beta)}[f_1]_{g_1} \Rightarrow \sigma_D^{(\alpha,\beta)}[f_1 \pm f_2]_{g_1} = \sigma_D^{(\alpha,\beta)}[f_1]_{g_1}$.

Similarly, if we consider $\rho^{(\alpha,\beta)}[f_1]_{g_1} < \rho^{(\alpha,\beta)}[f_2]_{g_1}$, then one can easily verify that $\sigma_D^{(\alpha,\beta)}[f_1 \pm f_2]_{g_1} = \sigma_D^{(\alpha,\beta)}[f_2]_{g_1}$.

**Case II.** Let us consider that $\rho^{(\alpha,\beta)}[f_1]_{g_1} > \rho^{(\alpha,\beta)}[f_2]_{g_1}$ hold. Also let $\varepsilon \ (> 0)$ are arbitrary. Then we get from (181) and (184) for a sequence of values $R$ tending to infinity that

$$M_{f_1 \pm f_2, D}(R) \leq M_{f_1, D}(R) + M_{f_2, D}(R)$$

$$\leq M_{g_1, D}(\alpha^{-1}(\log\{(\overline{\sigma}_D^{(\alpha,\beta)}[f_1]_{g_1} + \varepsilon)(\exp(\beta(R)))^{\rho^{(\alpha,\beta)}[f_1]_{g_1}}\}))(1 + B) \tag{187}$$

where $B = \dfrac{M_{g_1, D}(\alpha^{-1}(\log\{(\sigma_D^{(\alpha,\beta)}[f_2]_{g_1} + \varepsilon)(\exp(\beta(R)))^{\rho^{(\alpha,\beta)}[f_2]_{g_1}}\}))}{M_{g_1, D}(\alpha^{-1}(\log\{(\overline{\sigma}_D^{(\alpha,\beta)}[f_1]_{g_1} + \varepsilon)(\exp(\beta(R)))^{\rho^{(\alpha,\beta)}[f_1]_{g_1}}\}))}$ and in view of $\rho^{(\alpha,\beta)}[f_1]_{g_1} > \rho^{(\alpha,\beta)}[f_2]_{g_1}$, we can make the term $B$ sufficiently small by taking $R$ sufficiently large and therefore using the similar technique for as executed in the proof of Case I we get from (187) that $\overline{\sigma}_D^{(\alpha,\beta)}[f_1 \pm f_2]_{g_1} = \overline{\sigma}_D^{(\alpha,\beta)}[f_1]_{g_1}$ when $\rho^{(\alpha,\beta)}[f_1]_{g_1} > \rho^{(\alpha,\beta)}[f_2]_{g_1}$ holds. Likewise, if we consider $\rho^{(\alpha,\beta)}[f_1]_{g_1} < \rho^{(\alpha,\beta)}[f_2]_{g_1}$, then one can easily verify that $\overline{\sigma}_D^{(\alpha,\beta)}[f_1 \pm f_2]_{g_1} = \overline{\sigma}_D^{(\alpha,\beta)}[f_2]_{g_1}$.

Thus combining Case I and Case II, we obtain the first part of the theorem.

**Case III.** Let us consider that $\rho^{(\alpha,\beta)}[f_1]_{g_1} < \rho^{(\alpha,\beta)}[f_1]_{g_2}$ with at least $f_1(z)$ is of regular generalized relative Gol'dberg $(\alpha, \beta)$ growth with respect to $g_2(z)$. Therefore we obtain from (182) and (183) for a sequence of values $R$ tending to infinity that

$$M_{g_1 \pm g_2, D}(R) \leq M_{g_1, D}(R) + M_{g_2, D}(R),$$

$$\implies M_{g_1 \pm g_2, D}(\alpha^{-1}(\log\{(\sigma_D^{(\alpha,\beta)}[f_1]_{g_1} - \varepsilon)(\exp(\beta(R)))^{\rho^{(\alpha,\beta)}[f_1]_{g_1}}\}))$$

$$\leq M_{g_1, D}(\alpha^{-1}(\log\{(\sigma_D^{(\alpha,\beta)}[f_1]_{g_1} - \varepsilon)(\exp(\beta(R)))^{\rho^{(\alpha,\beta)}[f_1]_{g_1}}\}))$$

$$+ M_{g_2, D}(\alpha^{-1}(\log\{(\sigma_D^{(\alpha,\beta)}[f_1]_{g_1} - \varepsilon)(\exp(\beta(R)))^{\rho^{(\alpha,\beta)}[f_1]_{g_1}}\}))$$

$$i.e., M_{g_1 \pm g_2, D}(\alpha^{-1}(\log\{(\sigma_D^{(\alpha,\beta)}[f_1]_{g_1} - \varepsilon)(\exp(\beta(R)))^{\rho^{(\alpha,\beta)}[f_1]_{g_1}}\}))$$

$$\leq (1 + C) M_{f_1, D}(R) \qquad (188)$$

where $C = \dfrac{M_{g_2, D}(\alpha^{-1}(\log\{(\sigma_D^{(\alpha,\beta)}[f_1]_{g_1} - \varepsilon)(\exp(\beta(R)))^{\rho^{(\alpha,\beta)}[f_1]_{g_1}}\}))}{M_{g_2, D}(\alpha^{-1}(\log\{(\overline{\sigma}_D^{(\alpha,\beta)}[f_1]_{g_2} - \varepsilon)(\exp(\beta(R)))^{\rho^{(\alpha,\beta)}[f_1]_{g_2}}\}))}$. Since $\rho^{(\alpha,\beta)}[f_1]_{g_1} < \rho^{(\alpha,\beta)}[f_1]_{g_2}$, we can make the term $C$ sufficiently small ($< \varepsilon_1$ for $\varepsilon_1 > 0$) by taking $R$ sufficiently large. Therefore for any $\delta = 1 + \varepsilon_1$, we obtain in view of $C < \varepsilon_1$ from (188) and Theorem 8.2.4, for a sequence of values of $R$ tending to infinity that

$$M_{g_1 \pm g_2, D}(\alpha^{-1}(\log\{(\sigma_D^{(\alpha,\beta)}[f_1]_{g_1} - \varepsilon)(\exp(\beta(R)))^{\rho^{(\alpha,\beta)}[f_1]_{g_1}}\})) < \alpha M_{f_1, D}(R)$$

Therefore, making $\delta \to 1+$, we obtain from above for a sequence of values $R$ tending to infinity that

$$(\sigma_D^{(\alpha,\beta)}[f_1]_{g_1} - \varepsilon)(\exp(\beta(R)))^{\rho^{(\alpha,\beta)}[f_1]_{g_1}} < \exp(\alpha(M_{g_1 \pm g_2, D}^{-1}(M_{f_1, D}(R)))).$$

Since $\varepsilon > 0$ is arbitrary, we find that

$$\sigma_D^{(\alpha,\beta)}[f_1]_{g_1 \pm g_2} \geq \sigma_D^{(\alpha,\beta)}[f_1]_{g_1}. \qquad\qquad (189)$$

Now we may consider that $g(z) = g_1(z) \pm g_2(z)$. Also $\rho^{(\alpha,\beta)}[f_1]_{g_1} < \rho^{(\alpha,\beta)}[f_1]_{g_2}$ and at least $f_1(z)$ is of regular generalized relative Gol'dberg $(\alpha, \beta)$ growth with respect to $g_2(z)$. Then $\sigma_D^{(\alpha,\beta)}[f_1]_g = \sigma_D^{(\alpha,\beta)}[f_1]_{g_1 \pm g_2} \geq \sigma_D^{(\alpha,\beta)}[f_1]_{g_1}$. Further let $g_1(z) = (g(z) \pm g_2(z))$. Therefore in view of Theorem 8.2.4 and $\rho^{(\alpha,\beta)}[f_1]_{g_1} < \rho^{(\alpha,\beta)}[f_1]_{g_2}$, we obtain that $\rho^{(\alpha,\beta)}[f_1]_g < \rho^{(\alpha,\beta)}[f_1]_{g_2}$ as at least $f_1(z)$ is of regular generalized relative Gol'dberg $(\alpha, \beta)$ growth with respect to $g_2(z)$. Hence in view of (189), $\sigma_D^{(\alpha,\beta)}[f_1]_{g_1} \geq \sigma_D^{(\alpha,\beta)}[f_1]_g = \sigma_D^{(\alpha,\beta)}[f_1]_{g_1 \pm g_2}$. Therefore $\sigma_D^{(\alpha,\beta)}[f_1]_g = \sigma_D^{(\alpha,\beta)}[f_1]_{g_1} \Rightarrow \sigma_D^{(\alpha,\beta)}[f_1]_{g_1 \pm g_2} = \sigma_D^{(\alpha,\beta)}[f_1]_{g_1}$.

Similarly if we consider $\rho^{(\alpha,\beta)}[f_1]_{g_1} > \rho^{(\alpha,\beta)}[f_1]_{g_2}$ with at least $f_1(z)$ is of regular generalized relative Gol'dberg $(\alpha, \beta)$ growth with respect to $g_1(z)$, then $\sigma_D^{(\alpha,\beta)}[f_1]_{g_1 \pm g_2} = \sigma_D^{(\alpha,\beta)}[f_1]_{g_2}$.

**Case IV.** In this case suppose that $\rho^{(\alpha,\beta)}[f_1]_{g_1} < \rho^{(\alpha,\beta)}[f_1]_{g_2}$ with at least $f_1(z)$ is of regular generalized relative Gol'dberg $(\alpha, \beta)$ growth with respect to $g_2(z)$. Therefore from (182), we get for all sufficiently large values of $R$ that

$$M_{g_1 \pm g_2, D}(R) \leq M_{g_1, D}(R) + M_{g_2, D}(R),$$

$$\Longrightarrow M_{g_1 \pm g_2, D}(\alpha^{-1}(\log\{(\overline{\sigma}_D^{(\alpha,\beta)}[f_1]_{g_1} - \varepsilon)(\exp(\beta(R)))^{\rho^{(\alpha,\beta)}[f_1]_{g_1}}\}))$$

$$\leq M_{g_1, D}(\alpha^{-1}(\log\{(\overline{\sigma}_D^{(\alpha,\beta)}[f_1]_{g_1} - \varepsilon)(\exp(\beta(R)))^{\rho^{(\alpha,\beta)}[f_1]_{g_1}}\}))$$

$$+ M_{g_2, D}(\alpha^{-1}(\log\{(\overline{\sigma}_D^{(\alpha,\beta)}[f_1]_{g_1} - \varepsilon)(\exp(\beta(R)))^{\rho^{(\alpha,\beta)}[f_1]_{g_1}}\}))$$

$$i.e., M_{g_1 \pm g_2, D}(\alpha^{-1}(\log\{(\overline{\sigma}_D^{(\alpha,\beta)}[f_1]_{g_1} - \varepsilon)(\exp(\beta(R)))^{\rho^{(\alpha,\beta)}[f_1]_{g_1}}\}))$$

$$\leq (1 + D) M_{f_1, D}(R) \qquad (190)$$

where $D = \dfrac{M_{g_2, D}\left(\alpha^{-1}(\log\{(\overline{\sigma}_D^{(\alpha,\beta)}[f_1]_{g_1} - \varepsilon)(\exp(\beta(R)))^{\rho^{(\alpha,\beta)}[f_1]_{g_1}}\})\right)}{M_{g_2, D}\left(\alpha^{-1}(\log\{(\overline{\sigma}_D^{(\alpha,\beta)}[f_1]_{g_2} - \varepsilon)(\exp(\beta(R)))^{\rho^{(\alpha,\beta)}[f_1]_{g_2}}\})\right)}$ and now in view of $\rho^{(\alpha,\beta)}[f_1]_{g_1} <$

$\rho^{(\alpha,\beta)}[f_1]_{g_2}$, we can make the term $D$ sufficiently small by taking $R$ sufficiently large and therefore using the similar technique for as executed in the proof of Case III we get from (190) that $\overline{\sigma}_D^{(\alpha,\beta)}[f_1]_{g_1 \pm g_2} = \overline{\sigma}_D^{(\alpha,\beta)}[f_1]_{g_1}$ where $\rho^{(\alpha,\beta)}[f_1]_{g_1} < \rho^{(\alpha,\beta)}[f_1]_{g_2}$ and at least $f_1(z)$ is of regular generalized relative Gol'dberg $(\alpha, \beta)$ growth with respect to $g_2(z)$.

Similarly, if we consider $\rho^{(\alpha,\beta)}[f_1]_{g_1} > \rho^{(\alpha,\beta)}[f_1]_{g_2}$ with at least $f_1(z)$ is of regular generalized relative Gol'dberg $(\alpha, \beta)$ growth with respect to $g_1(z)$, then $\overline{\sigma}_D^{(\alpha,\beta)}[f_1]_{g_1 \pm g_2} = \overline{\sigma}_D^{(\alpha,\beta)}[f_1]_{g_2}$.

Thus combining Case III and Case IV, we obtain the second part of the theorem.

The third part of the theorem is a natural consequence of Theorem 8.2.5 and the first part and second part of the theorem. Hence its proof is omitted ∎

**Theorem 8.2.14** *Let $f_1(z)$, $f_2(z)$, $g_1(z)$ and $g_2(z)$ be any four entire functions of $n$ complex variables. Also let $\lambda^{(\alpha,\beta)}[f_1]_{g_1}$, $\lambda^{(\alpha,\beta)}[f_2]_{g_1}$, $\lambda^{(\alpha,\beta)}[f_1]_{g_2}$ and $\lambda^{(\alpha,\beta)}[f_2]_{g_2}$ are all non-zero and finite.*
*(A) If $\lambda^{(\alpha,\beta)}[f_i]_{g_1} > \lambda^{(\alpha,\beta)}[f_j]_{g_1}$ with at least $f_j(z)$ is of regular generalized relative Gol'dberg $(\alpha, \beta)$ growth with respect to $g_1(z)$ for $i = j = 1, 2$ and $i \neq j$, then*

$$\tau_D^{(\alpha,\beta)}[f_1 \pm f_2]_{g_1} = \tau_D^{(\alpha,\beta)}[f_i]_{g_1} \text{ and } \overline{\tau}_D^{(\alpha,\beta)}[f_1 \pm f_2]_{g_1} = \overline{\tau}_D^{(\alpha,\beta)}[f_i]_{g_1}.$$

*(B) If $\lambda^{(\alpha,\beta)}[f_1]_{g_i} < \lambda^{(\alpha,\beta)}[f_1]_{g_j}$ for $i = j = 1, 2$ and $i \neq j$, then*

$$\tau_D^{(\alpha,\beta)}[f_1]_{g_1 \pm g_2} = \tau_D^{(\alpha,\beta)}[f_1]_{g_i} \text{ and } \overline{\tau}_D^{(\alpha,\beta)}[f_1]_{g_1 \pm g_2} = \overline{\tau}_D^{(\alpha,\beta)}[f_1]_{g_i}.$$

*(C) Assume the functions $f_1(z), f_2(z), g_1(z)$ and $g_2(z)$ satisfy the following conditions:*
*(i) $\rho^{(\alpha,\beta)}[f_i]_{g_1} > \rho^{(\alpha,\beta)}[f_j]_{g_1}$ with at least $f_j(z)$ is of regular generalized relative Gol'dberg $(\alpha, \beta)$ growth with respect to $g_1(z)$ for $i = j = 1, 2$ and $i \neq j$;*
*(ii) $\rho^{(\alpha,\beta)}[f_i]_{g_2} > \rho^{(\alpha,\beta)}[f_j]_{g_2}$ with at least $f_j(z)$ is of regular generalized relative Gol'dberg $(\alpha, \beta)$ growth with respect to $g_2(z)$ for $i = j = 1, 2$ and $i \neq j$;*
*(iii) $\rho^{(\alpha,\beta)}[f_1]_{g_i} < \rho^{(\alpha,\beta)}[f_1]_{g_j}$ and $\rho^{(\alpha,\beta)}[f_2]_{g_i} < \rho^{(\alpha,\beta)}[f_2]_{g_j}$ holds simultaneously for $i = j = 1, 2$ and $i \neq j$;*
*(iv) $\lambda^{(\alpha,\beta)}[f_l]_{g_m} = \min[\max\{\lambda^{(\alpha,\beta)}[f_1]_{g_1}, \lambda^{(\alpha,\beta)}[f_2]_{g_1}\}, \max\{\lambda^{(\alpha,\beta)}[f_1]_{g_2}, \lambda^{(\alpha,\beta)}[f_2]_{g_2}\}] \mid l = m = 1, 2$;*
*then we have*

$$\tau_D^{(\alpha,\beta)}[f_1 \pm f_2]_{g_1 \pm g_2} = \tau_D^{(\alpha,\beta)}[f_l]_{g_m} \text{ and } \overline{\tau}_D^{(\alpha,\beta)}[f_1 \pm f_2]_{g_1 \pm g_2} = \overline{\tau}_D^{(\alpha,\beta)}[f_l]_{g_m}.$$

**Proof.** For any arbitrary positive number $\varepsilon(> 0)$, we have for all sufficiently large values of $R$ that

$$M_{f_k,D}(R) \leq M_{g_l,D}\left(\alpha^{-1}(\log\{(\overline{\tau}_D^{(\alpha,\beta)}[f_k]_{g_l} + \varepsilon)(\exp(\beta(R)))^{\lambda^{(\alpha,\beta)}[f_k]_{g_l}}\})\right), \tag{191}$$

$$M_{f_k,D}(R) \geq M_{g_l,D}\left(\alpha^{-1}(\log\{(\tau_D^{(\alpha,\beta)}[f_k]_{g_l} - \varepsilon)(\exp(\beta(R)))^{\lambda^{(\alpha,\beta)}[f_k]_{g_l}}\})\right) \tag{192}$$

$$i.e., \ M_{g_l,D}(R)$$

$$\leq M_{f_k,D}\left(\beta^{-1}\left(\log\left(\left(\frac{\exp(\alpha(R))}{\left(\tau_D^{(\alpha,\beta)}[f_k]_{g_l} - \varepsilon\right)}\right)^{\frac{1}{\lambda_{g_l}^{(\alpha,\beta)}(f_k)}}\right)\right)\right), \tag{193}$$

and for a sequence of values of $R$ tending to infinity we obtain that

$$M_{f_k,D}(R) \geq M_{g_l,D}\left(\alpha^{-1}(\log\{(\overline{\tau}_D^{(\alpha,\beta)}[f_k]_{g_l} - \varepsilon)(\exp(\beta(R)))^{\lambda^{(\alpha,\beta)}[f_k]_{g_l}}\})\right) \tag{194}$$

and

$$M_{f_k,D}(R) \leq M_{g_l,D}\left(\alpha^{-1}(\log\{(\tau_D^{(\alpha,\beta)}[f_k]_{g_l} + \varepsilon)(\exp(\beta(R)))^{\lambda^{(\alpha,\beta)}[f_k]_{g_l}}\})\right), \tag{195}$$

where $k = 1, 2$ and $l = 1, 2$.

**Case I.** Let $\lambda^{(\alpha,\beta)}[f_1]_{g_1} > \lambda^{(\alpha,\beta)}[f_2]_{g_1}$ with at least $f_2(z)$ is of regular generalized relative Gol'dberg $(\alpha, \beta)$ growth with respect to $g_1(z)$. Also let $\varepsilon(> 0)$ be arbitrary. Hence we get in view of (191) and (195) for a sequence of values $R$ tending to infinity that/

$$M_{f_1 \pm f_2,D}(R) \leq M_{f_1,D}(R) + M_{f_2,D}(R)$$

$$\leq M_{g_1,D}\left(\alpha^{-1}(\log\{(\tau_D^{(\alpha,\beta)}[f_1]_{g_1} + \varepsilon)(\exp(\beta(R)))^{\lambda^{(\alpha,\beta)}[f_1]_{g_1}}\})\right)$$

$$+ M_{g_1,D}\left(\alpha^{-1}(\log\{(\overline{\tau}_D^{(\alpha,\beta)}[f_2]_{g_1} + \varepsilon)(\exp(\beta(R)))^{\lambda^{(\alpha,\beta)}[f_2]_{g_1}}\})\right)$$

$$i.e., M_{f_1 \pm f_2,D}(R) \leq$$

$$M_{g_1,D}\left(\alpha^{-1}(\log\{(\tau_D^{(\alpha,\beta)}[f_1]_{g_1} + \varepsilon)(\exp(\beta(R)))^{\lambda^{(\alpha,\beta)}[f_1]_{g_1}}\})\right)(1 + E), \tag{196}$$

where $E = \dfrac{M_{g_1,D}\left(\alpha^{-1}(\log\{(\overline{\tau}_D^{(\alpha,\beta)}[f_2]_{g_1} + \varepsilon)(\exp(\beta(R)))^{\lambda^{(\alpha,\beta)}[f_2]_{g_1}}\right)}{M_{g_1,D}\left(\alpha^{-1}(\log\{(\tau_D^{(\alpha,\beta)}[f_1]_{g_1} + \varepsilon)(\exp(\beta(R)))^{\lambda^{(\alpha,\beta)}[f_1]_{g_1}}\})\right)}$ and in view of $\lambda^{(\alpha,\beta)}[f_1]_{g_1} >$

$\lambda^{(\alpha,\beta)}[f_2]_{g_1}$, we can make the term $E$ sufficiently small by taking $R$ sufficiently large. Therefore with the help of Theorem 8.2.1 and using the similar technique of Case I of Theorem 8.2.13, we get from (196) that

$$\tau_D^{(\alpha,\beta)}[f_1 \pm f_2]_{g_1} \leq \tau_D^{(\alpha,\beta)}[f_1]_{g_1}. \tag{197}$$

Further, we may consider that $f(z) = f_1(z) \pm f_2(z)$. Also suppose that $\lambda^{(\alpha,\beta)}[f_1]_{g_1} > \lambda^{(\alpha,\beta)}[f_2]_{g_1}$ and at least $f_2(z)$ is of regular generalized relative Gol'dberg $(\alpha,\beta)$ growth with respect to $g_1(z)$. Then $\tau_D^{(\alpha,\beta)}[f]_{g_1} = \tau_D^{(\alpha,\beta)}[f_1 \pm f_2]_{g_1} \leq \tau_D^{(\alpha,\beta)}[f_1]_{g_1}$. Now let $f_1(z) = (f(z) \pm f_2(z))$. Therefore in view of Theorem 8.2.1, $\lambda^{(\alpha,\beta)}[f_1]_{g_1} > \lambda^{(\alpha,\beta)}[f_2]_{g_1}$ and at least $f_2(z)$ is of regular generalized relative Gol'dberg $(\alpha,\beta)$ growth with respect to $g_1(z)$, we obtain that $\lambda^{(\alpha,\beta)}[f]_{g_1} > \lambda^{(\alpha,\beta)}[f_2]_{g_1}$ holds. Hence in view of (197), $\tau_D^{(\alpha,\beta)}[f_1]_{g_1} \leq \tau_D^{(\alpha,\beta)}[f]_{g_1} = \tau_D^{(\alpha,\beta)}[f_1 \pm f_2]_{g_1}$. Therefore $\tau_D^{(\alpha,\beta)}[f]_{g_1} = \tau_D^{(\alpha,\beta)}[f_1]_{g_1} \Rightarrow \tau_D^{(\alpha,\beta)}[f_1 \pm f_2]_{g_1} = \tau_D^{(\alpha,\beta)}[f_1]_{g_1}$.

Similarly, if we consider $\lambda^{(\alpha,\beta)}[f_1]_{g_1} < \lambda^{(\alpha,\beta)}[f_2]_{g_1}$ with at least $f_1(z)$ is of regular generalized relative Gol'dberg $(\alpha,\beta)$ growth with respect to $g_1(z)$ then one can easily verify that $\tau_D^{(\alpha,\beta)}[f_1 \pm f_2]_{g_1} = \tau_D^{(\alpha,\beta)}[f_2]_{g_1}$.

**Case II.** Let us consider that $\lambda^{(\alpha,\beta)}[f_1]_{g_1} > \lambda^{(\alpha,\beta)}[f_2]_{g_1}$ with at least $f_2(z)$ is of regular generalized relative Gol'dberg $(\alpha,\beta)$ growth with respect to $g_1(z)$. Also let $\varepsilon\,(>0)$ be arbitrary. Therefore we get in view of (191) for all sufficiently large values of $R$ that

$$M_{f_1 \pm f_2, D}(R) \leq M_{f_1, D}(R) + M_{f_2, D}(R)$$
$$\leq M_{g_1, D}\left(\exp^{[p-1]}\left\{\left(\overline{\tau}_D^{(\alpha,\beta)}[f_1]_{g_1} + \varepsilon\right)\left[\log^{[q-1]}\varphi(R)\right]^{\lambda^{(\alpha,\beta)}[f_1]_{g_1}}\right\}\right)$$
$$+ M_{g_1, D}\left(\exp^{[p-1]}\left\{\left(\overline{\tau}_D^{(\alpha,\beta)}[f_2]_{g_1} + \varepsilon\right)\left[\log^{[q-1]}\varphi(R)\right]^{\lambda^{(\alpha,\beta)}[f_2]_{g_1}}\right\}\right)$$

*i.e.,* $M_{f_1 \pm f_2, D}(R) \leq$

$$M_{g_1, D}\left(\exp^{[p-1]}\left\{\left(\overline{\tau}_D^{(\alpha,\beta)}[f_1]_{g_1} + \varepsilon\right)\left[\log^{[q-1]}\varphi(R)\right]^{\lambda^{(\alpha,\beta)}[f_1]_{g_1}}\right\}\right)(1 + F), \qquad (198)$$

where $F = \dfrac{M_{g_1, D}\left(\exp^{[p-1]}\left\{\left(\overline{\tau}_D^{(\alpha,\beta)}[f_2]_{g_1} + \varepsilon\right)\left[\log^{[q-1]}\varphi(R)\right]^{\lambda^{(\alpha,\beta)}[f_2]_{g_1}}\right\}\right)}{M_{g_1, D}\left(\exp^{[p-1]}\left\{\left(\overline{\tau}_D^{(\alpha,\beta)}[f_1]_{g_1} + \varepsilon\right)\left[\log^{[q-1]}\varphi(R)\right]^{\lambda^{(\alpha,\beta)}[f_1]_{g_1}}\right\}\right)}$ and in view of $\lambda^{(\alpha,\beta)}[f_1]_{g_1} > \lambda^{(\alpha,\beta)}[f_2]_{g_1}$, one can make the term $F$ sufficiently small by taking $R$ sufficiently large and therefore for similar reasoning of Case I we get from (198) that $\overline{\tau}_D^{(\alpha,\beta)}[f_1 \pm f_2]_{g_1} = \overline{\tau}_D^{(\alpha,\beta)}[f_1]_{g_1}$ when $\lambda^{(\alpha,\beta)}[f_1]_{g_1} > \lambda^{(\alpha,\beta)}[f_2]_{g_1}$ and at least $f_2(z)$ is of regular generalized relative Gol'dberg $(\alpha,\beta)$ growth with respect to $g_1(z)$.

Likewise, if we consider $\lambda^{(\alpha,\beta)}[f_1]_{g_1} < \lambda^{(\alpha,\beta)}[f_2]_{g_1}$ with at least $f_1(z)$ is of regular generalized relative Gol'dberg $(\alpha,\beta)$ growth with respect to $g_1(z)$ then one can easily verify that $\overline{\tau}_D^{(\alpha,\beta)}[f_1 \pm f_2]_{g_1} = \overline{\tau}_D^{(\alpha,\beta)}[f_2]_{g_1}$.

Thus combining Case I and Case II, we obtain the first part of the theorem.

**Case III.** Let us consider that $\lambda^{(\alpha,\beta)}[f_1]_{g_1} < \lambda^{(\alpha,\beta)}[f_1]_{g_2}$. Now we get from (192) for all sufficiently large values of $R$ that

$$M_{g_1 \pm g_2, D}(R) \leq M_{g_1, D}(R) + M_{g_2, D}(R)$$

$$\Rightarrow M_{g_1 \pm g_2, D}\left(\alpha^{-1}(\log\{(\tau_D^{(\alpha,\beta)}[f_1]_{g_1} - \varepsilon)(\exp(\beta(R)))^{\lambda^{(\alpha,\beta)}[f_1]_{g_1}}\right)$$

$$\leq M_{g_1, D}\left(\alpha^{-1}(\log\{(\tau_D^{(\alpha,\beta)}[f_1]_{g_1} - \varepsilon)(\exp(\beta(R)))^{\lambda^{(\alpha,\beta)}[f_1]_{g_1}}\right)$$

$$+ M_{g_2, D}\left(\alpha^{-1}(\log\{(\tau_D^{(\alpha,\beta)}[f_1]_{g_1} - \varepsilon)(\exp(\beta(R)))^{\lambda^{(\alpha,\beta)}[f_1]_{g_1}}\right)$$

$$M_{g_1 \pm g_2, D}\left(\alpha^{-1}(\log\{(\tau_D^{(\alpha,\beta)}[f_1]_{g_1} - \varepsilon)(\exp(\beta(R)))^{\lambda^{(\alpha,\beta)}[f_1]_{g_1}}\right)$$

$$\leq (1 + G)M_{f_1, D}(R) \quad (199)$$

where $G = \dfrac{M_{g_2, D}\left(\alpha^{-1}(\log\{(\tau_D^{(\alpha,\beta)}[f_1]_{g_1} - \varepsilon)(\exp(\beta(R)))^{\lambda^{(\alpha,\beta)}[f_1]_{g_1}}\right)}{M_{g_2, D}\left(\alpha^{-1}(\log\{(\tau_D^{(\alpha,\beta)}[f_1]_{g_2} - \varepsilon)(\exp(\beta(R)))^{\lambda^{(\alpha,\beta)}[f_1]_{g_2}}\right)}$, and as $\lambda^{(\alpha,\beta)}[f_1]_{g_1} < \lambda^{(\alpha,\beta)}[f_1]_{g_2}$,

we can make the term $G$ sufficiently small by taking $R$ sufficiently large. Now with the help of Theorem 8.2.3 and using the similar technique of Case III of Theorem 8.2.13, we get from (199) that

$$\tau_D^{(\alpha,\beta)}[f_1]_{g_1 \pm g_2} \geq \tau_D^{(\alpha,\beta)}[f_1]_{g_1}. \quad (200)$$

Again, we may consider that $g(z) = g_1(z) \pm g_2(z)$. As $\lambda^{(\alpha,\beta)}[f_1]_{g_1} < \lambda^{(\alpha,\beta)}[f_1]_{g_2}$, so $\tau_D^{(\alpha,\beta)}[f_1]_g = \tau_D^{(\alpha,\beta)}[f_1]_{g_1 \pm g_2} \geq \tau_D^{(\alpha,\beta)}[f_1]_{g_1}$. Also let $g_1(z) = (g(z) \pm g_2(z))$. Therefore in view of Theorem 8.2.3 and $\lambda^{(\alpha,\beta)}[f_1]_{g_1} < \lambda^{(\alpha,\beta)}[f_1]_{g_2}$ we obtain that $\lambda_g^{(p,q)}(f_1) < \lambda^{(\alpha,\beta)}[f_1]_{g_2}$ holds. Hence in view of (200) $\tau_D^{(\alpha,\beta)}[f_1]_{g_1} \geq \tau_D^{(\alpha,\beta)}[f_1]_g = \tau_D^{(\alpha,\beta)}[f_1]_{g_1 \pm g_2}$. Therefore $\tau_D^{(\alpha,\beta)}[f_1]_g = \tau_D^{(\alpha,\beta)}[f_1]_{g_1} \Rightarrow \tau_D^{(\alpha,\beta)}[f_1]_{g_1 \pm g_2} = \tau_D^{(\alpha,\beta)}[f_1]_{g_1}$.

Likewise, if we consider that $\lambda^{(\alpha,\beta)}[f_1]_{g_1} > \lambda^{(\alpha,\beta)}[f_1]_{g_2}$, then one can easily verify that $\tau_D^{(\alpha,\beta)}[f_1]_{g_1 \pm g_2} = \tau_D^{(\alpha,\beta)}[f_1]_{g_2}$.

**Case IV.** In this case further we consider $\lambda^{(\alpha,\beta)}[f_1]_{g_1} < \lambda^{(\alpha,\beta)}[f_1]_{g_2}$. Now we obtain from (193) and (194) for a sequence of values of $R$ tending to infinity that

$$M_{g_1 \pm g_2, D}(R) \leq M_{g_1, D}(R) + M_{g_2, D}(R)$$

$$M_{g_1 \pm g_2, D}\left(\alpha^{-1}(\log\{(\overline{\tau}_D^{(\alpha,\beta)}[f_1]_{g_1} - \varepsilon)(\exp(\beta(R)))^{\lambda^{(\alpha,\beta)}[f_1]_{g_1}}\})\right)$$

$$\leq M_{g_1, D}\left(\alpha^{-1}(\log\{(\overline{\tau}_D^{(\alpha,\beta)}[f_1]_{g_1} - \varepsilon)(\exp(\beta(R)))^{\lambda^{(\alpha,\beta)}[f_1]_{g_1}}\})\right)$$

$$+ M_{g_2, D}\left(\alpha^{-1}(\log\{(\overline{\tau}_D^{(\alpha,\beta)}[f_1]_{g_1} - \varepsilon)(\exp(\beta(R)))^{\lambda^{(\alpha,\beta)}[f_1]_{g_1}}\})\right)$$

*i.e.,* $M_{g_1 \pm g_2, D}\left(\alpha^{-1}(\log\{(\overline{\tau}_D^{(\alpha,\beta)}[f_1]_{g_1} - \varepsilon)(\exp(\beta(R)))^{\lambda^{(\alpha,\beta)}[f_1]_{g_1}}\})\right)$

$$\leq (1 + H)M_{f_1, D}(R) \quad (201)$$

where $H = \dfrac{M_{g_2,D}\left(\alpha^{-1}(\log\{(\overline{\tau}_D^{(\alpha,\beta)}[f_1]_{g_1}-\varepsilon)(\exp(\beta(R)))^{\lambda^{(\alpha,\beta)}[f_1]_{g_1}}\})\right)}{M_{g_2,D}\left(\alpha^{-1}(\log\{(\overline{\tau}_D^{(\alpha,\beta)}[f_1]_{g_2}-\varepsilon)(\exp(\beta(R)))^{\lambda^{(\alpha,\beta)}[f_1]_{g_2}}\})\right)}$, and in view of $\lambda^{(\alpha,\beta)}[f_1]_{g_1} <$

$\lambda^{(\alpha,\beta)}[f_1]_{g_2}$, we can make the term $H$ sufficiently small by taking $R$ sufficiently large and therefore using the similar technique as executed in the proof of Case IV of Theorem 8.2.13, we get from (201) that $\overline{\tau}_D^{(\alpha,\beta)}[f_1]_{g_1\pm g_2} = \overline{\tau}_D^{(\alpha,\beta)}[f_1]_{g_1}$ when $\lambda^{(\alpha,\beta)}[f_1]_{g_1} < \lambda^{(\alpha,\beta)}[f_1]_{g_2}$. Similarly, if we consider that $\lambda^{(\alpha,\beta)}[f_1]_{g_1} > \lambda^{(\alpha,\beta)}[f_1]_{g_2}$, then one can easily verify that $\overline{\tau}_D^{(\alpha,\beta)}[f_1]_{g_1\pm g_2} = \overline{\tau}_D^{(\alpha,\beta)}[f_1]_{g_2}$.

Thus combining Case III and Case IV, we obtain the second part of the theorem.

The proof of the third part of the Theorem is omitted as it can be carried out in view of Theorem 8.2.6 and the above cases. ∎

**Theorem 8.2.15** *Let $f_1(z)$, $f_2(z)$, $g_1(z)$ and $g_2(z)$ be any four entire functions of $n$ complex variables.*
*(A) The following condition is assumed to be satisfied:*
*(i) Either $\sigma_D^{(\alpha,\beta)}[f_1]_{g_1} \neq \sigma_D^{(\alpha,\beta)}[f_2]_{g_1}$ or $\overline{\sigma}_D^{(\alpha,\beta)}[f_1]_{g_1} \neq \overline{\sigma}_D^{(\alpha,\beta)}[f_2]_{g_1}$ holds, then*

$$\rho^{(\alpha,\beta)}[f_1 \pm f_2]_{g_1} = \rho^{(\alpha,\beta)}[f_1]_{g_1} = \rho^{(\alpha,\beta)}[f_2]_{g_1}.$$

*(B) The following conditions are assumed to be satisfied:*
*(i) Either $\sigma_D^{(\alpha,\beta)}[f_1]_{g_1} \neq \sigma_D^{(\alpha,\beta)}[f_1]_{g_2}$ or $\overline{\sigma}_D^{(\alpha,\beta)}[f_1]_{g_1} \neq \overline{\sigma}_D^{(\alpha,\beta)}[f_1]_{g_2}$ holds;*
*(ii) $f_1(z)$ is of regular generalized relative Gol'dberg growth $(\alpha,\beta)$ with respect to at least any one of $g_1(z)$ or $g_2(z)$, then*

$$\rho^{(\alpha,\beta)}[f_1]_{g_1\pm g_2} = \rho^{(\alpha,\beta)}[f_1]_{g_1} = \rho^{(\alpha,\beta)}[f_1]_{g_2}.$$

**Proof. Case I.** Suppose that $\rho^{(\alpha,\beta)}[f_1]_{g_1} = \rho^{(\alpha,\beta)}[f_2]_{g_1}$ $(0 < \rho^{(\alpha,\beta)}[f_1]_{g_1}, \rho^{(\alpha,\beta)}[f_2]_{g_1} < \infty)$. Now in view of Theorem 8.2.2 it is easy to see that $\rho^{(\alpha,\beta)}[f_1 \pm f_2]_{g_1} \leq \rho^{(\alpha,\beta)}[f_1]_{g_1} = \rho^{(\alpha,\beta)}[f_2]_{g_1}$. If possible let

$$\rho^{(\alpha,\beta)}[f_1 \pm f_2]_{g_1} < \rho^{(\alpha,\beta)}[f_1]_{g_1} = \rho^{(\alpha,\beta)}[f_2]_{g_1}. \tag{202}$$

Let $\sigma_D^{(\alpha,\beta)}[f_1]_{g_1} \neq \sigma_D^{(\alpha,\beta)}[f_2]_{g_1}$. Then in view of the first part of Theorem 8.2.13 and (202) we obtain that

$$\sigma_D^{(\alpha,\beta)}[f_1]_{g_1} = \sigma_{g_1,D}^{(\alpha,\beta)}[f_1 \pm f_2 \mp f_2]_{g_1} = \sigma_D^{(\alpha,\beta)}[f_2]_{g_1}$$

which is a contradiction. Hence

$$\rho^{(\alpha,\beta)}[f_1 \pm f_2]_{g_1} = \rho^{(\alpha,\beta)}[f_1]_{g_1} = \rho^{(\alpha,\beta)}[f_2]_{g_1}.$$

Similarly with the help of the first part of Theorem 8.2.13, one can obtain the same conclusion under the hypothesis $\overline{\sigma}_D^{(\alpha,\beta)}[f_1]_{g_1} \neq \overline{\sigma}_D^{(\alpha,\beta)}[f_2]_{g_1}$. This proves the first part of the theorem.

**Case II.** Let us consider that $\rho^{(\alpha,\beta)}[f_1]_{g_1} = \rho^{(\alpha,\beta)}[f_1]_{g_2}$ $(0 < \rho^{(\alpha,\beta)}[f_1]_{g_1}, \rho^{(\alpha,\beta)}[f_1]_{g_2} < \infty)$ and $f_1(z)$ is of regular generalized relative Gol'dberg growth $(\alpha,\beta)$ with respect to at least any one of $g_1(z)$ or $g_2(z)$ and $(g_1(z) \pm g_2(z))$. Therefore in view of Theorem 8.2.4, it follows that

$$\rho^{(\alpha,\beta)}[f_1]_{g_1 \pm g_2} \geq \rho^{(\alpha,\beta)}[f_1]_{g_1} = \rho^{(\alpha,\beta)}[f_1]_{g_2}$$

and if possible let

$$\rho^{(\alpha,\beta)}[f_1]_{g_1 \pm g_2} > \rho^{(\alpha,\beta)}[f_1]_{g_1} = \rho^{(\alpha,\beta)}[f_1]_{g_2}. \tag{203}$$

Let us consider that $\sigma_D^{(\alpha,\beta)}[f_1]_{g_1} \neq \sigma_D^{(\alpha,\beta)}[f_1]_{g_2}$. Then. in view of the proof of the second part of Theorem 8.2.13 and (203) we obtain that

$$\sigma_D^{(\alpha,\beta)}[f_1]_{g_1} = \sigma_D^{(\alpha,\beta)}[f_1]_{g_1 \pm g_2 \mp g_2} = \sigma_D^{(\alpha,\beta)}[f_1]_{g_2}$$

which is a contradiction. Hence

$$\rho^{(\alpha,\beta)}[f_1]_{g_1 \pm g_2} = \rho^{(\alpha,\beta)}[f_1]_{g_1} = \rho^{(\alpha,\beta)}[f_1]_{g_2}.$$

Again, in view of the proof of second part of Theorem 8.2.13 one can derive the same conclusion for the condition $\overline{\sigma}_D^{(\alpha,\beta)}[f_1]_{g_1} \neq \overline{\sigma}_D^{(\alpha,\beta)}[f_1]_{g_2}$ and therefore the second part of the theorem is established ∎

**Theorem 8.2.16** *Let* $f_1(z)$, $f_2(z)$, $g_1(z)$ *and* $g_2(z)$ *be any four entire functions of* $n$ *complex variables.*
*(A) The following conditions are assumed to be satisfied:*
*(i)* $f_1(z) \pm f_2(z)$ *is of regular generalized relative Gol'dberg growth* $(\alpha, \beta)$ *with respect to at least any one of* $g_1(z)$ *or* $g_2(z)$;
*(ii) Either* $\sigma_D^{(\alpha,\beta)}[f_1 \pm f_2]_{g_1} \neq \sigma_D^{(\alpha,\beta)}[f_1 \pm f_2]_{g_2}$ *or* $\overline{\sigma}_D^{(\alpha,\beta)}[f_1 \pm f_2]_{g_1} \neq \overline{\sigma}_D^{(\alpha,\beta)}[f_1 \pm f_2]_{g_2}$;
*(iii) Either* $\sigma_D^{(\alpha,\beta)}[f_1]_{g_1} \neq \sigma_D^{(\alpha,\beta)}[f_2]_{g_1}$ *or* $\overline{\sigma}_D^{(\alpha,\beta)}[f_1]_{g_1} \neq \overline{\sigma}_D^{(\alpha,\beta)}[f_2]_{g_1}$;
*(iv) Either* $\sigma_D^{(\alpha,\beta)}[f_1]_{g_2} \neq \sigma_D^{(\alpha,\beta)}[f_2]_{g_2}$ *or* $\overline{\sigma}_D^{(\alpha,\beta)}[f_1]_{g_2} \neq \overline{\sigma}_D^{(\alpha,\beta)}[f_2]_{g_2}$; *then*

$$\rho^{(\alpha,\beta)}[f_1 \pm f_2]_{g_1 \pm g_2} = \rho^{(\alpha,\beta)}[f_1]_{g_1} = \rho^{(\alpha,\beta)}[f_2]_{g_1} = \rho^{(\alpha,\beta)}[f_1]_{g_2} = \rho^{(\alpha,\beta)}[f_2]_{g_2}.$$

*(B) The following conditions are assumed to be satisfied:*
*(i)* $f_1(z)$ *and* $f_2(z)$ *are of regular generalized relative Gol'dberg growth* $(\alpha, \beta)$ *with respect to at least any one of* $g(z)$ *or* $g_2(z)$;
*(ii) Either* $\sigma_D^{(\alpha,\beta)}[f_1]_{g_1 \pm g_2} \neq \sigma_D^{(\alpha,\beta)}[f_2]_{g_1 \pm g_2}$ *or* $\overline{\sigma}_D^{(\alpha,\beta)}[f_1]_{g_1 \pm g_2} \neq \overline{\sigma}_D^{(\alpha,\beta)}[f_2]_{g_1 \pm g_2}$;
*(iii) Either* $\sigma_D^{(\alpha,\beta)}[f_1]_{g_1} \neq \sigma_D^{(\alpha,\beta)}[f_1]_{g_2}$ *or* $\sigma_D^{(\alpha,\beta)}[f_1]_{g_1} \neq \overline{\sigma}_D^{(\alpha,\beta)}[f_1]_{g_2}$;
*(iv) Either* $\sigma_D^{(\alpha,\beta)}[f_2]_{g_1} \neq \sigma_D^{(\alpha,\beta)}[f_2]_{g_2}$ *or* $\overline{\sigma}_D^{(\alpha,\beta)}[f_2]_{g_1} \neq \overline{\sigma}_D^{(\alpha,\beta)}[f_2]_{g_2}$; *then*

$$\rho^{(\alpha,\beta)}[f_1 \pm f_2]_{g_1 \pm g_2} = \rho^{(\alpha,\beta)}[f_1]_{g_1} = \rho^{(\alpha,\beta)}[f_2]_{g_1} = \rho^{(\alpha,\beta)}[f_1]_{g_2} = \rho^{(\alpha,\beta)}[f_2]_{g_2}.$$

We omit the proof of Theorem 8.2.16 as it is a natural consequence of Theorem 8.2.15.

**Theorem 8.2.17** *Let $f_1(z)$, $f_2(z)$, $g_1(z)$ and $g_2(z)$ be any four entire functions of $n$ complex variables.*

*(A) The following conditions are assumed to be satisfied:*

*(i) At least any one of $f_1(z)$ or $f_2(z)$ is of regular generalized relative Gol'dberg growth $(\alpha, \beta)$ with respect to $g_1(z)$;*

*(ii) Either $\tau_D^{(\alpha,\beta)}[f_1]_{g_1} \neq \tau_D^{(\alpha,\beta)}[f_2]_{g_1}$ or $\overline{\tau}_D^{(\alpha,\beta)}[f_1]_{g_1} \neq \overline{\tau}_D^{(\alpha,\beta)}[f_2]_{g_1}$ holds, then*

$$\lambda^{(\alpha,\beta)}[f_1 \pm f_2]_{g_1} = \lambda^{(\alpha,\beta)}[f_1]_{g_1} = \lambda^{(\alpha,\beta)}[f_2]_{g_1}.$$

*(B) The following conditions are assumed to be satisfied:*

*(i) $f_1(z)$, $g_1(z)$ and $g_2(z)$ be any three entire functions such that $\lambda^{(\alpha,\beta)}[f_1]_{g_1}$ and $\lambda^{(\alpha,\beta)}[f_1]_{g_2}$ exists;*

*(ii) Either $\tau_D^{(\alpha,\beta)}[f_1]_{g_1} \neq \tau_D^{(\alpha,\beta)}[f_1]_{g_2}$ or $\overline{\tau}_D^{(\alpha,\beta)}[f_1]_{g_1} \neq \overline{\tau}_D^{(\alpha,\beta)}[f_1]_{g_2}$ holds, then*

$$\lambda^{(\alpha,\beta)}[f_1]_{g_1 \pm g_2} = \lambda^{(\alpha,\beta)}[f_1]_{g_1} = \lambda^{(\alpha,\beta)}[f_1]_{g_2}.$$

**Proof. Case I.** Let $\lambda^{(\alpha,\beta)}[f_1]_{g_1} = \lambda^{(\alpha,\beta)}[f_2]_{g_1}$ $(0 < \lambda^{(\alpha,\beta)}[f_1]_{g_1}, \lambda^{(\alpha,\beta)}[f_2]_{g_1} < \infty)$ and at least $f_1(z)$ or $f_2(z)$ and $(f_1(z) \pm f_2(z))$ are of regular generalized relative Gol'dberg $(\alpha, \beta)$ growth with respect to $g_1(z)$. Now, in view of Theorem 8.2.1, it is easy to see that

$$\lambda^{(\alpha,\beta)}[f_1 \pm f_2]_{g_1} \leq \lambda^{(\alpha,\beta)}[f_1]_{g_1} = \lambda^{(\alpha,\beta)}[f_2]_{g_1}.$$

If possible let

$$\lambda^{(\alpha,\beta)}[f_1 \pm f_2]_{g_1} < \lambda^{(\alpha,\beta)}[f_1]_{g_1} = \lambda^{(\alpha,\beta)}[f_2]_{g_1}. \tag{204}$$

Let $\tau_D^{(\alpha,\beta)}[f_1]_{g_1} \neq \tau_D^{(\alpha,\beta)}[f_2]_{g_1}$. Then in view of the proof of the first part of Theorem 8.2.14 and (204) we obtain that

$$\tau_D^{(\alpha,\beta)}[f_1]_{g_1} = \tau_D^{(\alpha,\beta)}[f_1 \pm f_2 \mp f_2]_{g_1} = \tau_D^{(\alpha,\beta)}[f_2]_{g_1}$$

which is a contradiction. Hence

$$\lambda^{(\alpha,\beta)}[f_1 \pm f_2]_{g_1} = \lambda^{(\alpha,\beta)}[f_1]_{g_1} = \lambda^{(\alpha,\beta)}[f_2]_{g_1}.$$

Similarly in view of the proof of the first part of Theorem 8.2.14 , one can establish the same conclusion under the hypothesis $\overline{\tau}_D^{(\alpha,\beta)}[f_1]_{g_1} \neq \overline{\tau}_D^{(\alpha,\beta)}[f_2]_{g_1}$. This proves the first part of the theorem.

**Case II.** Let us consider that $\lambda^{(\alpha,\beta)}[f_1]_{g_1} = \lambda^{(\alpha,\beta)}[f_1]_{g_2}$ $(0 < \lambda^{(\alpha,\beta)}[f_1]_{g_1}, \lambda^{(\alpha,\beta)}[f_1]_{g_2} < \infty$. Therefore in view of Theorem 8.2.3, it follows that

$$\lambda^{(\alpha,\beta)}[f_1]_{g_1 \pm g_2} \geq \lambda^{(\alpha,\beta)}[f_1]_{g_1} = \lambda^{(\alpha,\beta)}[f_1]_{g_2}$$

and if possible let

$$\lambda^{(\alpha,\beta)}[f_1]_{g_1 \pm g_2} > \lambda^{(\alpha,\beta)}[f_1]_{g_1} = \lambda_{g_2}^{(\alpha,\beta)}(f_1). \tag{205}$$

Suppose $\tau_D^{(\alpha,\beta)} [f_1]_{g_1} \neq \tau_D^{(\alpha,\beta)} [f_1]_{g_2}$. Then in view of the second part of Theorem 8.2.14 and (205), we obtain that

$$\tau_D^{(\alpha,\beta)} [f_1]_{g_1} = \tau_D^{(\alpha,\beta)} [f_1]_{g_1 \pm g_2 \mp g_2} = \tau_D^{(\alpha,\beta)} [f_1]_{g_2}$$

which is a contradiction. Hence $\lambda^{(\alpha,\beta)} [f_1]_{g_1 \pm g_2} = \lambda^{(\alpha,\beta)} [f_1]_{g_1} = \lambda^{(\alpha,\beta)} [f_1]_{g_2}$. Analogously with the help of the second part of Theorem 8.2.14, the same conclusion can also be derived under the condition $\overline{\tau}_D^{(\alpha,\beta)} [f_1]_{g_1} \neq \overline{\tau}_D^{(\alpha,\beta)} [f_1]_{g_2}$ and therefore the second part of the theorem is established. ∎

**Theorem 8.2.18** *Let $f_1(z)$, $f_2(z)$, $g_1(z)$ and $g_2(z)$ be any four entire functions of $n$ complex variables.*
*(A) The following conditions are assumed to be satisfied:*
*(i) At least any one of $f_1(z)$ or $f_2(z)$ is of regular generalized relative Gol'dberg growth $(\alpha, \beta)$ with respect to $g_1(z)$ and $g_2(z)$.*
*(ii) Either $\tau_D^{(\alpha,\beta)} [f_1 \pm f_2]_{g_1} \neq \tau_D^{(\alpha,\beta)} [f_1 \pm f_2]_{g_2}$ or $\overline{\tau}_D^{(\alpha,\beta)} [f_1 \pm f_2]_{g_1} \neq \overline{\tau}_D^{(\alpha,\beta)} [f_1 \pm f_2]_{g_2}$ holds;*
*(iii) Either $\tau_D^{(\alpha,\beta)} [f_1]_{g_1} \neq \tau_D^{(\alpha,\beta)} [f_2]_{g_1}$ or $\overline{\tau}_D^{(\alpha,\beta)} [f_1]_{g_1} \neq \overline{\tau}_D^{(\alpha,\beta)} [f_2]_{g_1}$;*
*(iv) Either $\tau_D^{(\alpha,\beta)} [f_1]_{g_2} \neq \tau_D^{(\alpha,\beta)} [f_2]_{g_2}$ or $\overline{\tau}_D^{(\alpha,\beta)} [f_1]_{g_2} \neq \overline{\tau}_D^{(\alpha,\beta)} [f_2]_{g_2}$; then*

$$\lambda^{(\alpha,\beta)} [f_1 \pm f_2]_{g_1 \pm g_2} = \lambda^{(\alpha,\beta)} [f_1]_{g_1} = \lambda^{(\alpha,\beta)} [f_2]_{g_1} = \lambda^{(\alpha,\beta)} [f_1]_{g_2} = \lambda^{(\alpha,\beta)} [f_2]_{g_2}.$$

*(B) The following conditions are assumed to be satisfied:*
*(i) At least any one of $f_1(z)$ or $f_2(z)$ are of regular generalized relative Gol'dberg growth $(\alpha, \beta)$ with respect to $g_1(z) \pm g_2(z)$;*
*(ii) Either $\tau_D^{(\alpha,\beta)} [f_1]_{g_1 \pm g_2} \neq \tau_D^{(\alpha,\beta)} [f_2]_{g_1 \pm g_2}$ or $\overline{\tau}_D^{(\alpha,\beta)} [f_1]_{g_1 \pm g_2} \neq \overline{\tau}_D^{(\alpha,\beta)} [f_2]_{g_1 \pm g_2}$ holds;*
*(iii) Either $\tau_D^{(\alpha,\beta)} [f_1]_{g_1} \neq \tau_D^{(\alpha,\beta)} [f_1]_{g_2}$ or $\overline{\tau}_D^{(\alpha,\beta)} [f_1]_{g_1} \neq \overline{\tau}_D^{(\alpha,\beta)} [f_1]_{g_2}$ holds;*
*(iv) Either $\tau_D^{(\alpha,\beta)} [f_2]_{g_1} \neq \tau_{g_2,D}^{(\alpha,\beta)} (f_2)$ or $\overline{\tau}_D^{(\alpha,\beta)} [f_2]_{g_1} \neq \overline{\tau}_{g_2,D}^{(\alpha,\beta)} (f_2)$ holds, then*

$$\lambda^{(\alpha,\beta)} [f_1 \pm f_2]_{g_1 \pm g_2} = \lambda^{(\alpha,\beta)} [f_1]_{g_1} = \lambda^{(\alpha,\beta)} [f_2]_{g_1} = \lambda^{(\alpha,\beta)} [f_1]_{g_2} = \lambda^{(\alpha,\beta)} [f_2]_{g_2}.$$

We omit the proof of Theorem 8.2.18 as it is a natural consequence of Theorem 8.2.17.

**Theorem 8.2.19** *Let $f_1(z)$, $f_2(z)$, $g_1(z)$ and $g_2(z)$ be any four entire functions of $n$ complex variables.*
*Also let $\rho^{(\alpha,\beta)} [f_1]_{g_1}$, $\rho^{(\alpha,\beta)} [f_2]_{g_1}$, $\rho^{(\alpha,\beta)} [f_1]_{g_2}$ and $\rho^{(\alpha,\beta)} [f_2]_{g_2}$ are all non-zero.*
*(A) Assume the functions $f_1(z)$, $f_2(z)$ and $g_1(z)$ satisfy the following conditions:*
*(i) $g_1(z)$ satisfies the Property (G) and*
*(ii) $f_1(z)$ and $f_2(z)$ satisfy Property (X); then*

$$\sigma_D^{(\alpha,\beta)} [f_1 \cdot f_2]_{g_1} = \sigma_D^{(\alpha,\beta)} [f_i]_{g_1} \quad and \quad \overline{\sigma}_D^{(\alpha,\beta)} [f_1 \cdot f_2]_{g_1} = \overline{\sigma}_D^{(\alpha,\beta)} [f_i]_{g_1}.$$

*(B) Assume the functions $g_1(z)$, $g_2(z)$ and $f_1(z)$ satisfy the following conditions:*
*(i) $f_1(z)$ is of regular generalized relative Gol'dberg $(\alpha, \beta)$ growth with respect to at least*

*any one of $g_1(z)$ or $g_2(z)$ and $f_1(z)$ satisfy the Property (G) and*
*(ii) $g_1(z)$ and $g_2(z)$ satisfy Property (X); then*

$$\sigma_D^{(\alpha,\beta)}[f_1]_{g_1 \cdot g_2} = \sigma_D^{(\alpha,\beta)}[f_1]_{g_i} \quad \text{and} \quad \overline{\sigma}_D^{(\alpha,\beta)}[f_1]_{g_1 \cdot g_2} = \overline{\sigma}_D^{(\alpha,\beta)}[f_1]_{g_i}.$$

*(C) Assume the functions $f_1(z)$, $f_2(z)$, $g_1(z)$ and $g_2(z)$ satisfy the following conditions:*
*(i) $g_1(z) \cdot g_2(z)$, $f_1(z)$ and $f_2(z)$ satisfy the Property (G);*
*(ii) $f_1(z)$ and $f_2(z)$ satisfy Property (X);*
*(iii) $g_1(z)$ and $g_2(z)$ satisfy Property (X);*
*(iv) $f_1(z)$ is of regular generalized relative Gol'dberg $(\alpha,\beta)$ growth with respect to at least any one of $g_1(z)$ or $g_2(z)$;*
*(v) $f_2(z)$ is of regular generalized relative Gol'dberg $(\alpha,\beta)$ growth with respect to at least any one of $g_1(z)$ or $g_2(z)$;*
*(vi) $\rho^{(\alpha,\beta)}[f_l]_{g_m} = \max[\min\{\rho^{(\alpha,\beta)}[f_1]_{g_1}, \rho^{(\alpha,\beta)}[f_1]_{g_2}\}, \min\{\rho^{(\alpha,\beta)}[f_2]_{g_1}, \rho^{(\alpha,\beta)}[f_2]_{g_2}\}] \mid l, m = 1, 2;$ then*

$$\sigma_D^{(\alpha,\beta)}[f_1 \cdot f_2]_{g_1 \cdot g_2} = \sigma_D^{(\alpha,\beta)}[f_l]_{g_m} \quad \text{and} \quad \overline{\sigma}_D^{(\alpha,\beta)}[f_1 \cdot f_2]_{g_1 \cdot g_2} = \overline{\sigma}_D^{(\alpha,\beta)}[f_l]_{g_m}.$$

**Proof. Case I.** Let $g_1(z)$ satisfies the Property (G) and $f_1(z)$ and $f_2(z)$ satisfy Property (X). We have using (181) for all sufficiently large values of $R$ and for any arbitrary $\varepsilon > 0$ that

$$\begin{aligned}
M_{f_1 \cdot f_2, D}(R) &\leq M_{f_1, D}(R) \cdot M_{f_2, D}(R) \\
&\leq M_{g_1, D}(\alpha^{-1}(\log\{(\sigma_D^{(\alpha,\beta)}[f_1]_{g_1} + \varepsilon)(\exp(\beta(R)))^{\rho^{(\alpha,\beta)}[f_1]_{g_1}}\})) \\
&\quad \cdot M_{g_1, D}(\alpha^{-1}(\log\{(\sigma_D^{(\alpha,\beta)}[f_2]_{g_1} + \varepsilon)(\exp(\beta(R)))^{\rho^{(\alpha,\beta)}[f_2]_{g_1}}\}))
\end{aligned}$$

Suppose that $\max\left\{\rho^{(\alpha,\beta)}[f_1]_{g_1}, \rho^{(\alpha,\beta)}[f_2]_{g_1}\right\} = \rho^{(\alpha,\beta)}[f_1]_{g_1}$, then

$$M_{f_1 \cdot f_2, D}(R) < (M_{g_1, D}(\alpha^{-1}(\log\{(\sigma_D^{(\alpha,\beta)}[f_1]_{g_1} + \varepsilon)(\exp(\beta(R)))^{\rho^{(\alpha,\beta)}[f_1]_{g_1}}\})))^2$$

Now in view of Theorem 8.2.8,

$$i.e. M_{f_1 \cdot f_2, D}(R) < (M_{g_1, D}(\alpha^{-1}(\log\{(\sigma_D^{(\alpha,\beta)}[f_1]_{g_1} + \varepsilon)(\exp(\beta(R)))^{\rho^{(\alpha,\beta)}[f_1 \cdot f_2]_{g_1} + \varepsilon_1}\})\}))^2$$

for $\varepsilon_1 > 0$.

Since $g_1(z)$ satisfies the Property (G), we have

$$M_{f_1 \cdot f_2, D}(R) < M_{g_1, D}((\alpha^{-1}(\log\{(\sigma_D^{(\alpha,\beta)}[f_1]_{g_1} + \varepsilon)(\exp(\beta(R)))^{\rho^{(\alpha,\beta)}[f_1 \cdot f_2]_{g_1} + \varepsilon_1}\})\}))^\delta) \quad (206)$$

where $\delta > 1$.

Since $\varepsilon, \varepsilon_1 > 0$ is arbitrary, we obtain from (206) by letting $\delta \longrightarrow 1+$ that

$$\sigma_D^{(\alpha,\beta)}[f_1 \cdot f_2]_{g_1} \leq \sigma_D^{(\alpha,\beta)}[f_1]_{g_1}. \quad (207)$$

Again since $f_1(z)$ and $f_2(z)$ are satisfy Property (X), then using (181) for all sufficiently large values of $R$ and for any arbitrary $\varepsilon > 0$, we have

$$M_{f_1,D}(R) < M_{f_1 \cdot f_2, D}(R)$$
$$\leq M_{g_1,D}\left(\alpha^{-1}(\log\{(\sigma_D^{(\alpha,\beta)}[f_1 \cdot f_2]_{g_1} + \varepsilon)(\exp(\beta(R)))^{\rho^{(\alpha,\beta)}[f_1 \cdot f_2]_{g_1}}\})\right)$$

In view of Theorem 8.2.8,

$$M_{f_1,D}(R) \leq M_{g_1,D}\left(\alpha^{-1}(\log\{(\sigma_D^{(\alpha,\beta)}[f_1 \cdot f_2]_{g_1} + \varepsilon)(\exp(\beta(R)))^{\rho^{(\alpha,\beta)}[f_1]_{g_1}}\})\right)$$

which implies

$$\sigma_D^{(\alpha,\beta)}[f_1 \cdot f_2]_{g_1} \geq \sigma_D^{(\alpha,\beta)}[f_1]_{g_1} \tag{208}$$

Hence from from (207) and (208), we conclude that

$$\sigma_D^{(\alpha,\beta)}[f_1 \cdot f_2]_{g_1} = \sigma_D^{(\alpha,\beta)}[f_1]_{g_1}.$$

Similarly, if we consider $\rho^{(\alpha,\beta)}[f_1]_{g_1} < \rho^{(\alpha,\beta)}[f_2]_{g_1}$, then one can verify that

$$\sigma_D^{(\alpha,\beta)}[f_1 \cdot f_2]_{g_1} = \sigma_D^{(\alpha,\beta)}[f_2]_{g_1}.$$

**Case II.** Let $\rho^{(\alpha,\beta)}[f_1]_{g_1} > \rho^{(\alpha,\beta)}[f_2]_{g_1}$ and $g_1$ satisfy the Property (G). Now for any arbitrary $\varepsilon > 0$, like Case I, we have from (184) for a sequence of values of $R$ tending to infinity that

$$M_{f_1 \cdot f_2, D}(R) \leq M_{f_1,D}(R) \cdot M_{f_2,D}(R)$$
$$\leq M_{g_1,D}\left(\alpha^{-1}(\log\{(\overline{\sigma}_D^{(\alpha,\beta)}[f_1]_{g_1} + \varepsilon)(\exp(\beta(R)))^{\rho^{(\alpha,\beta)}[f_1]_{g_1}}\})\right)$$
$$\cdot M_{g_1,D}\left(\alpha^{-1}(\log\{(\overline{\sigma}_D^{(\alpha,\beta)}[f_2]_{g_1} + \varepsilon)(\exp(\beta(R)))^{\rho^{(\alpha,\beta)}[f_2]_{g_1}}\})\right)^{\Large/}$$

Now using the similar technique for a sequence of values of $R$ tending to infinity as explored in the proof of Case I, one can easily verify from above that $\overline{\sigma}_D^{(\alpha,\beta)}[f_1 \cdot f_2]_{g_1} = \overline{\sigma}_D^{(\alpha,\beta)}[f_1]_{g_1}$ under the conditions specified in the theorem. Similarly, if we consider $\rho^{(\alpha,\beta)}[f_1]_{g_1} < \rho^{(\alpha,\beta)}[f_2]_{g_1}$, then one can also verify that $\overline{\sigma}_D^{(\alpha,\beta)}[f_1 \cdot f_2]_{g_1} = \overline{\sigma}_D^{(\alpha,\beta)}[f_2]_{g_1}$. Therefore the first part of theorem follows from Case I and Case II.

**Case III.** Let $f_1$ satisfy the Property (G) and $\rho^{(\alpha,\beta)}[f_1]_{g_1} < \rho^{(\alpha,\beta)}[f_1]_{g_2}$ with $f_1(z)$ is of regular generalized relative Gol'dberg $(\alpha, \beta)$ growth with respect to at least any one of $g_1(z)$ or $g_2(z)$. For a sequence of values of $R$ tending to infinity that

$$M_{g_1 \cdot g_2, D}(R) \leq M_{g_1,D}(R) \cdot M_{g_2,D}(R),$$

$$i.e., M_{g_1 \cdot g_2, D}\left(\alpha^{-1}(\log\{(\sigma_D^{(\alpha,\beta)}[f_1]_{g_1} - \varepsilon)(\exp(\beta(R)))^{\rho^{(\alpha,\beta)}[f_1]_{g_1}}\})\right)$$
$$\leq M_{g_1,D}\left(\alpha^{-1}(\log\{(\sigma_D^{(\alpha,\beta)}[f_1]_{g_1} - \varepsilon)(\exp(\beta(R)))^{\rho^{(\alpha,\beta)}[f_1]_{g_1}}\})\right)$$
$$\cdot M_{g_2,D}\left(\alpha^{-1}(\log\{(\sigma_D^{(\alpha,\beta)}[f_1]_{g_1} - \varepsilon)(\exp(\beta(R)))^{\rho^{(\alpha,\beta)}[f_1]_{g_1}}\})\right),$$

Now $\rho^{(\alpha,\beta)}[f_1]_{g_1} < \rho^{(\alpha,\beta)}[f_1]_{g_2}$ implies $\dfrac{M_{g_2,D}(\alpha^{-1}(\log\{(\sigma_D^{(\alpha,\beta)}[f_1]_{g_1}-\varepsilon)(\exp(\beta(R)))^{\rho^{(\alpha,\beta)}[f_1]_{g_1}}\}))}{M_{g_2,D}(\alpha^{-1}(\log\{(\overline{\sigma}_D^{(\alpha,\beta)}[f_1]_{g_2}-\varepsilon)(\exp(\beta(R)))^{\rho^{(\alpha,\beta)}[f_1]_{g_2}}\}))} <$

$1$, so

$$M_{g_1\cdot g_2,D}\left(\alpha^{-1}(\log\{(\sigma_D^{(\alpha,\beta)}[f_1]_{g_1}-\varepsilon)(\exp(\beta(R)))^{\rho^{(\alpha,\beta)}[f_1]_{g_1}}\})\right)$$
$$< M_{g_1,D}\left(\alpha^{-1}(\log\{(\sigma_D^{(\alpha,\beta)}[f_1]_{g_1}-\varepsilon)(\exp(\beta(R)))^{\rho^{(\alpha,\beta)}[f_1]_{g_1}}\})\right)$$
$$\cdot M_{g_2,D}\left(\alpha^{-1}(\log\{(\overline{\sigma}_D^{(\alpha,\beta)}[f_1]_{g_2}-\varepsilon)(\exp(\beta(R)))^{\rho^{(\alpha,\beta)}[f_1]_{g_2}}\})\right)$$

Now we have in view of (182) and (183) for a sequence of values of $R$ tending to infinity that

$$M_{g_1\cdot g_2,D}\left(\alpha^{-1}(\log\{(\sigma_D^{(\alpha,\beta)}[f_1]_{g_1}-\varepsilon)(\exp(\beta(R)))^{\rho^{(\alpha,\beta)}[f_1]_{g_1}}\})\right) < [M_{f_1,D}(R)]^2 \qquad (209)$$

Since $f_1(z)$ satisfy the Property (G), we have from (209) for any $\delta > 1$,

$$M_{g_1\cdot g_2,D}\left(\alpha^{-1}(\log\{(\sigma_D^{(\alpha,\beta)}[f_1]_{g_1}-\varepsilon)(\exp(\beta(R)))^{\rho^{(\alpha,\beta)}[f_1]_{g_1}}\})\right) < \left[M_{f_1,D}(R^\delta)\right] \qquad (210)$$

Since $\varepsilon > 0$ is arbitrary, it follows from (210) by taking $\delta \to 1+$ and for a sequence of values $R$ tending to infinity that

$$\sigma_D^{(\alpha,\beta)}[f_1]_{g_1\cdot g_2} \geq \sigma_D^{(\alpha,\beta)}[f_1]_{g_1}. \qquad (211)$$

Next since $g_1(z)$ and $g_2(z)$ are satisfy Property (X), then for all sufficiently large values of $R$, we have from(181) and Theorem8.2.10

$$M_{g_1,D}(R) < M_{g_1\cdot g_2,D}(R)$$
$$i.e.,\ M_{g_1\cdot g_2,D}\left(\alpha^{-1}(\log\{(\sigma_D^{(\alpha,\beta)}[f_1]_{g_1}+\varepsilon)(\exp(\beta(R)))^{\rho^{(\alpha,\beta)}[f_1]_{g_1}}\})\right)$$
$$> M_{g_1,D}\left(\alpha^{-1}(\log\{(\sigma_D^{(\alpha,\beta)}[f_1]_{g_1}+\varepsilon)(\exp(\beta(R)))^{\rho^{(\alpha,\beta)}[f_1]_{g_1}}\})\right)$$
$$\geq M_{f_1,D}(R)$$

So

$$\sigma_D^{(\alpha,\beta)}[f_1]_{g_1} \geq \sigma_D^{(\alpha,\beta)}[f_1]_{g_1\cdot g_2,} \qquad (212)$$

Hence from (211) and (212), we conclude that

$$\sigma_D^{(\alpha,\beta)}[f_1]_{g_1\cdot g_2,} = \sigma_D^{(\alpha,\beta)}[f_1]_{g_1}.$$

Similarly, if we consider $\rho^{(\alpha,\beta)}[f_1]_{g_1} > \rho^{(\alpha,\beta)}[f_1]_{g_2}$, then one can verify that $\sigma_D^{(\alpha,\beta)}[f_1]_{g_1\cdot g_2,} = \sigma_D^{(\alpha,\beta)}[f_1]_{g_2}$.

**Case IV.** Suppose $f_1(z)$ satisfy the Property (G). Also let $\rho^{(\alpha,\beta)}[f_1]_{g_1} < \rho^{(\alpha,\beta)}[f_1]_{g_2}$ with $f_1(z)$ is of regular generalized relative Gol'dberg $(\alpha,\beta)$ growth with respect to at least any one of $g_1(z)$ or $g_2(z)$. Therefore like Case I and in view of (182), we obtain for all sufficiently large values of $R$ that

$$M_{g_1 \cdot g_2, D}(R) \le M_{g_1, D}(R) \cdot M_{g_2, D}(R)$$

$$\Rightarrow \quad M_{g_1 \cdot g_2, D}\left(\alpha^{-1}(\log\{(\overline{\sigma}_D^{(\alpha,\beta)}[f_1]_{g_1} - \varepsilon)(\exp(\beta(R)))^{\rho^{(\alpha,\beta)}[f_1]_{g_1}}\})\right)$$

$$\le M_{g_1, D}\left(\alpha^{-1}(\log\{(\overline{\sigma}_D^{(\alpha,\beta)}[f_1]_{g_1} - \varepsilon)(\exp(\beta(R)))^{\rho^{(\alpha,\beta)}[f_1]_{g_1}}\})\right)$$

$$\cdot M_{g_2, D}\left(\alpha^{-1}(\log\{(\overline{\sigma}_D^{(\alpha,\beta)}[f_1]_{g_1} - \varepsilon)(\exp(\beta(R)))^{\rho^{(\alpha,\beta)}[f_1]_{g_1}}\})\right)$$

Now using the similar technique for all sufficiently large values of $R$ as explored in the proof of Case III, one can easily verify that $\overline{\sigma}_D^{(\alpha,\beta)}[f_1]_{g_1 \cdot g_2} = \overline{\sigma}_D^{(\alpha,\beta)}[f_1]_{g_1}$ under the conditions specified in the theorem. Likewise, if we consider $\rho^{(\alpha,\beta)}[f_1]_{g_1} > \rho^{(\alpha,\beta)}[f_1]_{g_2}$ with at least $f_1$ is of regular generalized relative Gol'dberg growth $(\alpha,\beta)$ with respect to $g_1$, then one can verify that $\overline{\sigma}_D^{(\alpha,\beta)}[f_1]_{g_1 \cdot g_2} = \overline{\sigma}_D^{(\alpha,\beta)}[f_1]_{g_2}$. Therefore the second part of theorem follows from Case III and Case IV.

Proof of the third part of the Theorem is omitted as it can be carried out in view of Theorem 8.2.11 and the above cases. ■

**Theorem 8.2.20** *Let $f_1(z)$, $f_2(z)$, $g_1(z)$ and $g_2(z)$ be any four entire functions of $n$ complex variables. Also let $\lambda^{(\alpha,\beta)}[f_1]_{g_1}$, $\lambda^{(\alpha,\beta)}[f_2]_{g_1}$, $\lambda^{(\alpha,\beta)}[f_1]_{g_2}$ and $\lambda^{(\alpha,\beta)}[f_2]_{g_2}$ are all non-zero and finite.*
*(A) Assume the functions $f_1(z)$, $f_2(z)$ and $g_1(z)$ satisfy the following conditions:*
*(i) At least $f_1(z)$ or $f_2(z)$ is of regular generalized relative Gol'dberg $(\alpha,\beta)$ growth with respect to $g_1(z)$ and $g_1(z)$ satisfy the Property (G) and*
*(ii) $f_1(z)$ and $f_2(z)$ satisfy Property (X); then*

$$\tau_D^{(\alpha,\beta)}[f_1 \cdot f_2]_{g_1} = \tau_D^{(\alpha,\beta)}[f_i]_{g_1} \quad and \quad \overline{\tau}_D^{(\alpha,\beta)}[f_1 \cdot f_2]_{g_1} = \overline{\tau}_D^{(\alpha,\beta)}[f_i]_{g_1}.$$

*(B) Assume the functions $g_1(z)$, $g_2(z)$ and $f_1(z)$ satisfy the following conditions:*
*(i) $f_1(z)$ satisfy the Property (G) and*
*(ii) $g_1(z)$ and $g_2(z)$ satisfy Property (X); then*

$$\tau_D^{(\alpha,\beta)}[f_1]_{g_1 \cdot g_2} = \tau_D^{(\alpha,\beta)}[f_1]_{g_i} \quad and \quad \overline{\tau}_D^{(\alpha,\beta)}[f_1]_{g_1 \cdot g_2} = \overline{\tau}_D^{(\alpha,\beta)}[f_1]_{g_i}.$$

*(C) Assume the functions $f_1(z)$, $f_2(z)$, $g_1(z)$ and $g_2(z)$ satisfy the following conditions:*
*(i) $g_1(z) \cdot g_2(z)$, $f_1(z)$ and $f_2(z)$ are satisfy the Property (G);*
*(ii) $f_1(z)$ and $f_2(z)$ satisfy Property (X);*
*(iii) $g_1(z)$ and $g_2(z)$ satisfy Property (X);*
*(iv) At least $f_1(z)$ or $f_2(z)$ is of regular generalized relative Gol'dberg $(\alpha,\beta)$ growth with respect to $g_1(z)$ for $i = 1, 2$, $j = 1, 2$ and $i \neq j$;*
*(v) At least $f_1(z)$ or $f_2(z)$ is of regular generalized relative Gol'dberg $(\alpha,\beta)$ growth with respect to $g_2(z)$ for $i = 1, 2$, $j = 1, 2$ and $i \neq j$;*

$(vi)$ $\lambda^{(\alpha,\beta)}\left[f_l\right]_{g_m} = \min[\max\{\lambda^{(\alpha,\beta)}\left[f_1\right]_{g_1}, \lambda^{(\alpha,\beta)}\left[f_2\right]_{g_1}\}, \max\{\lambda^{(\alpha,\beta)}\left[f_1\right]_{g_2}, \lambda^{(\alpha,\beta)}\left[f_2\right]_{g_2}\}] \mid l, m = 1, 2;$ *then*

$$\tau_D^{(\alpha,\beta)}\left[f_1 \cdot f_2\right]_{g_1 \cdot g_2} = \tau_D^{(\alpha,\beta)}\left[f_l\right]_{g_m} \quad and \quad \overline{\tau}_D^{(\alpha,\beta)}\left[f_1 \cdot f_2\right]_{g_1 \cdot g_2} = \overline{\tau}_D^{(\alpha,\beta)}\left[f_l\right]_{g_m}.$$

We omit the proof of Theorem 8.2.20 as it is a natural consequence of Theorem 8.2.19 and Theorem 8.2.12 .

# 8.3   Conclusion.

From the property of generalized relative Gol'dberg order $(\alpha, \beta)$ and generalized relative Gol'dberg lower order $(\alpha, \beta)$, the following propositions are immediate:

**Proposition 8.3.1** *Let $f_i(z)$, $f_j(z)$ and $g_i(z)$ be any three entire functions of $n$ complex variables with $0 < \rho^{(\alpha_i,\beta_i)}\left[f_i\right]_{g_i} < \infty$ and $0 < \rho^{(\alpha_i,\beta_j)}\left[f_j\right]_{g_i} < \infty$. Then the following relations may occurs:*

$$(i) \quad \beta_i(R) = \beta_j(R) \ and \ \rho^{(\alpha_i,\beta_i)}\left[f_i\right]_{g_i} > \rho^{(\alpha_i,\beta_j)}\left[f_j\right]_{g_i},$$

$$(ii) \quad \beta_i(R) = \beta_j(R) \ and \ \rho^{(\alpha_i,\beta_i)}\left[f_i\right]_{g_i} = \rho^{(\alpha_i,\beta_j)}\left[f_j\right]_{g_i},$$

$$(iii) \quad \beta_i(R) > \beta_j(R) \ and \ \rho^{(\alpha_i,\beta_i)}\left[f_i\right]_{g_i} > \rho^{(\alpha_i,\beta_j)}\left[f_j\right]_{g_i},$$

$$(iv) \quad \beta_i(R) > \beta_j(R) \ and \ \rho^{(\alpha_i,\beta_i)}\left[f_i\right]_{g_i} = \rho^{(\alpha_i,\beta_j)}\left[f_j\right]_{g_i},$$

*and*

$$(v) \quad \beta_i(R) > \beta_j(R) \ and \ \rho^{(\alpha_i,\beta_i)}\left[f_i\right]_{g_i} > \rho^{(\alpha_i,\beta_j)}\left[f_j\right]_{g_i}.$$

**Proposition 8.3.2** *Let $f_i(z)$, $g_i(z)$ and $g_j(z)$ be any three entire functions of $n$ complex variables with $0 < \lambda^{(\alpha_i,\beta_i)}\left[f_i\right]_{g_i} < \infty$ and $0 < \lambda^{(\alpha_j,\beta_i)}\left[f_i\right]_{g_j} < \infty$. Then the following relations may occurs:*

$$(i) \quad \alpha_i(R) = \alpha_j(R) \ and \ \lambda^{(\alpha_i,\beta_i)}\left[f_i\right]_{g_i} < \lambda^{(\alpha_j,\beta_i)}\left[f_i\right]_{g_j},$$

$$(ii) \quad \alpha_i(R) = \alpha_j(R) \ and \ \lambda^{(\alpha_i,\beta_i)}\left[f_i\right]_{g_i} = \lambda^{(\alpha_j,\beta_i)}\left[f_i\right]_{g_j},$$

$$(iii) \quad \alpha_i(R) > \alpha_j(R) \ and \ \lambda^{(\alpha_i,\beta_i)}\left[f_i\right]_{g_i} < \lambda^{(\alpha_j,\beta_i)}\left[f_i\right]_{g_j},$$

$$(iv) \quad \alpha_i(R) > \alpha_j(R) \ and \ \lambda^{(\alpha_i,\beta_i)}\left[f_i\right]_{g_i} = \lambda^{(\alpha_j,\beta_i)}\left[f_i\right]_{g_j},$$

*and*

$$(v) \quad \alpha_i(R) > \alpha_j(R) \ and \ \lambda^{(\alpha_i,\beta_i)}\left[f_i\right]_{g_i} > \lambda^{(\alpha_j,\beta_i)}\left[f_i\right]_{g_j}.$$

In this chapter, we investigate certain properties and find out the limiting value of generalized relative Gol'dberg order $(\alpha, \beta)$ ( respectively generalized relative Gol'dberg lower order $(\alpha, \beta)$), generalized relative Gol'dberg type $(\alpha, \beta)$ ( respectively generalized relative Gol'dberg lower type $(\alpha, \beta)$) and generalized relative Gol'dberg weak type $(\alpha, \beta)$ ( respectively generalized relative Gol'dberg upper weak type $(\alpha, \beta)$) of entire functions of $n$ complex variables under some different conditions considering the first two conditions of Proposition 8.3.1 and Proposition 8.3.2 for $\rho$, $\lambda$ and $\rho = \lambda$ whichever is applicable. Now question may arise about the limiting values of generalized relative Gol'dberg order $(\alpha, \beta)$ ( respectively generalized relative Gol'dberg lower order $(\alpha, \beta)$) as well as generalized relative Gol'dberg type $(\alpha, \beta)$ ( respectively generalized relative Gol'dberg lower type $(\alpha, \beta)$) and generalized relative Gol'dberg weak type $(\alpha, \beta)$ ( respectively generalized relative Gol'dberg upper weak type $(\alpha, \beta)$) when any one of the last three cases of Proposition 8.3.1 and Proposition 8.3.2 are considered and this may be a further scope of study for the future researchers in this branch.

# References

[1] T. Biswas and R. Biswas: Sum and product theorems relating to (p,q)-$\varphi$ relative Gol'dberg order and (p,q)-$\varphi$ relative Gol'dberg lower order of entire functions of several variables, Uzbek Math. J., 2018(4) (2018), 160-169.

[2] C. Roy: Some properties of entire functions in one and several complex variables, Ph.D. Thesis ( 2010), University of Calcutta.

CONFLICT OF INTEREST

The authors confirm that this book contents have no conflict of interest.

# Conclusion

This book is mainly focused on some growth properties of entire functions of several complex variables, which covers the important branch of complex analysis specially the theory of analytic functions of several variables. All the Chapters of this book deals with some growth properties of entire functions of $n$ complex variables, with the generalization of Gol'dberg order, relative Gol'dberg order, Gol'dberg type and relative Gol'dberg type etc. after introducing non-negative continuous functions $\alpha$ and $\beta$ defined on $(-\infty, +\infty)$. This book opens the new era of future research. Also the concept of generalized Gol'dberg order and generalized Gol'dberg type should have a broad range of applications in complex dynamics, factorization theory of entire functions of several complex variables, the solution of complex differential equations etc.

During previous decades, several authors made closed investigations on the growth properties of entire functions of several complex variables using different growth indicators such as Gol'dberg order, $(p, q)$-th Gol'dberg order, relative Gol'dberg order etc. In this book we wish to establish some basic growth properties of entire functions of several complex variables on the basis of their generalized Gol'dberg order $(\alpha, \beta)$, generalized relative Gol'dberg order $(\alpha, \beta)$, generalized Gol'dberg type $(\alpha, \beta)$, generalized relative Gol'dberg type $(\alpha, \beta)$ where $\alpha$ and $\beta$ continuous non-negative functions defined on $(-\infty, +\infty)$. We have also discussed about the particular cases when it coincide with present definitions. Integral representations of some definitions are given in some Chapters with some comparative studies.

So, this book (Monograph) will enrich some parts of Pure Mathematics and will give some scopes of study for the future researchers in this branch of complex analysis.

# Bibliography

[1] B. A. Fuks: Introduction to the theory of analytic functions of several complex variables, American Mathematical Soci., Providence, R. I., 1963.

[2] A. A. Gol'dberg: Elementary remarks on the formulas defining order and type of functions of several variables, Dokl. Akad. Nauk Arm. SSR, 29 (1959), 145-151 (Russian).

[3] S. K. Datta and A. R. Maji: Study of Growth properties on the basis of generalised Gol'dberg order of composite entire functions of several complex variables,International J. of Math.Sci.& Engg.Appls, 5(V) (2011), 297-311.

[4] X. Shen, J. Tu and H.Y. Xu: Complex oscillation of a second-order linear differential equation with entire coefficients of $[p, q] - \varphi$ order, Adv. Difference Equ., 2014, 2014:200, 14 pages.

[5] T. Biswas and R. Biswas: Sum and product theorems relating to (p,q)-$\varphi$ relative Gol'dberg order and (p,q)-$\varphi$ relative Gol'dberg lower order of entire functions of several variables,Uzbek Math. J., 2018(4) (2018), 160-169.

[6] B. C. Mondal and C. Roy: Relative gol'dberg order of an entire function of several variables, Bull Cal. Math. Soc., 102(4) (2010), 371-380.

[7] B. Prajapati and A. Rastogi: Some results o $p^{th}$ gol'dberg relative order, International Journal of Applied Mathematics and Statistical Sciences, 5(2) (2016), 147-154.

[8] T. Biswas: Some growth analysis of entire functions of several variables on the basis of their (p,q)-th relative Gol'dberg order and (p,q)-th relative Gol'dberg type, Palest. J. Math., 9(1) (2020), 149-158.

[9] T. Biswas: Sum and product theorems relating to relative (p,q)-th Gol'dberg order, relative (p,q)-th Gol'dberg type and relative (p,q)-th Gol'dberg weak type of entire functions of several variables, J. Interdiscip. Math., 22(1) (2019), 53-63.

[10] T. Biswas: Some results relating to (p,q)-th relative Gol'dberg order and (p,q)-relative Gol'dberg type of entire functions of several variables, J. Fract. Calc. Appl., 10(2) (2019), 249-272.

[11] T. Biswas and R. Biswas: Some growth properties of entire functions of several complex variables on the basis of their (p,q)-$\varphi$ relative Gol'dberg order and (p,q)-$\varphi$ relative Gol'dberg lower order, Electron. J. Math. Anal. Appl., 8(1) (2020), 229-236.

[12] T. Biswas and R. Biswas: Some growth estimations based on (p,q)-$\varphi$ relative Gol'dberg type and (p,q)-$\varphi$ relative Gol'dberg weak type of entire functions of several complex variables, Korean J. Math., 28(3) (2020), 489-507.

[13] D. Banerjee and S. Sarkar: A note on (p,q)$^{th}$ relative Gol'dberg order of entire functions of several variables. Bull. Allahabad Math. Soc., 34(1) (2019), 25-37.

[14] D. Banerjee and S. Sarkar: On (p,q)$^{th}$ Gol'dberg order and (p,q)$^{th}$ Gol'dberg type of an entire function of several complex variables represented by multiple dirichlet series, South East Asian J. Math. Math. Sci., 15(1) (2019), 15-24.

[15] D. Banerjee and S. Sarkar: (p,q)$^{th}$ relative Gol'dberg order of entire functions of several variables. International J. of Math. Sci. & Engg. Appls., 11(III) (2017), 185-201.

[16] T. Biswas and R. Biswas: On some (p,q)-$\varphi$ relative Gol'dberg type and (p,q)-$\varphi$ relative Gol'dberg weak type based growth properties of entire functions of several complex variables, Ital. J. Pure Appl. Math., N. 44 (2020), 403-414.

[17] S.K. Datta and A.R. Maji: Some study of the comparative growth rates on the basis of generalised relative Gol'dberg order of composite entire functions of several complex variables, International J. of Math. Sci. & Engg. Appls, 5(V) (2011), 335-344.

[18] S.K. Datta and A.R. Maji: Some study of the comparative growth properties on the basis of relative Gol'dberg order of composite entire functions of several complex variables, Int. J. Contemp. Math. Sci., 6(42) (2011), 2075-2082.

[19] A. Feruj: Gol'dberg order and Gol'dberg type of entire functions represented by multiple Dirichlet series, Ganit J. Bangladesh Math. Soc., 29, (2009), 63-70.

[20] C. Roy: Some properties of entire functions in one and several complex vaiables, Ph.D. Thesis ( 2010), University of Calcutta.

[21] P. K. Sarkar: On Gol'dberg order and Gol'dberg type of an entire function of several complex variables represented by multiple Dirichlet series, Indian J. of pure and App. Math. 13(10) (1982), 1221-1229.

[22] U. V. Singh and A. Rastogi: On Gol'dberg qth Order and Gol'dberg qth Type of an Entire Function Represented by Multiple Dirichlet Series, International Journal of Mathematics and its Applications, 3(3-D) (2015), 51-56.

# SUBJECT INDEX

## A

Assessment, quantitative 43, 71

## B

Bounded complete n-circular domain 1, 2, 3, 4, 5, 8, 9, 10, 11, 16, 17, 18, 28, 30, 33

## C

Complex analysis 147
Complex differential equations 147
Complex dynamics 147

## D

Different 42, 43, 48, 71, 97, 116, 144
  conditions 42, 48, 97, 116, 144
  functions scale 43, 71
Dimensions 28, 42, 70, 97
Divergence 34, 36, 53, 56, 57, 60
Domain, arbitrary bounded complex n-circular 1

## E

Equality, sign of 123, 124, 125
Equivalence relations 50

## F

Factorization theory 147
Functions 8, 11, 16, 23, 27, 28, 30, 31, 32, 37, 40, 41, 43, 47, 49, 50, 61, 68, 69, 71, 72, 81, 82, 98, 104, 111, 113, 117, 118, 147
  analytic 147
  increasing 16, 30, 32, 37, 43, 49, 71, 98, 117
  of n-complex variables 8, 11, 23, 27, 28, 31, 40, 41, 47, 50, 61, 68, 69, 72, 81, 82, 104, 111, 113, 118

## G

Generalized Gol'dberg 2, 3, 9, 16, 17, 18, 26, 42, 70, 121
  irregular 3, 9, 17
  definitions of 2, 16, 17
  notion of 26, 42, 70
  positive 17, 18
  regular 3, 9, 17
  regular relative 121
Generalized 2, 3, 4, 13, 16, 17, 18, 28, 30, 40, 42, 43, 147
  Gol'dberg order 2, 3, 4, 13, 16, 17, 18, 28, 42, 43, 147
  relative hyper Gol'dberg order 30, 40
  relative logarithmic Gol'dberg lower order 30, 40
Gol'dberg order 2, 4, 6, 7, 10, 13, 16, 17, 18, 19, 28, 30, 31, 40, 43, 49, 50, 98, 147, 149
Growth 2, 13, 17, 30, 31, 42, 47, 50, 64, 70, 71, 121, 122, 124, 125, 126, 128, 129, 131, 132, 134, 140, 143, 147
  analysis 30
  different 13
  irregular 2
  properties, basic 13, 147
  ratios 42
  scale 17, 50
Growth indicators 1, 2, 4, 6, 7, 9, 10, 12, 13, 16, 17, 18, 30, 31, 39, 50, 51, 60, 71, 147
  different 1, 147
  relative 39, 60, 71
Growth rates 43, 72
  relative 72

## I

Index-pair 4, 5, 6, 7, 10, 11
  lower 5
  relative 10, 11

9 789814 998055